HPLC of Peptides and Proteins

METHODS IN MOLECULAR BIOLOGY™

John M. Walker, SERIES EDITOR

276. **Capillary Electrophoresis of Proteins and Peptides,** edited by *Mark A. Strege and Avinash L. Lagu, 2004*
275. **Chemoinformatics,** edited by *Jürgen Bajorath, 2004*
274. **Photosynthesis Research Protocols,** edited by *Robert Carpentier, 2004*
273. **Platelets and Megakaryocytes,** *Volume 2: Perspectives and Techniques,* edited by *Jonathan M. Gibbns and Martyn P. Mahaut-Smith, 2004*
272. **Platelets and Megakaryocytes,** *Volume 1: Functional Assays,* edited by *Jonathan M. Gibbns and Martyn P. Mahaut-Smith, 2004*
271. **B Cell Protocols,** edited by *Hua Gu and Klaus Rajewsky, 2004*
270. **Parasite Genomics Protocols,** edited by *Stuart N. Isaacs, 2004*
269. **Vaccina Virus and Poxvirology:** *Methods and Protocols,* edited by *Stuart N. Isaacs, 2004*
268. **Public Health Microbiology:** *Methods and Protocols,* edited by *John F. T. Spencer and Alicia L. Ragout de Spencer, 2004*
267. **Recombinant Gene Expression:** *Reviews and Protocols, Second Edition,* edited by *Paulina Balbas and Argelia Johnson, 2004*
266. **Genomics, Proteomics, and Clinical Bacteriology:** *Methods and Reviews,* edited by *Neil Woodford and Alan Johnson, 2004*
265. **RNA Interference, Editing, and Modification:** *Methods and Protocols,* edited by *Jonatha M. Gott, 2004*
264. **Protein Arrays:** *Methods and Protocols,* edited by *Eric Fung, 2004*
263. **Flow Cytometry, Second Edition**, edited by *Teresa S. Hawley and Robert G. Hawley, 2004*
262. **Genetic Recombination Protocols**, edited by *Alan S. Waldman, 2004*
261. **Protein–Protein Interactions:** *Methods and Applications,* edited by *Haian Fu, 2004*
260. **Mobile Genetic Elements:** *Protocols and Genomic Applications,* edited by *Wolfgang J. Miller and Pierre Capy, 2004*
259. **Receptor Signal Transduction Protocols, Second Edition,** edited by *Gary B. Willars and R. A. John Challiss, 2004*
258. **Gene Expression Profiling:** *Methods and Protocols,* edited by *Richard A. Shimkets, 2004*
257. **mRNA Processing and Metabolism:** *Methods and Protocols,* edited by *Daniel R. Schoenberg, 2004*
256. **Bacterial Artifical Chromosomes,** *Volume 2: Functional Studies,* edited by *Shaying Zhao and Marvin Stodolsky, 2004*
255. **Bacterial Artifical Chromosomes,** *Volume 1: Library Construction, Physical Mapping, and Sequencing,* edited by *Shaying Zhao and Marvin Stodolsky, 2004*
254. **Germ Cell Protocols,** *Volume 2: Molecular Embryo Analysis, Live Imaging, Transgenesis, and Cloning,* edited by *Heide Schatten, 2004*
253. **Germ Cell Protocols,** *Volume 1: Sperm and Oocyte Analysis,* edited by *Heide Schatten, 2004*
252. **Ribozymes and siRNA Protocols, Second Edition,** edited by *Mouldy Sioud, 2004*
251. **HPLC of Peptides and Proteins:** *Methods and Protocols,* edited by *Marie-Isabel Aguilar, 2004*
250. **MAP Kinase Signaling Protocols,** edited by *Rony Seger, 2004*
249. **Cytokine Protocols,** edited by *Marc De Ley, 2004*
248. **Antibody Engineering:** *Methods and Protocols,* edited by *Benny K. C. Lo, 2004*
247. ***Drosophila*** **Cytogenetics Protocols,** edited by *Daryl S. Henderson, 2004*
246. **Gene Delivery to Mammalian Cells:** *Volume 2: Viral Gene Transfer Techniques,* edited by *William C. Heiser, 2004*
245. **Gene Delivery to Mammalian Cells:** *Volume 1: Nonviral Gene Transfer Techniques,* edited by *William C. Heiser, 2004*
244. **Protein Purification Protocols, Second Edition,** edited by *Paul Cutler, 2004*
243. **Chiral Separations:** *Methods and Protocols,* edited by *Gerald Gübitz and Martin G. Schmid, 2004*
242. **Atomic Force Microscopy:** *Biomedical Methods and Applications,* edited by *Pier Carlo Braga and Davide Ricci, 2004*
241. **Cell Cycle Checkpoint Control Protocols,** edited by *Howard B. Lieberman, 2004*
240. **Mammalian Artificial Chromosomes:** *Methods and Protocols,* edited by *Vittorio Sgaramella and Sandro Eridani, 2004*
239. **Cell Migration in Inflammation and Immunity:** *Methods and Protocols,* edited by *Daniele D'Ambrosio and Francesco Sinigaglia, 2004*
238. **Biopolymer Methods in Tissue Engineering,** edited by *Anthony P. Hollander and Paul V. Hatton, 2004*
237. **G Protein Signaling:** *Methods and Protocols,* edited by *Alan V. Smrcka, 2004*
236. **Plant Functional Genomics**: *Methods and Protocols*, edited by *Erich Grotewold, 2004*
235. ***E. coli*** **Plasmid Vectors**: *Methods and Applications*, edited by *Nicola Casali and Andrew Preston, 2003*
234. **p53 Protocols**, edited by *Sumitra Deb and Swati Palit Deb, 2003*
233. **Protein Kinase C Protocols**, edited by *Alexandra C. Newton, 2003*
232. **Protein Misfolding and Disease:** *Principles and Protocols*, edited by *Peter Bross and Niels Gregersen, 2003*
231. **Directed Evolution Library Creation**: *Methods and Protocols*, edited by *Frances H. Arnold and George Georgiou, 2003*
230. **Directed Enzyme Evolution**: *Screening and Selection Methods*, edited by *Frances H. Arnold and George Georgiou, 2003*
229. **Lentivirus Gene Engineering Protocols**, edited by *Maurizio Federico, 2003*
228. **Membrane Protein Protocols**: *Expression, Purification, and Characterization*, edited by *Barry S. Selinsky, 2003*
227. **Membrane Transporters:** *Methods and Protocols,* edited by *Qing Yan, 2003*
226. **PCR Protocols, Second Edition,** edited by *John M. S. Bartlett and David Stirling, 2003*
225. **Inflammation Protocols,** edited by *Paul G. Winyard and Derek A. Willoughby, 2003*
224. **Functional Genomics:** *Methods and Protocols,* edited by *Michael J. Brownstein and Arkady B. Khodursky, 2003*
223. **Tumor Suppressor Genes:** *Volume 2: Regulation, Function, and Medicinal Applications,* edited by *Wafik S. El-Deiry, 2003*
222. **Tumor Suppressor Genes:** *Volume 1: Pathways and Isolation Strategies,* edited by *Wafik S. El-Deiry, 2003*

METHODS IN MOLECULAR BIOLOGY™

HPLC of Peptides and Proteins

Methods and Protocols

Edited by

Marie-Isabel Aguilar

Department of Biochemistry and Molecular Biology
Monash University, Clayton, Victoria, Australia

HUMANA PRESS ✻ TOTOWA, NEW JERSEY

999 Riverview Drive, Suite 208
Totowa, New Jersey 07512

www.humanapress.com

This publication is printed on acid-free paper. ∞
ANSI Z39.48-1984 (American Standards Institute Permanence of Paper for Printed Library Materials.)

Cover design by Patricia F. Cleary.

Cover illustration: Cartoon depicting the selective retention of a peptide from a mixture of peptides and proteins. Illustration by Lee Tzong-Hsien, William Farrugia, and Marie-Isabel Aguilar.

For additional copies, pricing for bulk purchases, and/or information about other Humana titles, contact Humana at the above address or at any of the following numbers: Tel.: 973-256-1699; Fax: 973-256-8341; E-mail: humana@humanapr.com; or visit our Website: www.humanapress.com

Printed in the United States of America. 10 9 8 7 6 5 4 3 2 1

ISSN: 1064-3745

E-ISBN: 1-59259-742-4

Library of Congress Cataloging-in-Publication Data

HPLC of peptides and proteins : methods and protocols / edited by
Marie-Isabel Aguilar.
p. ; cm. -- (Methods in molecular biology ; 251)
Includes bibliographical references and index.
ISBN 0-89603-977-3 (alk. paper)
1. Peptides--Analysis. 2. Proteins--Analysis. 3. High performance liquid chromatography.
[DNLM: 1. Peptides--chemistry. 2. Chromatography, High Pressure Liquid--methods. 3. Proteins--chemistry. QU 68 H872 2004] I. Aguilar, Marie-Isabel. II. Methods in molecular biology (Totowa, N.J.); v. 251.
QP552.P4H73 2004
572'.6536--dc22

2003017464

Preface

The introduction of high-performance liquid chromatography (HPLC) to the analysis of peptides and proteins some 25 years ago revolutionized the biological sciences by enabling the rapid and sensitive analysis of peptide and protein structure through the exquisite speed, sensitivity, and resolution that can be easily obtained. Today, HPLC in its various modes has become the pivotal technique in the characterization of peptides and proteins and currently plays a critical role in both our understanding of biological processes and in the development of peptide- and protein-based pharmaceuticals.

The number of applications of HPLC in peptide and protein purification continues to expand at an extremely rapid rate. Solid-phase peptide synthesis and recombinant DNA techniques have allowed the production of large quantities of peptides and proteins that need to be highly purified. HPLC techniques are also used extensively in the isolation and characterization of novel proteins that will become increasingly important in the postgenomic age. The design of multidimensional purification schemes to achieve high levels of product purity further demonstrates the power of HPLC techniques not only in the characterization of cellular events, but also in the production of peptide- and protein-based therapeutics. HPLC continues to be at the heart of the analytical techniques with which scientists in both academia and in industry must arm themselves to be able to fully characterize the identity, purity, and potency of peptides and proteins.

The aim of *HPLC of Peptides and Proteins: Methods and Protocols* is to provide the beginner with a sufficiency of the practical information needed to develop separation and analytical protocols for peptide and protein analysis. This volume opens with an overview of the basic theory and general methodology of HPLC, with particular reference to the key separation parameters that can be manipulated to achieve high resolution. Each of the commonly used HPLC techniques are covered in Chapters 2–9, whereas methods for capillary to large-scale preparative isolation are described in Chapters 10–15. Chapters 16–27 provide those already experienced in HPLC with a number of specific applications, as in case studies to illustrate the analytical approaches to a particular separation or assay challenge, with examples drawn from contemporary fields in biochemistry and biotechnology. These applications include proteolytic mapping, posttranslational modifications, neuropeptide processing, glycopeptides and glycoproteins,

MHC-binding peptides, toxins/venoms, membrane proteins, antibodies, combinatorial and proteome analysis, and enzymatic activity.

HPLC of Peptides and Proteins: Methods and Protocols will be a valuable resource for a wide range of scientists, including biochemists, molecular biologists, pharmacologists, and microbiologists, who work with peptides and/or proteins in both academic and biotechnology laboratories.

Finally, I would like to thank all of the authors for their enthusiastic participation and excellent contributions.

Marie-Isabel Aguilar

Contents

Contributors

JOAQUIN ABIAN • *Structural and Biological Mass Spectrometry Unit, Department of Medical Bioanalysis, IIBB-CSIC, IDIBAPS, Barcelona, Spain*
MARIE-ISABEL AGUILAR • *Department of Biochemistry and Molecular Biology, Monash University, Clayton, Victoria, Australia*
PAUL ALEWOOD • *Institute for Molecular Bioscience, University of Queensland, Brisbane, Australia*
JEAN-LOUIS AUBAGNAC • *CNRS-UMR 5810, Laboratoire des Aminoacides, Peptides et Protéines, Universités Montpellier I et II, UM II, Montpellier, France*
KÁLMÁN BENEDEK • *Director of Analytical Technologies, iGORI, Thousand Oaks, CA*
JOSEPH BERTOLINI • *CSL Bioplasma, Broadmeadows, Victoria, Australia*
ISAAC BLANCO • *Neurosciences Institute, Department of Biochemistry and Molecular Biology, School of Medicine, Universitat Autònoma de Barcelona, Bellaterra, Spain*
MALCOLM BUCKLE • *Enzymologie et Cinétique Structurale, LBPA (UMR 8113 du CNRS), Ecole Normale Supérieur de Cachan, Cachan, France*
MONTSERRAT CARRASCAL • *Structural and Biological Mass Spectrometry Unit, Department of Medical Bioanalysis, IIBB-CSIC, IDIBAPS, Barcelona, Spain*
CHRISTINE ENJALBAL • *CNRS-UMR 5810, Laboratoire des Aminoacides, Peptides et Protéines, Universités Montpellier I et II, UM II, Montpellier, France*
JAMES FINLAYSON • *Senior Chromatographer, Auspep Pty Ltd., West Melbourne, Victoria, Australia*
STUART R. GALLANT • *Process Development Department, Cell Genesys Inc., San Francisco, CA*
PETER GOMME • *CSL Bioplasma, Broadmeadows, Victoria, Australia*
JORDI GÓMEZ-RAMIREZ • *Neurosciences Institute, Department of Biochemistry et Molecular Biology, School of Medicine, Universitat Autònoma de Barcelona, Bellaterra, Spain*
TAKAO HAYAKAWA • *Division of Biological Chemistry and Biologicals, National Institute of Health Sciences, Tokyo, Japan*

WILFRIED HELLIGER • *Institute of Medical Chemistry and Biochemistry, University of Innsbrück, Innsbrück, Austria*
PETER HØJRUP • *Department of Biochemistry and Molecular Biology, University of Southern Denmark, Odense, Denmark*
SATSUKI ITOH • *Division of Biological Chemistry and Biologicals, National Institute of Health Sciences, Tokyo, Japan*
NANA KAWASAKI • *Division of Biological Chemistry and Biologicals, National Institute of Health Sciences, Tokyo, Japan*
MARTIN R. LARSEN • *Protein Research Group, Department of Molecular Biology and Biochemistry, University of Southern Denmark, Odense, Denmark*
REBECCA A. LEW • *Peptide Biology Laboratory, Baker Heart Research Institute, Melbourne, Victoria, Australia*
HERBERT LINDNER • *Institute of Medical Chemistry and Biochemistry, University of Innsbrück, Innsbrück, Austria*
JEAN MARTINEZ • *CNRS-UMR 5810, Laboratoire des Aminoacides, Peptides et Protéines, Universités Montpellier I et II, UM I, Montpellier, France*
EDOUARD C. NICE • *Ludwig Institute for Cancer Research, Melbourne, Victoria, Australia*
KATHLEEN R. NOON • *Department of Biological Chemistry, University of Michigan, Ann Arbor, MI*
MIYAKO OHTA • *Division of Biological Chemistry and Biologicals, National Institute of Health Sciences, Tokyo, Japan*
JORDI ORTIZ • *Neurosciences Institute, Department of Biochemistry and Molecular Biology, School of Medicine, Universitat Autònoma de Barcelona, Bellaterra, Spain*
MATTHEW J. POWELL • *Department of Chemistry, West Virginia University, Morgantown WV*
MAREE S. POWELL • *The Helen M. Schutt Trust Laboratory, The Austin Research Institute, The Austin And Repatriation Medical Centre, Heidelberg, Victoria, Australia*
ANTHONY W. PURCELL • *Department of Immunology and Microbiology, University of Melbourne, Parkville, Victoria, Australia*
PIERRE SANCHEZ • *CNRS-UMR 5810, Laboratoire des Aminoacides, Peptides et Protéines, Universités Montpellier I et II, UM II, Montpellier, France*
DENIS B. SCANLON • *R&D Manager, Auspep Pty Ltd, West Melbourne, Victoria, Australia*
TUN-LI SHEN • *Department of Chemistry, Brown University, Providence, RI*
PETER STANTON • *Prince Henry's Institute of Medical Research, Clayton Victoria, Australia*

Gilles Subra • *CNRS-UMR 5810, Laboratoire des Aminoacides, Peptides et Protéines, Universités Montpellier I et II, UM I, Montpellier, France*
Patrick Thomas • *CSL Bioplasma, Broadmeadows, Victoria, Australia*
Aaron T. Timperman • *Department of Chemistry, West Virginia University, Morgantown, WV*
Julian P. Whitelegge • *The Pasarow Mass Spectrometry Laboratory, Departments of Psychiatry and Biobehavioral Sciences and Biochemistry and The Neuropsychiatric Institute, University of California, Los Angeles, CA*
David Wilson • *Xenome Ltd., Indooroopilly, Queensland, Australia*
Bruce D. Wines • *The Helen M. Schutt Trust Laboratory, The Austin Research Institute, The Austin and Repatriation Medical Centre, Heidelberg, Victoria, Australia*
Michael Zachariou • *School of Engineering and Science, Swinburne University, Hawthorn, Victoria, Australia*

I

Modes of HPLC

1

HPLC of Peptides and Proteins

Basic Theory and Methodology

Marie-Isabel Aguilar

1. Introduction

High-performance liquid chromatography (HPLC) is now firmly established as the premier technique for the analysis and purification of a wide range of molecules. In particular, HPLC in its various modes has become the central technique in the characterization of peptides and proteins and has, therefore, played a critical role in the rapid advances in the biological and biomedical sciences over the last 10 years.

The enormous success of HPLC can be attributed to a number of inherent features associated with reproducibility, ease of selectivity manipulation, and generally high recoveries. The most significant feature is the excellent resolution that can be achieved under a wide range of conditions for very closely related molecules, as well as structurally quite distinct molecules. This arises from the fact that all interactive modes of chromatography are based on recognition forces that can be subtly manipulated through changes in the elution conditions that are specific for the particular mode of chromatography. Peptides and proteins interact with the chromatographic surface in an orientation-specific manner, in which their retention time is determined by the molecular composition of specific contact regions. For larger polypeptides and proteins that adopt a significant degree of secondary and tertiary structure, the chromatographic contact region comprises a small proportion of the total molecular surface. Hence, the unique orientation of a peptide or protein at a particular stationary phase surface forms the basis of the exquisite selectivity that can be achieved with HPLC techniques. All biological processes depend on specific

From: *Methods in Molecular Biology, vol. 251, HPLC of Peptides and Proteins: Methods and Protocols*
Edited by: M.-I. Aguilar © Humana Press Inc., Totowa, NJ

interactions between molecules and affinity chromatography exploits these specific interactions to allow the purification of a biomolecule on the basis of its biological function or individual chemical structure. In contrast reversed phase HPLC, ion-exchange and hydrophobic interaction chromatography separate peptides and proteins on the basis of differences in surface hydrophobicity or surface charge. These techniques therefore allow the separation of complex mixtures whereas affinity chromatography normally results in the purification of one or a small number of closely related components of a mixture.

Reversed-phase chromatography (RPC) is arguably the most commonly used mode of separation for peptides, although ion-exchange (IEC) and size exclusion (SEC) chromatography also find application. The three-dimensional structure of proteins can be sensitive to the often harsh conditions employed in RPC, and as a consequence, RPC is employed less for the isolation of proteins where it is important to recover the protein in a biologically active form. IEC, SEC, and affinity chromatography are therefore the most commonly used modes for proteins, but RPC and hydrophobic interaction (HIC) chromatography are also employed.

HPLC is extremely versatile for the isolation of peptides and proteins from a wide variety of synthetic or biological sources. The number of applications of HPLC in peptide and protein purification continue to expand at an extremely rapid rate. Solid-phase peptide synthesis and recombinant DNA techniques have allowed the production of large quantities of peptides and proteins which need to be highly purified. The design of multidimensional purification schemes to achieve high levels of product purity further highlight the power of HPLC techniques in the analysis and isolation of peptide and proteins samples. The complexity of the mixture to be chromatographed depends on the nature of the source and the degree of preliminary clean-up that can be performed. In the case of synthetic peptides, RPC is generally employed both for the initial analysis and the final large scale purification. The isolation of proteins from a biological cocktail however, often requires a combination of techniques to produce a homogenous sample. HPLC techniques are then introduced at the later stages following initial precipitation, clarification and preliminary separations using soft gel. Purification protocols therefore need to be tailored to the specific target molecule. The key factor that underpins the development of a successful separation protocol is the ability to manipulate the retention of the target molecule so that it can be resolved from other contaminating components. This chapter thus provides an outline of the general theory of chromatography and the factors that control both the retention time and peakwidth of solutes undergoing separation in terms of the parameters that control resolution. This information can then be used to understand the approaches used to perform separations with specific modes of chromatography as outlined in the remaining chapters in this book.

2. The Molecular Basis of Separation

The separation of a mixture of peptides and proteins in interactive modes of chromatography arises from the differential adsorption of each solute according to their respective affinity for the immobilized stationary phase. Thus, when a particular molecule has a very high affinity for a specific stationary phase, i.e., when the equilibrium distribution coefficient K is high, then that solute is retained to a greater extent than another molecule with a lower affinity for the stationary phase. The degree and nature of the binding affinity is clearly dependent on the structure of the solute and the immobilized ligands. For example, in the case of RPC and HIC, binding is mediated predominantly through hydrophobic interactions between the solute and the immobilized *n*-alkyl ligands. In IEC, the binding is through electrostatic interactions, whereas in different modes of affinity chromatography, binding involves a mixture of hydrophobic, electrostatic, and polar forces. In the case of size exclusion chromatography, the differential movement along the column is a result of the extent to which each solute can permeate the porous structure of the stationary phase.

An additional factor that influences the appearance and relative separation of a peak is the degree of bandbroadening of the solute band during migration through the column. Thus, as it moves down the column, the solute band broadens as a consequence of a number of factors including longitudinal diffusion, brownian motion, eddy diffusion, and mobile phase and stagnant phase mass transfer. These effects result in bandbroadening that generally increases with increasing residence time in the column. The resulting degree of separation or selectivity between constituent solutes in a mixture is thus a subtle interplay between the relative affinity of the molecules for the stationary phase and the degree of diffusive processes that occur during separation.

3. Retention and Bandwidth Relationships

The time taken for a solute to pass though a chromatographic column is referred to as the retention time t_r. This retention time is measured as the time taken by the solute, following injection, to emerge from the column and to be detected as illustrated in **Fig. 1**. In order to allow retention times to be compared to different columns or under different conditions, the retention time of a solute is normally compared with the retention time of a molecule which is not retained on the specific column of interest. This allows the unitless capacity factor k' of a solute to be expressed in terms of the retention time t_r, through the relationship

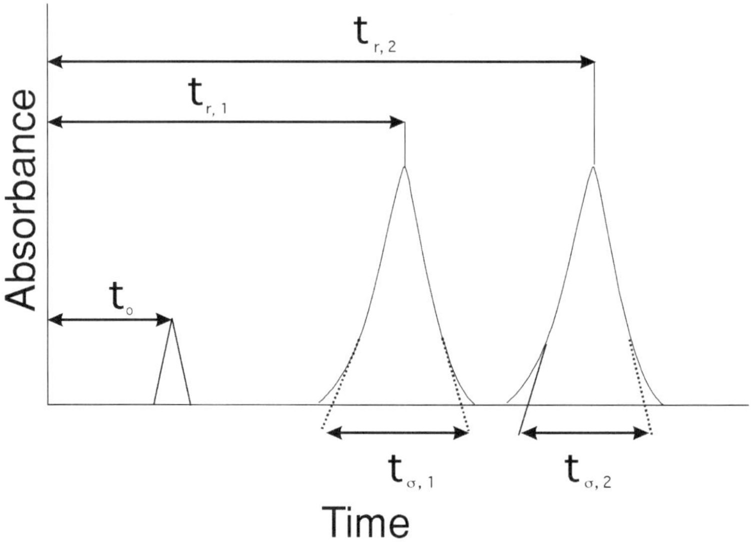

Fig. 1. Diagram of the retention parameters that describe a chromatographic separation. The retention time of a nonretained solute is denoted by t_0, while the retention times of two retained solutes, 1 and 2, are given by $t_{r,1}$ and $t_{r,2}$. The corresponding peakwidths for solutes 1 and 2 are denoted σ,1 and σ,2, and together with the retention times are they used to detrmine the resolution of the separation according to **Eq. 9.**

$$k' = (t_r - t_o) / t_o \tag{1}$$

where t_o is the retention time of a nonretained solute. The capacity factor k' can also be defined as the ratio n_s/n_m where n_s and n_m are the number of moles of solute in the stationary phase and mobile phase respectively as follows:

$$k' = n_s / n_m \tag{2}$$

or alternatively as

$$k' = [X]_s V_s / [X]_m V_m \tag{3}$$

where $[X]_s$ and $[X]_m$ refer to the concentrations of the solute in the stationary and mobile phases, respectively, and V_s and V_m are the corresponding volumes of the stationary and mobile phases. Since the ratio $[X]_s / [X]_m$ is the equilibrium distribution coefficient K and the ratio V_s / V_m defines the phase ratio Φ of the chromatographic system, the capacity factor can also be expressed as follows:

$$k' = \Phi[X]_s / [X]_m \tag{4}$$

or

$$k' = \Phi K \tag{5}$$

Equation 5 thus formerly describes the direct thermodynamic relationship between the retention of a peptide or protein and its affinity for the stationary phase material.

The practical significance of k' can be related to the selectivity parameter α, defined as the ratio of the capacity factors of two adjacent peaks as follows:

$$\alpha = k'_i / k'_j \tag{6}$$

which allows the definition of a chromatographic elution window in which retention times can be manipulated to maximise the separation of components within a mixture. Clearly, the aim is to obtain as high a value of α as possible, which reflects a high degree of separation between two peaks. The second factor involved in defining the quality of a separation is the peak width σ_t. The degree of peak broadening is directly related to the efficiency of the column and can be expressed in terms of the number of theoretical plates N as follows:

$$N = (t_r)^2 / \sigma_r^2. \tag{7}$$

N can also be expressed in terms of the reduced plate height equivalent h, the column length L, and the particle diameter of the stationary phase material d_p, as

$$N = hL / d_p. \tag{8}$$

The resolution R_s between two components of a mixture, therefore, depends on both selectivity and bandwidth according to

$$R_s = 1 / 4 \sqrt{N} (\alpha - 1)[1 / (1 + k')]. \tag{9}$$

This equation describes the relationship between the quality of a separation and the relative retention, selectivity, and the bandwidth. It also provides the formal basis upon which resolution can be manipulated to achieve a particular level of separation. Thus, when faced with an unsatisfactory separation, the aim is to improve resolution by one of three possible strategies. The first is to increase α as previously and the second, but related, approach is to vary k' within a defined range normally $1 < k' < 10$ through variation in the experimental elution conditions such as solvent strength, separation time, or nature of the immobilized ligand. Third, one can increase N, for example, by using very small particles in microbore or narrow bore columns.

An appreciation of the factors that control the resolution of peptides and proteins in interactive modes of chromatography can assist in the development and manipulation of separation protocols to obtain the desired separation. The optimization of high-resolution separations of peptides and proteins involves the separation of sample components through manipulation of both retention times and solute peak shape. For example, inspection of the schematic separation shown in **Fig. 1** demonstrates baseline separation between the two components which corresponds to a high value of both selectivity α, and resolution. A scenario can be envisaged where it may be desirable to decrease the retention times of the solutes to allow more rapid analysis times. However, resolution may be sacrificed and the final separation conditions are often likely to be a tradeoff between rate of analysis and quality of separation.

An enormous range of different separation techniques are available for peptide and protein analysis. The challenge facing the scientist who wishes to analyze and/or purify their peptide or protein sample is the selection of the initial separation conditions and subsequent optimisation of the appropriate experimental parameters. The following chapters thus provide a practical guide to performing peptide and protein analyses under a range of different separation modes. In addition, the reader is guided through the experimental options available to achieve a high-resolution separation of a peptide or protein mixture, an exercise which is underpinned by the theoretical relationships provided in this chapter.

2

Reversed-Phase High-Performance Liquid Chromatography

Marie-Isabel Aguilar

1. Introduction

Reversed-phase high-performance liquid chromatography (RP-HPLC) involves the separation of molecules on the basis of hydrophobicity. The separation depends on the hydrophobic binding of the solute molecule from the mobile phase to the immobilized hydrophobic ligands attached to the stationary phase, i.e., the sorbent. A schematic diagram showing the binding of a peptide or a protein to a reversed-phase surface is shown in **Fig. 1**. The solute mixture is initially applied to the sorbent in the presence of aqueous buffers, and the solutes are eluted by the addition of organic solvent to the mobile phase. Elution can proceed either by isocratic conditions where the concentration of organic solvent is constant, or by gradient elution whereby the amount of organic solvent is increased over a period of time. The solutes are, therefore, eluted in order of increasing molecular hydrophobicity. RP-HPLC is a very powerful technique for the analysis of peptides and proteins because of a number of factors that include: (1) the excellent resolution that can be achieved under a wide range of chromatographic conditions for very closely related molecules as well as structurally quite distinct molecules; (2) the experimental ease with which chromatographic selectivity can be manipulated through changes in mobile phase characteristics; (3) the generally high recoveries and, hence, high productivity; and (4) the excellent reproducibility of repetitive separations carried out over a long period of time, which is caused partly by the stability of the sorbent materials under a wide range of mobile phase conditions *(1,2)*. However, RP-HPLC can cause the irreversible denaturation of protein samples thereby reducing the potential recovery of material in a biologically active form.

From: *Methods in Molecular Biology, vol. 251, HPLC of Peptides and Proteins: Methods and Protocols*
Edited by: M.-I. Aguilar © Humana Press Inc., Totowa, NJ

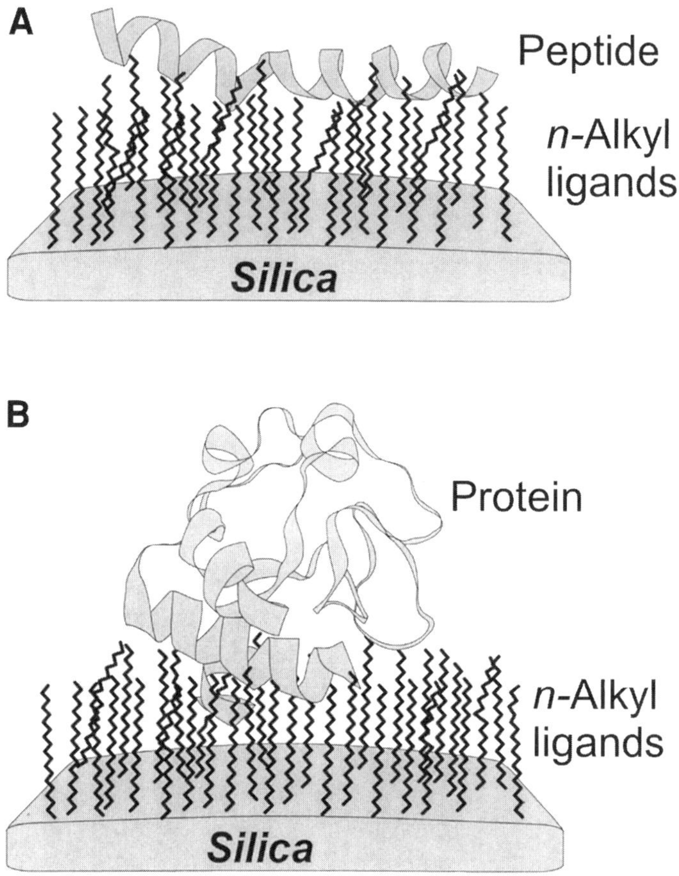

Fig. 1. Schematic representation of the binding of **(A)** a peptide and **(B)** a protein, to an RP-HPLC silica-based sorbent. The peptide or protein interacts with the immobilized hydrophobic ligands through the hydrophobic chromatographic contact region.

The RP-HPLC experimental system for the analysis of peptides and proteins usually consists of an *n*-alkylsilica-based sorbent from which the solutes are eluted with gradients of increasing concentrations of organic solvent such as acetonitrile containing an ionic modifier such as trifluoroacetic acid (TFA) *(1,2)*. Complex mixtures of peptides and proteins can be routinely separated and low picomolar—femtomolar amounts of material can be collected for further characterization. Separations can be easily manipulated by changing the gradient slope, the operating temperature, the ionic modifier, or the organic solvent composition.

The extensive use of RP-HPLC for the purification of peptides, small polypeptides with molecular weights up to 10,000, and related compounds of pharmaceutical interest has not been replicated to the same extent for larger polypeptides (molecular mass > 10 KDa) and globular proteins. The combination of the traditionally used acidic buffering systems and the hydrophobicity of the *n*-alkylsilica supports which can result in low mass yields or the loss of biological activity of larger polypeptides and proteins have often discouraged practitioners from using RP-HPLC methods for large-scale protein separations. The loss of enzymatic activity, the formation of multiple peaks for compositionally pure samples, and poor yields of protein can all be attributed to the denaturation of protein solutes during the separation process using RP-HPLC *(3–6)*.

RP-HPLC is extremely versatile for the isolation of peptides and proteins from a wide variety of synthetic or biological sources and is used for both analytical and preparative applications (***1–2***, *see also* Chs. 10–21). Analytical applications range from the assessment of purity of peptides following solid-phase peptide synthesis (*see* Ch. 14), to the analysis of tryptic maps of proteins. Preparative RP-HPLC is also used for the micropurification of protein fragments for sequencing to large-scale purification of synthetic peptides and recombinant proteins. The complexity of the mixture to be chromatographed will depend on the nature of the source and the degree of preliminary clean-up that can be performed. In the case of synthetic peptides, RP-HPLC is generally employed both for the initial analysis and the final large-scale purification. The purification of synthetic peptides usually involves an initial separation on an analytical scale to assess the complexity of the mixture followed by large-scale purification and collection of the target product. A sample of the purified material can then be subjected to RP-HPLC analysis under the same or different elution conditions to check for purity. The isolation of proteins from a biological cocktail derived from a tissue extract or biological fluid for example, often requires a combination of techniques to produce a homogenous sample. HPLC techniques are then introduced at the later stages following initial precipitation, clarification, and preliminary separations using soft gels.

The challenge facing the scientist who wishes to analyze and/or purify their peptide or protein sample by RP-HPLC is the selection of the initial separation conditions and subsequent optimization of the appropriate experimental parameters. This chapter describes a standard method that can be used as an initial procedure for the RP-HPLC analysis of a peptide sample, and then different experimental options available to achieve a high-resolution separation of a peptide or protein mixture using RP-HPLC are outlined in **Subheading 4.**

2. Materials

2.1. Chemicals

1. Acetonitrile (CH_3CN), HPLC grade.
2. Milli-Q water.
3. Trifluoroacetic acid (TFA).

2.2. Equipment and Supplies

1. HPLC solvent delivery system with binary gradient capability and a UV detector.
2. Reversed-phase octadecylsilica (C18) column (*see* **Note 1**) (4.6 mm id (internal diameter) × 250 mm length (*see* **Note 2**), 5 μm particle size, 300 Å pore size (*see* **Note 3**).
3. C18 guard column.
4. Solvent filtration apparatus equipped with a 0.22-μm Teflon filter.
5. Sample filters, 0.22 μm porosity.
6. Buffer A: 0.1% (v/v) TFA in water (*see* **Note 4**).
7. Buffer B: 100% CH_3CN containing 0.1% (v/v) TFA (*see* **Note 5**).

3. Methods

3.1. Sample Preparation

Dissolve 1 mg of sample in 1 mL of Buffer A. If there is some undissolved material, filter the sample through a 0.22-μm filter.

3.2. Solvent Preparation

Filter all solvents through a 0.22-μm filter before use. This removes particulates that could block solvent lines or the column and also serves to degass the solvent. If the HPLC instrument is not installed with on-line degassing capability, check with your instrument requirements to assess whether further degassing is required.

3.3. Column Equilibration and Blank Run

1. Connect the guard and the column to the solvent delivery system according to the HPLC system requirements and equilibrate under the following initial conditions.
   a. Solvent: 100% Buffer A
   b. Flow rate: 1 mL/min (*see* **Note 6**)
   c. Detection wavelength: 215 nm (*see* **Note 7**)
   D. Temperature: Ambient (*see* **Note 8**)
2. Once a stable baseline is obtained, inject 10 μL of Milli-Q water (either manually or via an automatic injector). It is generally advisable to perform two blank runs to ensure proper equilibration of the column.

3.4. Sample Injection and Analysis

Once a stable baseline is obtained, inject 10 μL of the sample (either manually or via an automatic injector) and use a linear gradient from 0 to 100% buffer B over 30 min to elute the sample (*see* **Note 9**).

Figure 2 shows a typical chromatogram of a crude synthetic peptide. The large majority of components should normally elute within the gradient time. Thus, each individual method is relatively straightforward to perform. The

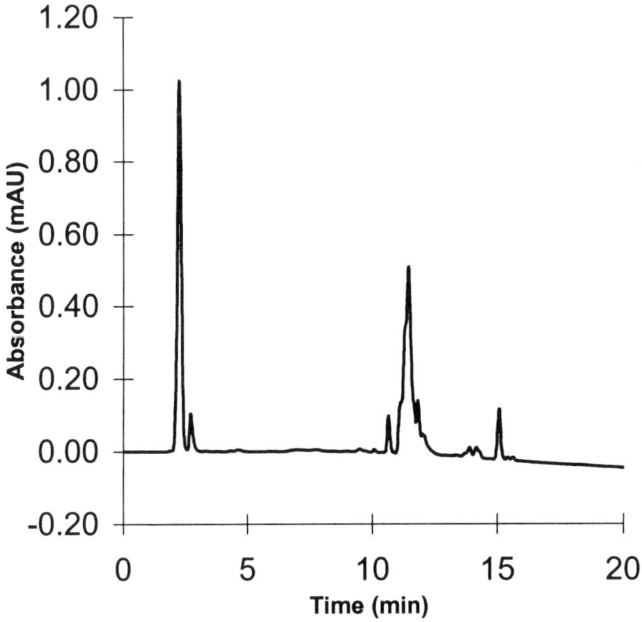

Fig. 2. RP-HPLC elution profile illustrating the purification of a synthetic peptide. Analytical profile (1 mg) of a crude peptide mixture from solid phase peptide synthesis. Column: Zorbax 300 RP-C18, 25 cm × 4.6 mm id, 5-µm particle size, 30 nm pore size. Conditions, linear gradient from 0–60% acetonitrile with 0.1%TFA over 30 min, flow rate of 1 mL/min, 25°C.

scope lies in the wide range of operating parameters that can be changed in order to manipulate the resolution of peptide and protein mixtures in RP-HPLC. These parameters include the immobilized ligand (*see* **Note 1**), the column packing geometry (*see* **Note 3**), the column dimensions (*see* **Note 2**), the ionic additive (*see* **Note 4**), the organic solvent (*see* **Note 5**), the mobile phase flow rate (*see* **Note 6**), the gradient time and gradient shape (*see* **Note 9**), and the operating temperature (*see* **Note 8**).

4. Notes

1. The most commonly employed experimental procedure for the RP-HPLC analysis of peptides and proteins generally involves the use of a C18-based sorbent and a mobile phase. The chromatographic packing materials that are generally used are based on microparticulate porous silica which allows the use of high linear flow velocities resulting in favorable mass transfer properties and rapid analysis times *(7,8)*. The silica is chemically modified by a derivatized silane bearing an *n*-alkyl hydrophobic ligand. The most commonly used ligand is C18, whereas

n-butyl (C4) and *n*-octyl (C8) also find important application and phenyl and cyanopropyl ligands can provide different selectivity ***(9)***. The process of chemical immobilization of the silica surface results in approx half of the surface silanol group being modified. The sorbents are, therefore, generally subjected to further silanization with a small reactive silane to produce an end-capped packing material. The type of *n*-alkyl ligand significantly influences the retention of peptides and proteins and can therefore be used to manipulate the selectivity of peptide and protein separations. Although the detailed molecular basis of the effect of ligand structure is not fully understood, a number of factors including the relative hydrophobicity and ligand chain length, flexibility, and the degree of exposure of surface silanols all play a role in the retention process. An example of the effect of chain length on peptide separations can be seen in **Fig. 3** ***(1)***. It can be seen that the peaks labeled T_3 and T_{13} are fully resolved on the C4 packing but cannot be separated on the C18 material. In contrast, the peptides T_5 and T_{18} are unresolved on the C4 column but fully resolved on the C18 material. In addition to effects on peptide selectivity, the choice of ligand type can also influence protein recovery and conformational integrity of protein samples. Generally higher protein recoveries are obtained with the shorter and less hydrophobic *n*-butyl ligands. However, proteins have also been obtained in high yield with *n*-octadecyl silica ***(10–12)***. Silica-based packings are also susceptible to cleavage at pH values greater than 7. This limitation can severely restrict the utility of these materials for separations which require basic pH conditions to effect resolution. In these cases, alternative stationary phases have been developed including cross-linked polystyrene divinylbenzene ***(13,14)*** and porous zirconia ***(15,16)***, which are all stable to hydrolysis at alkaline pHs.

2. The desired level of efficiency and sample loading size determines the dimension of the column to be used. For small peptides and proteins, increased resolution will be obtained with increases in column length. Thus, for applications such as tryptic mapping, column lengths between 15–25 cm and id of 4.6 mm are generally employed. However, for larger proteins, low mass recovery and loss of biological activity may result with these columns as a result of irreversible binding and/or denaturation. In these cases, shorter columns of between 2 and 20 cm in length can be used. For preparative applications in the 1–500 mg scale, such as the purification of synthetic peptides, so-called semipreparative columns of dimensions 30 cm × 1 cm id and preparative columns of 30 cm × 2 cm id can be used.

   The selection of the internal diameter of the column is based on the sample capacity and detection sensitivity. Whereas most analytical applications are carried out with columns of internal diameter of 4.6 mm id, for samples derived from previously unknown proteins where there is a limited supply of material, the task is to maximize the detection sensitivity. In these cases, the use of narrow bore columns of 1 or 2 mm id can be used that allow the elution and recovery of samples in much smaller volumes of solvent (*see* Chapter 11). Capillary chromatography is also finding increasing application where capillary columns of internal

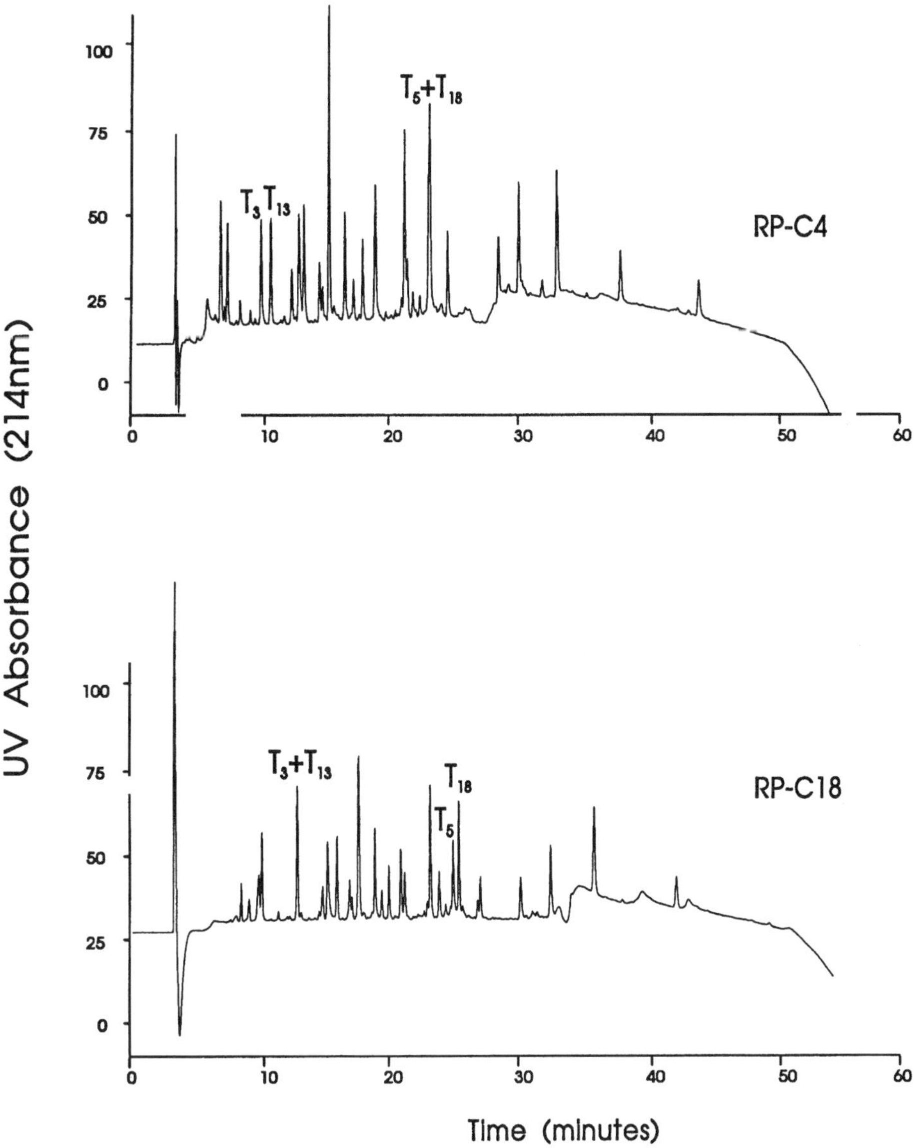

Fig. 3. The influence of *n*-alkyl chain length on the separation of tryptic peptides derived from porcine growth hormone. **Top:** Bakerbond (J. T. Baker, Phillipsburg, NJ) RP-C4, 25 cm × 4.6 mm id, 5 µm particle size, 30 nm pore size. **Bottom:** Bakerbond (J. T. Baker) RP-C18, 25 cm × 4.6 mm id, 5 µm particle size, 30 nm pore size. Conditions, linear gradient from 0–90% acetonitrile with 0.1%TFA over 60 min, flow rate of 1 mL/min, 25°C (Reproduced from **ref. *1*** by permission of Academic).

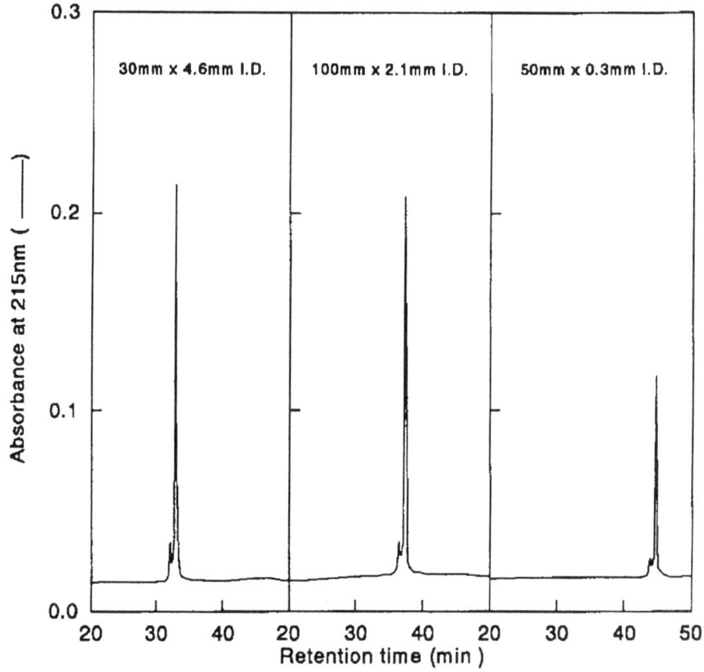

Fig. 4. Effect of column internal diameter on detector sensitivity. Column: Brownlee RP-300 C8 (7 μm particle size, 30 nm pore size), 3 cm × 4.6 mm id and 10 cm × 2.1 mm id (Applied Biosystems) and 5 cm × 0.32 mm id. Conditions: linear gradient from 0–60% acetonitrile with 0.1% TFA over 60 min, 45°C. Flow rates, 1 mL/min, 200 μLl/min, and 4 μL/min for the 4.6, 2.1, and 0.32 mm id columns, respectively. Sample loadings, lysozyme, 10 μg, 4 μg, and 0.04 μg for the 4.6, 2.1, and 0.32 mm id columns, respectively. (Reproduced from **ref.** ***17*** by permission of Elsevier Science, copyright 1992.)

diameter between 0.2–0.4 mm and column length of 15 cm result in the analysis of femtomole of sample (*see* Chapter 10). The effect of decreasing column internal diameter on detection sensitivity is shown in **Fig. 4** for the analysis of lysozyme on a C18 material packed into columns of 4.6, 2.1, and 0.3 mm id ***(17)***.

3. The geometry of the particle in terms of the particle diameter and pore size, is also an important feature of the packing material. Improved resolution can be achieved by decreasing the particle diameter and the most commonly used range of particle diameters for analytical scale RP-HPLC is 3–5 μm. There are also examples of the use of nonporous particles of smaller diameter ***(18)***. For preparative scale separations, 10–20 μm particles are utilized. The pore size of RP-HPLC sorbents is also an important factor that must be considered. For peptides, the pore size generally ranges between 100–300 Å depending on the size of the peptides. Porous materials of ≥300 Å pore size are necessary for the separation of proteins, as the

solute molecular diameter must be at least one-tenth the size of the pore diameter to avoid restricted diffusion of the solute and to allow the total surface area of the sorbent material to be accessible. The development of particles with 6000–8000 Å pores with a network of smaller pores of 500–1000 Å has also allowed very rapid peptide and protein separations to be achieved ***(19,20)***.

4. RP-HPLC is generally carried out with an acidic mobile phase, with TFA the most commonly used additive because of its volatility. However, for high sensitivity applications, the amount of TFA in buffer B can be adjusted downward because phosphoric acid, perchloric acid, formic acid, hydrochloric acid, acetic acid, and heptaflourobutyric acid have also been used ***(21–23)***. Alternative additives such as nonionic detergents can be used for the isolation of more hydrophobic proteins such as membrane proteins (***24***, *see* Chapter 22).
5. One of the most powerful characteristics of RP-HPLC is the ability to manipulate solute retention and resolution through changes in the composition of the mobile phase. In RP-HPLC, peptide and protein retention is a result of multisite interactions with the ligands. The practical consequence of this is that high resolution isocratic elution of peptides and proteins can rarely be achieved as the experimental window of solvent concentration required for their elution is very narrow. Mixtures of peptides and proteins are therefore routinely eluted by the application of a gradient of increasing organic solvent concentration. The three most commonly employed organic solvents in RP-HPLC are acetonitrile, methanol, and 2-propanol, which all exhibit high optical transparency in the detection wavelengths used for peptide and protein analysis. Acetonitrile provides the lowest viscosity solvent mixtures and 2-propanol is the strongest eluent. An example of the influence of organic solvent is shown in **Fig. 5** where changes in selectivity can be observed for a number of peptide peaks in the tryptic map. In addition to the eluotropic effects, the nature of the organic solvent can also influence the conformation of both peptides and proteins and will, therefore, have an additional effect on selectivity through changes in the structure of the hydrophobic contact region. In the case of proteins, this may also impact on the level of recovery of biologically active material.
6. The typical experiment with an analytical scale column would utilize flow rates ranging between 0.5–2.0 mL/min. With microbore columns (1–2 mm id) flow rates of 50–250 μL/min are used, whereas for capillary columns of 0.2–0.4 mm id, flow rates of 1–4 μL/min are applied. At the preparative end of the scale with columns of 10–20 mm id, flow rates ranging between 5–20 mL/min are required.
7. Detection of peptides and proteins in RP-HPLC, generally involves detection between 210 and 220 nm, which is specific for the peptide bond, or at 280 nm, which corresponds to the aromatic amino acids tryptophan and tyrosine. The use of photodiode array detectors can enhance the detection capabilities by the on-line accumulation of complete solute spectra. The spectra can then be used to identify peaks specifically on the basis of spectral characteristics and for the assessment of peak purity ***(24–26)***. In addition, second derivative spectroscopy can provide information on the conformational integrity of proteins following elution ***(27,28)***

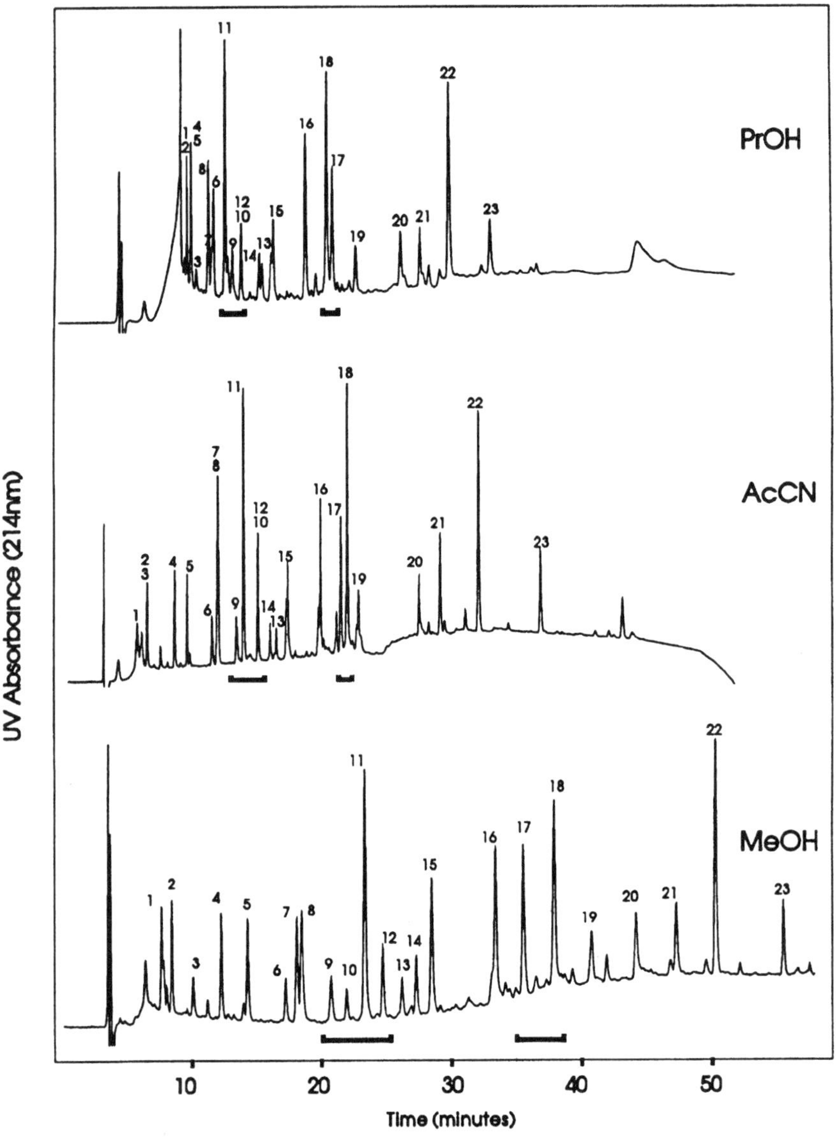

Fig. 5. The influence of organic solvent on the RPC of tryptic peptides derived from porcine growth hormone. Column: Bakerbond (J. T. Baker) RP-C4, 25 cm × 4.6 mm id, 5 µm particle size, 30 nm pore size. Conditions, linear gradient from 0–90% 2-propanol (top), acetonitrile (middle) or methanol (bottom) with 0.1%TFA over 60 min, flow rate of 1 mL/min, 25°C. (Reproduced from **ref. *1*** by permission of Academic.)

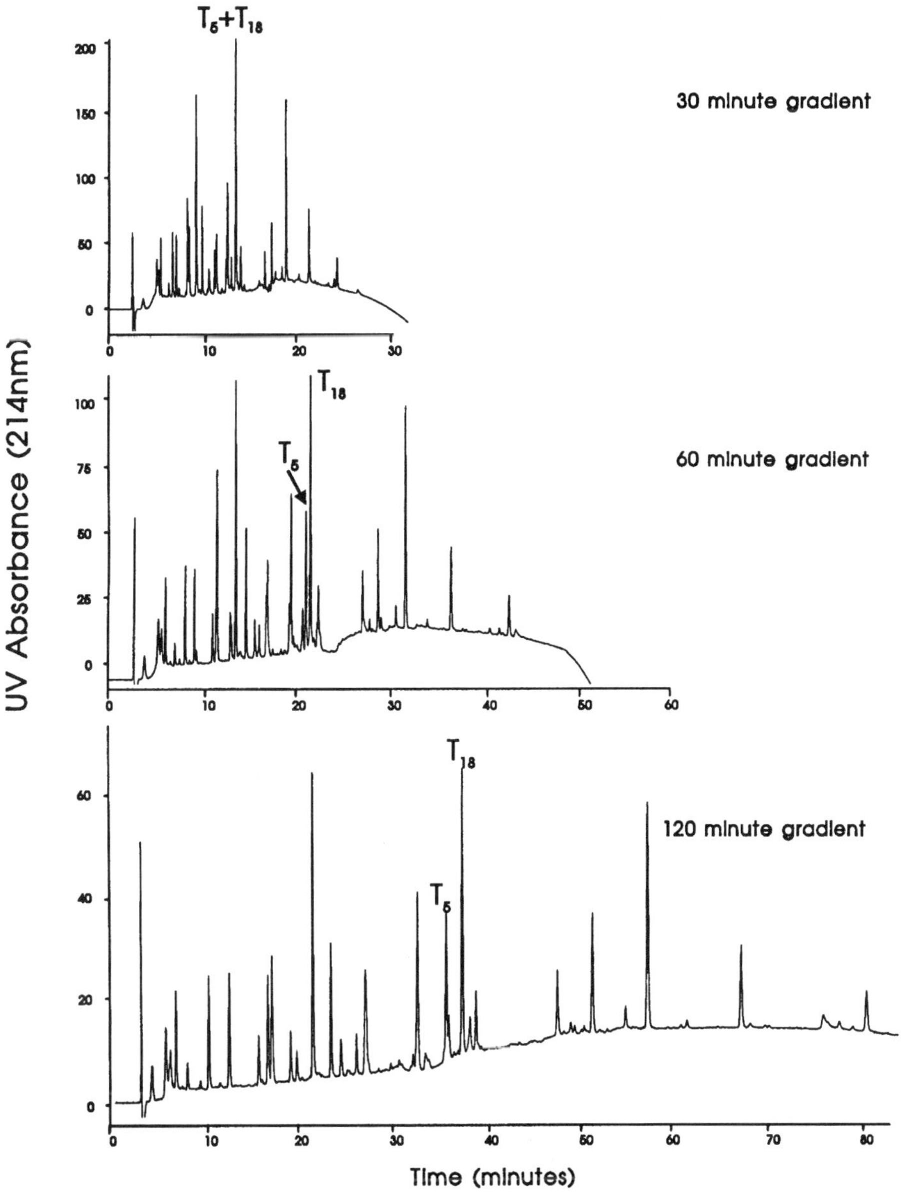

Fig. 6. Effect of gradient time on the reversed phase chromatography of tryptic peptides of porcine growth hormone. Column: Bakerbond (J. T. Baker) RP-C4, 25 cm × 4.6 mm id, 5 µm particle size, 30 nm pore size. Conditions, linear gradient from 0–90% acetonitrile with 0.1%TFA over 30 min (top), 60 min (middle), or 120 min (bottom) at a flow rate of 1 mL/min, 25°C. (Reproduced from **ref.** ***1*** by permission of Academic.)

8. The operating temperature can also be used to manipulate resolution. Although the separation of peptides and proteins is normally carried out at ambient temperature, solute retention in RP-HPLC is influenced by temperature through changes in solvent viscosity. In addition to this, peptide and protein conformation can also be manipulated by temperature. Changes in temperature can therefore also be used to manipulate the structure and retention of peptide mixtures. For peptides, it has been shown that secondary structure can actually be enhanced through binding to the hydrophobic sorbent ***(29)***. In the case of proteins that are to be subjected to further chemical analysis and thus where recovery of a biologically active protein is not essential, increasing temperature can be used to modulate retention via denaturation of the protein structure ***(3–6)***. However, if the efficient recovery of both mass and biological activity is of paramount importance, the use of elevated temperatures is not an option.
9. The choice of gradient conditions will depend on the nature of the molecules of interest. The influence of gradient time on the separation of a series of tryptic peptides proteins is shown in **Fig. 6** *(1)*. Generally the use of longer gradient times provides improved separation. However, these conditions also increase the residence time of the peptide or protein solute at the sorbent surface, which may then result in an increase in the degree of denaturation.

References

1. Aguilar, M. I. and Hearn, M. T. W. (1996) High resolution reversed phase high performance liquid chromatography of peptides and proteins. *Meth. Enzymol.* **270,** 3–26.
2. Mant, C. T. and Hodges, R. S. (1996) Analysis of peptides by high performance liquid chromatography. *Meth. Enzymol.* **271,** 3–50.
3. Purcell, A. W., Aguilar, M. I., and Hearn, M. T. W. (1995) Conformational effects in the RP-HPLC of polypeptides. II: The role of insulin A and B chains in the chromatographic behaviour of insulin. *J. Chromatogr.* **711,** 71–79.
4. Richards, K. L., Aguilar, M. I., and Hearn, M. T. W. (1994) The effect of protein conformation on experimental bandwidths in RP-HPLC. *J. Chromatogr.* **676,** 33–41.
5. Oroszlan, P., Wicar, S., Teshima, G., Wu, S.-L., Hancock, W. S., and Karger, B. L. (1992) Conformational effects in the reversed phase chromatographic behaviour of recombinant human growth hormone (rhGH) and N-methionyl recombinant human growth hormone (met-hGH). *Anal. Chem.* **64,** 1623–1631.
6. Lin S. and Karger B. L. (1990) Reversed phase chromatographic behaviour of proteins in different unfolded states. *J. Chromatogr.* **499,** 89–102.
7. Unger, K. K. (1990) Silica as a support, in *HPLC of Biological Macromolecules: Methods and Applications* (Gooding, K. M. and Regnier, F. E., eds.), Dekker, New York, pp. 3–24.
8. Henry, M. (1991) Design requirements of silica-based matrices for biopolymer chromatography. *J. Chromatogr.* **544,** 413–443.

9. Zhou, N. E., Mant, C. T., Kirkland J. J., and Hodges R. S. (1991) Comparison of silica-based cyanopropyl and octyl reversed phase packings for the separation of peptides and proteins. *J. Chromatogr.* **548,** 179–193.
10. Moy, F. J., Li, Y.-C., Rauenbeuhler, P., Winkler, M. E., Scheraga, H. A., and Montelione, G. T. (1993) Solution structure of human type-α transforming growth factor determined by heteronuclear NMR spectroscopy and refined by energy minimisation with restraints. *Biochemistry* **32,** 7334–7353.
11. Chlenov, M. A., Kandyba, E. I., Nagornaya, L. V., Orlova, I. L., and Volgin, Y. V. (1993) High performance liquid chromatography of human glycoprotein hormones. *J. Chromatogr.* **631,** 261–267.
12. Welinder, B. S., Sorenson, H. H., and Hansen, B. (1986) Reversed-phase high performance liquid chromatography of insulin. Resolution and recovery in relation to column geometry and buffer components. *J. Chromatogr.* **361,** 357–363.
13. Welinder, B. S. (1991) Use of polymeric reversed-phase columns for the characterisation of polypeptides extracted from human pancreata. II. Effect of the stationary phase. *J. Chromatogr.* **542,** 83–99.
14. Amersham Pharmacia Biotech. Source 5RPC ST 4.6/150 High Performance Reversed Phase Chromatgoraphy Data File 18-1132-36, 1999; Amersham Pharmacia Biotech. Purification of Amyloid-β 1-42 at High pH Using a Polymer Stationary Phase Application Note 18-1130-92, 1998.
15. Wirth, H.-J., Eriksson, K.-O., Holt, P., Aguilar, M. I., and Hearn, M. T. W. (1993) Ceramic based particles as chemically stable chromatographic supports. *J. Chromatogr.* **646,** 129–141.
16. Sun, L. and Carr, P. W. (1995) Chromatography of proteins using polybutadiene-coated zirconia. *Anal. Chem.* **67,** 3717–3721.
17. Moritz, R. L. and Simpson, R. J. (1992) Application of capillary reversed phase high performance liquid chromatography to high sensitivity protein sequence analysis. *J. Chromatogr.* **599,** 119–130.
18. Jilge, G., Janzen, R., Unger, K. K., Kinkel, K. N., and Hearn, M. T. W. (1987) Evaluation of advanced silica packings for the separation of biopolymers by high performance liquid chromatography. III. Retention and selectivity of proteins and peptides in gradient elution on non-porous monodisperse 1.5 μm reversed phase silicas. *J. Chromatogr.* **397,** 71–80.
19. Paliwal, S. K., deFrutos, M., and Regnier, F. E. (1996) Rapid separations of proteins by liquid chromatography. *Meth. Enzymol.* **270,** 133–151.
20. Premstaller, A., Oberacher, H., Walcher, W., et al. (2001) High-performance liquid chromatography-electrospray ionization mass spectrometry using monolithic capillary columns for proteomic studies. *Anal. Chem.* **73,** 2390–2396.
21. Young, P. M. and Wheat, T. E. (1990) Optimisation of high performance liquid chromatographic peptide separations with alternative mobile and stationary phases. *J. Chromatogr.* **512,** 273–281.
22. Poll, D. J. and Harding, D. R. K. (1989) Formic acid as a milder alternative to trifluoroacetic acid and phosphoric acid in two dimensional peptide mapping. *J. Chromatogr.* **469,** 231–239.

23. Thevenon, G. and Regnier, F. E. (1989) Reversed-phase liquid chromatography of proteins with strong acids. *J. Chromatogr.* **476,** 499–511.
24. Mant, C. T. and Hodges, R. S. (1991) The effects of anionic ion-pairing reagents on peptide retention in reversed phase chromatography, in *High Performance Liquid Chromatography of Peptides and Proteins: Separation, Analysis and Conformation* (Mant, C. T. and Hodges, R. S., eds.), CRC, Boca Raton, FL, pp. 327–341.
25. Welling, G. W., Van der Zee, R., and Welling-Wester, S. (1987) Column liquid chromatography of integral membrane proteins. *J. Chromatogr.* **418,** 223–243.
26. Frank, J., Braat, A. and Duine, J. A. (1987) Assessment of protein purity by chromatography and multiwavelength detection. *Anal. Biochem.* **162,** 65–73.
27. Nyberg, F., Pernow, C., Moberg, U., and Eriksson, R. B. (1986) High performance liquid chromatography and diode array detection for the identification of peptides containing aromatic amino acids in studies of endorphin-degrading activity in human cerebrospinal fluid. *J. Chromatogr.* **359,** 541–551.
28. Rozing, G. P. (1996) Diode array detection. *Meth. Enzymol.* **270,** 201–234.
29. Lazoura, E., Maidonis, J., Bayer, E., Hearn, M. T. W., and Aguilar, M. I. (1997) The conformational analysis of NPY-[18-36] analogues at hydrophobic surfaces. *Biophys. J.* **72,** 238–246.

3

Ion-Exchange Chromatography

Peter Stanton

1. Introduction

Ion-exchange chromatography (IEX) separates biomolecules (proteins, polypeptides, nucleic acids, polynucleotides, charged carbohydrates, and polysaccharides) based on differences in their charge. IEX can be a highly selective chromatographic technique, being able to resolve, for example, proteins which differ by only a single charged group *(1)*. The process relies upon the formation of ionic bonds between the charged groups on biomolecules (typically, $-NH_3^+$, $=NH_2^+$, $\geq NH^+$, $-COO^-$, PO_4^-, SO_3^{2-}), and an ion-exchange gel/support carrying the opposite charge. Non-bound biomolecules (i.e., neutral molecules which do not carry any electrical charge, or molecules carrying the same charge as the ion-exchange support) are removed by washing, and bound biomolecules are recovered by elution with a buffer of either higher ionic strength, or altered pH. The advantages of IEX are 1) high resolving power, 2) separations can be fast, 3) in general, recoveries are high, 4) buffer components are nondenaturing and frequently compatible with further downstream chromatographic separation or assay systems, 5) process can be used as a concentration step, to recover proteins from a dilute solution. The disadvantages of IEX are few, but include 1) the sample must be applied to the IEX support under conditions of low ionic strength and controlled pH, which sometimes requires an extra buffer exchange step to be inserted, 2) chromatographic instrumentation should be resistant to salt-induced corrosion, and 3) postchromatographic concentration of dilute solutions of recovered proteins can result in high salt concentrations (>1 *M*), unsuitable, for example, in biological assays unless buffer exchange is carried out. Applications for IEX are numerous, from analytical scale column chromatographic separations in the research laboratory

From: *Methods in Molecular Biology, vol. 251, HPLC of Peptides and Proteins: Methods and Protocols*
Edited by: M.-I. Aguilar © Humana Press Inc., Totowa, NJ

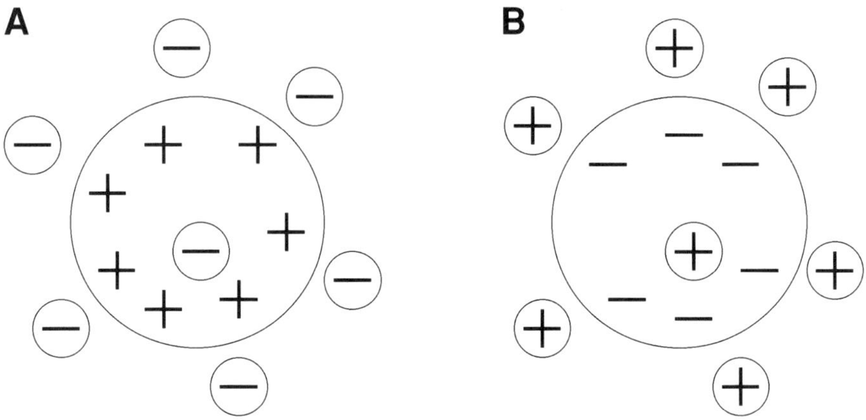

Fig. 1. Types of ion-exchangers. For an anion-exchanger **(A)**, the gel matrix is positively charged, with negatively charged counter-ions (anions) in solution. These anions are reversibly exchanged with other anions in the process of ion-exchange chromatography. For a cation-exchanger **(B)**, the gel matrix is negatively charged, with positively charged counter-ions (cations) in solution.

through to preparative scale column separations at the industrial level. In selected cases, IEX can also be applied successfully in a batch-mode, which has advantages in simplicity and lack of requirement for expensive chromatographic equipment.

There are two basic types of ion-exchangers: 1) anion-exchanger and 2) cation exchanger. The anion exchanger has positively charged groups which have been immobilized onto a chromatographic support, and will therefore bind and exchange negatively charged ions (anions) (*see* **Fig. 1A**), while a cation-exchanger has negatively charged immobilized groups which will bind and exchange positively charged ions (cations) (*see* **Fig. 1B**). In solution, proteins and other biomolecules are ionized, and the extent of ionization is dependent on the pH of the solution. For any given protein, the pH at which the total positive charge is equal to the total negative charge is known as the isoelectric point (p*I*). Hence, when pH = p*I*, the total *net* charge on the protein is zero (*see* **Fig. 2**). At pHs less than the p*I*, the total net charge on the protein will be positive, thus the protein should bind to a cation-exchange column. At pHs greater than the p*I*, the total net charge on the protein will now be negative, and the protein should bind to an anion-exchange column (*see* **Fig. 2**). As a rule of thumb, proteins with p*I* < 6 (i.e., acidic proteins) are chromatographed on an anion-exchange column, while proteins of p*I* > 8 (basic proteins) are chro-

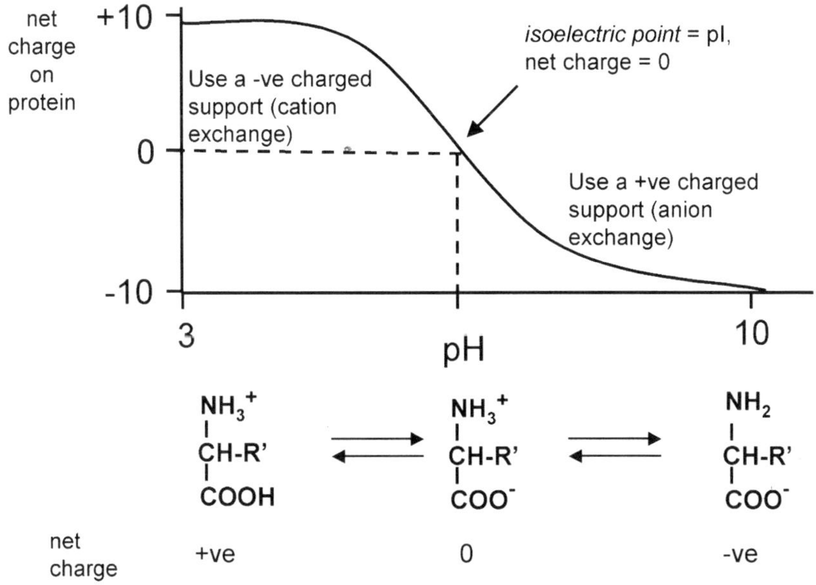

Fig. 2. Generic titration curve for proteins. In solution, charged groups on proteins are ionized, and the extent of ionisation is dependent on solution pH. At a particular pH known as the isoelectric point (p*I*), the total positive charge on the protein is equivalent to the total negative charge, hence the *net* protein charge is zero. At pHs more acidic than the p*I*, the net protein charge is positive as carboxyl and other acidic groups are protonated, and amino groups are ionised. Under these conditions, a negatively charged ion-exchanger (cation exchange) should be used. Conversely, at pHs more basic than the p*I*, the net protein charge is negative, and a positively charged ion-exchanger (anion exchange) should be employed.

matographed on a cation-exchange column, and proteins with p*I* between 6 and 8 can be chromatographed on either type *(5)*. In the lab, if the p*I* of the protein(s) of interest is unknown, it will be necessary to conduct a pilot experiment to determine the best ion exchanger to use (*see* **Subheading 3.2.**). A list of the p*I*s of a series of standard proteins is shown in **Table 1**.

Despite the fact that the net charge on a protein is zero at its p*I*, it is not unusual that some binding to ion-exchangers will occur. This is owing to 1) the nonuniform distribution of charged groups on the surface of the protein and 2) potential for differences between the pH of the microenvironment inside the pores of the ion-exchange support and the pH of the bulk eluent *(3)*. Regardless, as the solubility of many proteins is lowest at or near their p*I* *(4)*, it is suggested to avoid chromatography at this pH value to prevent potential on-column precipitation (*see* **Note 1**).

Table 1
Isoelectric Points and Molecular Weights of Some Standard Proteins[a]

Protein	p*I*	M. Wt
Pepsin	~1	35,500
Ovalbumin	4.6	43,000
Thyroglobulin	4.6	660,000
Albumin (bovine serum)	4.9	68,000
Urease	5.0	480,000
β-Lactoglobulin	5.2	20,000
Insulin	5.3	5,700
Hemoglobin (horse)	6.8	64,000
Myoglobin (horse)	7.0	17,500
Carbonic anhydrase	7.3	29,000
Chymotrypsinogen	9.5	25,700
Cytochrome-*c*	10.7	12,000
Lysozyme (hen egg white)	11.0	14,300

[a]Data from **refs.** ***2*** and ***5***.

The types of charged groups commonly immobilized to chromatographic supports are shown in **Table 2**. These groups can be categorized into *weak* or *strong* ion-exchangers, depending on the pH range over which the exchanger remains charged. Strong ion exchangers, typically containing sulphonic acid groups (cation-exchange) or quaternary amino groups (anion-exchange) (*see* **Table 2**), remain ionized over a wide pH range, whereas weak ion-exchangers are ionized in a narrower pH range. Hence, an advantage for the use of a strong ion-exchanger is that the charge on the exchanger is independent of pH over the pH range 2–12, and therefore, the interaction between the solute and the exchanger follows a simple mechanism ***(1)***. In addition, the sample loading capacity of the matrix is not altered at high or low pH. In contrast, the sample loading capacity of a weak ion-exchanger varies considerably with pH, therefore both the charge on the support and the amount of sample which can be loaded can be less predictable ***(3)***. It is important to realize that the terms "weak" and "strong" *do not* refer to the strength of attraction between the charged exchanger and the solute/molecule of interest.

It is beyond the scope of this chapter to consider the theoretical mechanisms that affect resolution, selectivity, and capacity in ion-exchange chromatography. For detailed considerations of the retention behavior of biomolecules on IEX supports, the reader is referred to reviews elsewhere ***(1,3)***.

Table 2
Types of Functional Groups Immobilized on Ion-Exchange Gels[a]

Functional group	Name	Type of exchanger
$-O-CH_2CH_2N^+H(CH_2CH_3)_2$	Diethylaminoethyl (DEAE)	Weak anion
$-O-CH_2CH_2N^+H(CH_2CH_3)_2-CH_2-CHOH-CH_3$	Quaternary aminoethyl (QAE)	Strong anion
$-O-CH_2OH-CH_2-O-CH_2-CHOH-CH_2-N^+(CH_3)_3$	Quaternary ammonium (Q)	Strong anion
$-O-CH_2-COO^-$	Carboxymethyl (CM)	Weak cation
$-O-CH_2-CHOH-CH_2-O-CH_2-CH_2-CH_2SO_3^-$	Sulphopropyl (SP)	Strong cation
$-O-CH_2-CHOH-CH_2-O-CH_2-CHOH-CH_2SO_3^-$	Methyl sulphonate (S)	Strong cation

[a]Adapted from **ref.** ***1***.

2. Materials

2.1. Gel Media and Selection Criteria

The selection of appropriate ion-exchange gel media depends on the characteristics of the biomolecule in question (including p*I*, molecular size, and pH stability), as well as on the type of separation being considered (analytical or preparative scale), the column capacity required (*see* **Subheading 2.2.**), the stability of the various supports at the chosen pH, and finally the cost of the support. IEX supports for low-pressure applications are commonly based on polysaccharides (agarose, dextran, cellulose), whereas for HPLC, supports are based on silica particles or polymers (cross-linked polystyrene/divinyl benzene, or methacrylate copolymer) (*see* **Table 3** for a list of some of the commercially available IEX supports).

If the p*I* of the protein is unknown, then one should either set up a pilot experiment to choose the optimal exchanger (*see* **Subheading 3.2.**), or use an anion-exchange support with a running buffer of pH 8.0–8.5 as a starting point. This latter alternative recognizes that the pI of many proteins is less than 7–7.5 ***(3)***, hence, these will bind to an anion-exchanger at pH 8.5. If the pI of the protein is known, choose either an anion-exchange support with running buffers of pH at least one pH unit greater than the pI, or a cation exchanger with running buffers at least one pH unit below the pI ***(1)***. Include at this selection stage any knowledge about the pH stability of the protein of interest, as some proteins readily denature or lose biological activity at extremes of pH.

As a starting point, choose either a strong anion or strong cation exchanger (*see* **Table 2**), as these supports give a more consistent performance and loading capacity over a wide pH range. After the initial experiments are completed, chromatography on the weak ion-exchangers can be investigated, as in some cases, these can yield improved resolution although this is sample dependent and must be empirically established.

IEX supports differ in their mechanical stability. Exchangers based on "soft" supports (agarose, dextran, or cellulose, *see* **Table 3**) are suitable for low pressure, open column chromatography of proteins, but these supports are limited to lower linear flow rates, and can swell and shrink to differing extents depending on the pH and ionic strength of the eluent *(1)*. Alternatively, high-performance media with small, controlled bead sizes (5–10 μm) manufactured from silica, or rigid crosslinked polystyrene/divinyl benzene derivatives (for example, Mono-Q, Mono-S, **Table 3**), do not swell or shrink and are stable at greater pressures making them suitable for HPLC *(5)*. Silica-based HPLC packings have a disadvantage in that they must be used in the pH range 2–8, as more basic pHs cause the packing to hydrolyze and disintegrate. The mechanical stability of these supports makes them an attractive choice for analytical-scale ion-exchange chromatography, however, commercial cost can become an issue for preparative scale purification. This problem has been addressed by the availability of similar monodisperse packings with particle sizes of 15, 25, or 30 μm (e.g., Source 15S, MacroPrep 25S) suitable for larger scale applications *(1)*.

Pore-size varies between the various IEX supports (*see* **Table 3**). In general, supports with smaller pore sizes (~125 Å) are best suited for low molecular weight solutes such as nucleotides and small peptides, whereas larger pore sizes (300 Å or greater) will work better for larger peptides, proteins, and nucleic acids *(3)*. Detailed information about the exclusion limits of particular IEX supports should be checked in the product literature. As examples, Sephadex ion exchangers are based on either the Sephadex G-50 or Sephadex G-25 particles, of which the G-50 support is more porous and better suited for the chromatography of proteins of molecular weights greater than 30,000 *(1)*. Similarly, the TSK DEAE-2SW prepacked HPLC column (*see* **Table 3**) has a pore size of 125 Å and an exclusion limit of 10,000 Daltons, whereas the TSK DEAE-5PW HPLC column has a pore size of 1000 Å and an exclusion limit of 1,000,000 Daltons.

Invariably, the choice of IEX support will involve some level of compromise between the main factors that influence resolution (particle type and size, porosity) and capacity and cost. Small particle sizes (10 μm) give higher resolution, but also give higher backpressure, which limits flow rates and requires specialist pumps and associated hardware. Larger particles (25–50 μm) allow faster flow rates and lower backpressures, but do not have the same resolution as smaller particles. Note also that the capacity of ion-exchange supports tends to increase with particle size (*see* **Subheading 2.2.**), therefore choose a support that has the smallest particle size (*see* **Table 3**) consistent with the scale of the purification.

2.2. Capacity of IEX Supports

The capacity of an ion-exchange support for a particular sample is dependent on a number of variables, including charge and molecular size of the sample, experimental conditions including buffer pH and ionic strength, and flow rate. Product information from the manufacturers can include various measures of capacity, which are 1) total ionic capacity, 2) available capacity, and 3) dynamic capacity ***(4)***. The total ionic capacity of an exchanger refers to the total amount of immobilized charged groups on the support. In practice, most ion-exchangers are porous (*see* **Table 3**), and the ability of large sample ions such as proteins to access all immobilized charged groups is limited by their molecular size and shape. Hence, *available capacity* refers to the amount of a specific protein that will bind to the ion-exchanger under defined conditions of pH, ionic strength, and sample concentration. This parameter is determined in a static or batch type of experiment, and does not take into account the effects of flow rate and column dimensions that contribute to capacity when the ion-exchanger is used in a column chromatography format, which is known as *dynamic capacity*. The dynamic capacity of an ion-exchanger is normally less than the available capacity under the same experimental conditions ***(4)***.

In practice, the available or dynamic capacities of ion-exchangers (*see* **Table 3**) will be of most use in deciding how much of the support is needed for a particular experiment. Where possible, consult the product literature and use the reported capacity as determined with a protein of similar molecular weight to that being purified, and under similar conditions of pH and ionic strength. It is generally recommended not to exceed ~10–20% of the stated capacity if maximum resolution is to be maintained ***(1)***. If column capacity is expected to be an important parameter in the success of your purification scheme, (e.g., for increased sample throughput, shorter overall purification times, and better economy) consideration should be given to pilot experiments aimed at determining ion-exchange capacity with your target protein.

2.3. Choice of Appropriate Buffers

As the mechanism of ion-exchange chromatography involves binding of a charged sample to the ion-exchange support, it is preferable that the buffering ions not compete in this process ***(1,4)***. Therefore, the buffering ions should be of the same charge as the ion-exchanger, meaning that anionic buffers (e.g., phosphate, acetate, citrate) should be used for cation-exchange, and cationic buffers (e.g., piperazine, Tris, bis-Tris) for anion-exchange. A list of commonly used buffers and recommended concentrations is given in **Table 4**. For the same reason, any buffer additives (e.g., detergents, protease inhibitors) should also be either noncharged, or carry the same charge as the ion-exchanger. Keep

Table 3
List of Some of the Commercially Available Ion-Exchangers[a,b]

Product by IEX type[c]	Functional group	Matrix type	Particle size (μm)	Pore size (Å)	Total ionic capacity (μmoles/mL gel)	Available capacity[d] (mg/mL gel)	Dynamic capacity[d] (mg/mL gel)	pH range[e]	Form[f]	Company[g]
SCX										
Mono S	Sulphonic acid	Crosslinked PS/DVB[i]	10	–	140–180	–	75 (human IgG)	3–11	pp	AB
Mini S	Sulphonic acid	Crosslinked PS/DVB	3	–	16–30	–	1.3[j] (ribonuclease)	3–11	pp	AB
Source 15S	Sulphonic acid	Crosslinked PS/DVB	15	–	na	–	80 (lysozyme)	2–12	pp,b	AB
Source 30S	Sulphonic acid	Crosslinked PS/DVB	30	–	na	–	80 (lysozyme)	2–12	pp,b	AB
SP Sepharose HP	Sulphopropyl	Crosslinked 6% agarose	24–44	–	140–200	–	55 (RNase)	4–13	pp,b	AB
SP Sepharose FF	Sulphopropyl	Crosslinked 6% agarose	45–165	–	180–250	–	50 (human IgG)	4–13	pp,b	AB
SP Sephadex	Sulphopropyl	Crosslinked dextran	40–125	–	90–300	–	1.6–8.0 (human IgG)	na	b	AB
TSK SP-5PW	Sulphopropyl	hydrophilic resin	10	1000	na	40 (hemoglobin)	–	2–12[h]	pp	S, A
TSK SP-NPR	Sulphopropyl	Hydrophilic resin	2.5	–		5 (hemoglobin)		2–12[h]	pp	S, A
Macro-Prep 25S	Sulphonic acid	Hydrophilic resin	25	725	110±30	–	–	1–14	b	BR
Macro-Prep High S	Sulphonic acid	Hydrophilic resin	50	1000	160±40	–	60 (human IgG)	1–14	b	BR
PL-SCX 1000 Å	Sulphonic acid	Macroporous PS/DVB	8,10	1000	na	–	–	1–14	pp	P
PL-SCX 4000 Å	Sulphonic acid	Macroporous PS/DVB	8,10	4000	na	–	–	1–14	pp	P
SynChropak SCX	Sulphonic acid	Silica	6	300	na	–	–	2–8	pp	A
WCX										
CM Sepharose FF	Carboxymethyl	Crosslinked 6% agarose	45–165	–	90–130	–	50 (RNase)	6–10	pp,b	AB
CM Sephadex	Carboxymethyl	Crosslinked dextran	40–125	–	170–550	–	1.6–7.0 (human IgG)	6–10	b	AB
TSK CM-5PW	Carboxymethyl	Hydrophilic resin	10	1000	na	45 (hemoglobin)	–	2–12[h]	pp	S, A
TSK CM-2SW	Carboxymethyl	Silica	5	125	na	na	–	2–7.58	pp	S, A
TSK CM-3SW	Carboxymethyl	Silica	10	250	na	110 (haemoglobin)	–	2–7.58	pp	S, A
Macro-Prep CM	Carboxymethyl	Hydrophilic resin	50	1000	210±40	–	–	1–14[h]	b	BR
SynChropak WCX	Carboxymethyl	Silica	6	300	na	–	–	2-8	pp	A

SAX										
Mono Q	Quaternary amine	Crosslinked PS/DVB	10	–	270–370	–	65 (BSA)	3–11	pp	AB
Mini Q	Quaternary amine	Crosslinked PS/DVB	3	–	60–90	–	1.4[j] (α-amylase)	3–11	pp	AB
Source 15Q	Quaternary amine	Crosslinked PS/DVB	15	–	na	–	45 (BSA)	2–12	pp,b	AB
Source 30Q	Quaternary amine	Crosslinked PS/DVB	30	–	na	–	45 (BSA)	2–12	pp,b	AB
Q Sepharose HP	Quaternary amine	Crosslinked 6% agarose	24–44	–	140–200	–	70 (BSA)	2–12	pp,b	AB
Q Sepharose FF	Quaternary amine	Crosslinked 6% agarose	45–165	–	180–250	–	120 (HSA)	2–12	pp,b	AB
QAE Sephadex	Diethyl-(2-hydroxy propyl)aminoethyl	Crosslinked dextran	40–125	–	100–500	–	10–80 (HSA)	na	b	AB
Macro-Prep 25 Q	Quaternary amine	Hydrophilic resin	25	725	220±40	–		1–14	b	BR
Macro-Prep High Q	Quaternary amine	Hydrophilic resin	50	1000	400±75	–	40 (BSA)	1–14	b	BR
PL-SAX 1000 Å	Quaternary amine	Macroporous PS/DVB	8,10	1000	na	–		1–14	pp	P
PL-SAX 4000 Å	Quaternary amine	Macroporous PS/DVB	8,10	4000	na	–		1–14	pp	P
SynChropak SAX	quaternary amine	silica	6	300	na	–		2–8	pp	A
WAX										
DEAE Sepharose FF	Diethylaminoethyl	Crosslinked 6% agarose	45–165	–	110–160	–	110 (HSA)	2–9	pp,b	AB
DEAE Sephacel	Diethylaminoethyl	Bead-formed cellulose	–	–	100–140	–	160 (HSA)	2–9	b	AB
DEAE Sephadex	Diethylaminoethyl	Crosslinked dextran	40–125	–	175–500	–	30–110 (HSA)	2–9	b	AB
TSK DEAE-5PW	Diethylaminoethyl	Hydrophilic resin	10	1000	na	30 (BSA)	–	2–12[h]	pp	S, A
TSK DEAE-NPR	Diethylaminoethyl	Hydrophilic resin	2.5	–	na	5 (BSA)	–	2–12[h]	pp	S, A
TSK DEAE-2SW	Diethylaminoethyl	Silica	5	125	na	na	–	2–7.5[h]	pp	S, A
TSK DEAE-3SW	Diethylaminoethyl	Silica	10	250	na	120 (BSA)	–	2–7.5[h]	pp	S, A
Macro-Prep DEAE	diethylaminoethyl	hydrophilic resin	50	1000	175±75	–	–	1–14[h]	pp	BR
SynChropak WAX	polyethyleneimine	silica	6	300	na	–	–	2–8	pp	A

[a]Details as provided by the manufacturer.

[b]For a complete listing of commercially available ion-exchangers, *see* **ref. *3*** and ***5***.

[c]SCX, strong cation-exchange; WCX, weak cation-exchange; SAX, strong anion exchange; WAX, weak anion exchange.

[d]Capacity determined with the protein shown in parentheses.

[e]Refers to the pH range over which the ion-exchange groups remain charged, and a high sample capacity is maintained.

[f]Available as either pre-packed columns (pp), or bulk ion-exchanger (b).

[g]Further information available at: Amersham Biosciences (AB) http://www.chromatography.amershambiosciences.com, Bio-Rad (BR) http://www.bio-rad.com/, Sigma-Aldrich (SA) http://www.sigmaaldrich.com, Supelco (S) http://www.sigmaaldrich.com, Agilent (A) http://www.chem.agilent.com, Polymerlabs (P) http://polymerlabs.com.

[h]Refers to the stability of the matrix, as given by the manufacturer. For details of the working pH range of this support, see other exchangers with the same functional group.

[i]PS/DVB: polystyrene/divinyl benzene.

[j]mg/column.

Table 4
Buffers for Ion-Exchange Chromatography[a]

pK_a (25°C)	pH range	Buffer	Concentration (m*M*)	Counter-Ion
Cation exchange				
2.00	1.5–2.5	Maleic acid	20	Na^+
2.88	2.4–3.4	Malonic acid	20	Na^+/Li^+
3.13	2.6–3.6	Citric acid	20	Na^+
3.81	3.6–4.3	Lactic acid	50	Na^+
3.75	3.8–4.3	Formic acid	50	Na^+/Li^+
4.21	4.3–4.8	Butanedioic acid	50	Na^+
4.76	4.8–5.2	Acetic acid	50	Na^+/Li^+
5.68	5.0–6.0	Malonic acid	50	Na^+/Li^+
7.20	6.7–7.6	Phosphate	50	Na^+
7.55	7.6–8.2	HEPES	50	Na^+/Li^+
8.35	8.2–8.7	BICINE	50	Na^+
Anion exchange				
4.75	4.5–5.0	*N*-methylpiperazine	20	Cl^-
5.68	5.0–6.0	Piperazine	20	$Cl^-/HCOO^-$
5.96	5.5–6.0	*L*-histidine	20	Cl^-
6.46	5.8–6.4	*bis*-Tris	20	Cl^-
6.80	6.4–7.3	*bis*-Tris propane	20	Cl^-
7.76	7.3–7.7	Triethanolamine	20	Cl^-/CH_3COO^-
8.06	7.6–8.5	Tris-Cl	20	Cl^-
8.52	8.0–8.5	*N*-methyldiethanolamine	50	$SO_2^-/Cl^-/CH_3COO^-$
8.88	8.4–8.8	Diethanolamine	20	Cl^-
8.64	8.5–9.0	1,3-Diaminopropane	20	Cl^-
9.50	9.0–9.5	Ethanolamine	20	Cl^-
9.73	9.5–9.8	Piperazine	20	Cl^-
10.47	9.8–10.3	1,3-Diaminopropane	20	Cl^-
11.12	10.6–11.6	Piperidine	20	Cl^-

[a]Adapted from **refs. *1*** and ***4***.

in mind that buffer pK_a's vary with temperature, and that the pK_a data listed in **Table 4** are for 25°C. The extent to which these pK_a's change with temperature is given elsewhere ***(1,4)***.

In most cases, cation-exchangers are supplied with Na^+ as the counter-ion, whereas Cl^- is commonly used as the counter-ion for anion-exchangers. If required, the counter-ion can be changed by washing the IEX support in an excess concentration (0.5–1.0 *M*) of a salt of the new counter-ion (*see* **Subheading 3.1.**).

2.4. Preparation of Buffers

Buffers should be prepared using high quality deionized or distilled water, and filtered and degassed as appropriate if HPLC columns are being used. In order to ensure run-to-run reproducibility, it is essential that buffer pH be accurately set using a quality pH electrode which is routinely calibrated using two pH standards. Monitor and record the volume of acid or base used to set the pH for any particular buffer, and as a quality control step make sure that a similar volume is always added. In some cases, buffers may include a proportion of an organic solvent (e.g., 10–20% (v/v) acetonitrile) to limit hydrophobic interactions with the exchanger, and the solvent should be added after the pH of the buffer has been set.

2.5. Batch vs Column Purification

IEX separations can be carried out using either a batch technique, or with the exchanger packed into a chromatographic column (or with prepacked columns supplied by the manufacturer). Each approach has its advantages. Batch processing can readily be applied to selectively and rapidly separate proteins into two groups, consisting of a group bound to the exchanger and subsequently eluted with elution buffer, and the nonbound group. If the desired protein is present exclusively in either group, a useful purification can be achieved using a minimum of specialist equipment. In contrast, much greater resolution between proteins can be gained using column chromatography, hence, this is the recommended approach if IEX is to be used as part of a purification strategy.

2.6. Mode of Elution

Elution of bound components from an ion-exchange support is normally achieved by increasing the salt concentration of the eluent, although changing the buffer pH can also be used. The salt concentration can be increased in either a step-wise fashion (i.e., discontinuous elution) as normally used in batchwise ion-exchange chromatography, or by the application of a linear gradient (continuous elution), which is the method of choice for column chromatography as higher resolution can be more readily achieved.

The number of increments to include in a stepwise elution system must be determined empirically, although a minimum of three steps is commonly employed. These will include an initial step to remove all weakly bound molecules, followed by elution with a higher salt concentration to desorb the more strongly bound molecule of interest, and a final wash with the maximum salt concentration to strip the column of remaining bound substances. Additional steps can be included to suit the particular application.

Elution with linear gradients will require a suitable pumping system and a gradient controller (*see* **Subheading 2.7.**). Nonlinear gradients can also be employed if available, however, it is suggested that these be included into a separation only after initial experiments with linear gradients have indicated their suitability. Use of a linear gradient will require a decision about the gradient steepness parameter, which is a measure of the rate of change of counter ion on the column during a separation. The steepness of the gradient is dictated by the column volume, flow rate, the initial and final salt concentrations of salt in the eluent, and the gradient time ***(3)***. Shallow gradients will result in optimal resolution, but this occurs at the expense of separation time and increased band-broadening ***(3)***. In contrast, fast gradients yield short separation times and minimal band-broadening, but at the expense of resolution. Where available, use the manufacturers recommendations for the gradient steepness parameter as a starting point, and vary this according to the desired resolution. For example, a rate of change of 17.5 m*M*/mL of Cl^- as the counter-ion is suggested to give optimal resolution for analytical separations on the strong anion exchanger Mono Q (5 cm length, 0.5 cm internal diameter) ***(1)***, but a slower rate (8.75 m*M* Cl^-/min) may give better results. If no data is available about the gradient steepness for an ion-exchanger, set up an experiment with a gradient from 0.05 *M* to 1 *M* of the counter-ion over 10–20 column volumes as a starting point ***(3)***, and compare the resolution achieved with similar data obtained at faster and slower rates of change.

The concentration of salt needed to elute all bound substances from an ion-exchange column is dependent on the type of exchanger, the pH of the buffer, and the p*I*s of the molecules loaded onto the gel. Hence, no general recommendation can be given, although the normal maximum salt concentration needed is in the range 0.5 *M*–1 *M*. This concentration can readily be adjusted after the results of the first chromatographic experiment become available. A series of empirical tests can also be undertaken to determine the maximum salt concentration needed (*see* **Subheading 3.2.**).

As mentioned earlier (*see* **Subheading 1.** and **Fig. 1**), the charge on a protein is pH dependent. Hence, once bound to the IEX support, alteration of the pH of the eluent can be used to change the charge on the protein and bring it closer to its p*I*, causing elution of the protein from the ion-exchanger. However, it is difficult to produce a linear pH gradient at a constant salt concentration, and for this reason, pH gradients are not as common as salt gradients as a means to elute proteins in ion-exchange chromatography ***(4)***. In contrast, steps in pH are easier to control and reproduce, and for anion-exchangers this requires a drop in eluent pH, while an increase in pH is needed for cation exchangers (*see* **Fig. 1**). An in-line pH electrode can be included in the flow path immediately postcolumn to monitor eluent pH, or alternatively this can be measured in collected fractions manually. A combination of a change in pH with an increase in salt concentration can also be considered as a means of elution in ion-exchange chromatography *(4)*.

2.7. Recommended Hardware Configuration

A typical system for low-pressure ion-exchange chromatography for protein purification will require a pumping system capable of generating gradients, plus an injector (or a mechanism to introduce the sample into the buffer flow), the chromatographic column, an in-line UV (280 nm) detector and chart recorder, and a fraction collector. Access to a stand-alone conductivity meter to monitor the conductivity of the column eluent is also very useful. A "bare-bones" system without the UV detector and chart recorder can also be used, although this has the disadvantage that UV measurements on eluted fractions must be done separately. For low-pressure systems a peristaltic pump and a simple gradient formerly made up of two interconnected vessels of identical shape can be used to generate a linear gradient. If using this approach, make sure that the gradient maker is level, and place the two buffers (the eluting buffer containing NaCl, and the initial equilibration buffer) into the two vessels, with the initial buffer vessel placed over a magnetic stirrer and connected by a second stopcock to the pump. Start the mixer, and open the stopcock between the vessels before starting the pump (*see* **Note 2**). Nonlinear gradients can also be generated by this approach by altering the shape of the nonmixing vessel.

IEX chromatography hardware for FPLC and HPLC should be matched to the expected flow rates and back-pressures of the columns to be used. Several systems for low-pressure and medium pressure ion-exchange chromatography [e.g., ÄKTA-FPLC (Amersham BioSciences), BioLogic HR system (Bio-Rad)] are available from the major manufacturers which are specifically designed for use with halide-containing buffers (*see* **Note 3**).

If columns are to be packed with bulk ion-exchanger, this should be carried out as specified by the manufacturer, and may require initial swelling of the gel in some cases, or use of a preswollen support. In either case, the support has to be pre-equilibrated before packing into the column (*see* **Subheading 3.1.**). Alternatively, prepacked columns may be used (*see* **Table 3**). As IEX is an adsorptive technique, columns are normally short, with a bed height of 5–15 cm *(1)*. Scaling-up of IEX columns is readily achieved by increasing the column diameter *(1)*.

3. Method

The following procedures describe ion-exchange chromatography on an anion-exchange support, using as an example the strong anion-exchanger

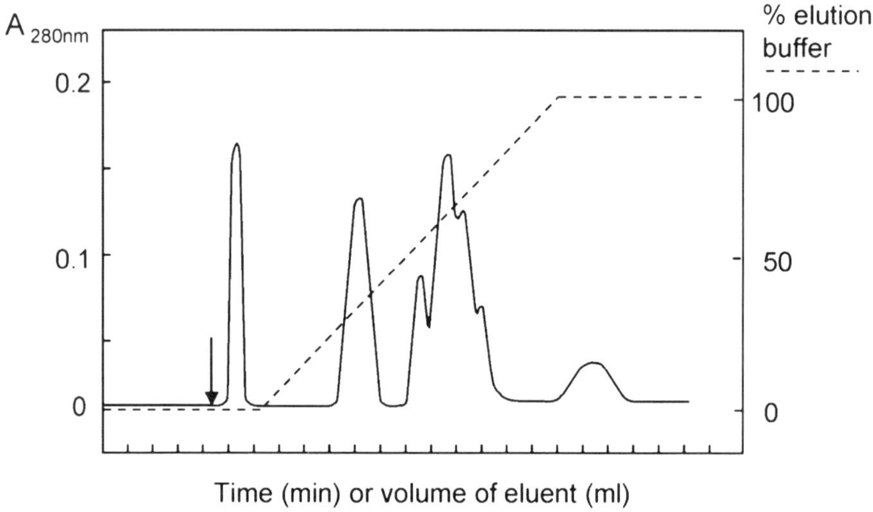

Fig. 3. Example of a typical ion-exchange column chromatogram. Initially, the ion-exchange column is equilibrated in the starting buffer (low ionic strength) until a stable baseline of UV absorbance at 280 nm (filled line) is achieved. The pH and conductivity of the column eluent at this point should be the same as the starting buffer, and confirms adequate equilibration. The sample is then injected (indicated by arrow), and the column is washed with starting buffer until all nonbound proteins are eluted. A linear gradient of elution buffer (high ionic strength) (dashed line) is then initiated to elute bound proteins, after which the column is washed with 100% elution buffer to remove strongly bound material. At the completion of the run, the column is stripped with stripping solution (1–2 *M* NaCl) (not shown) before reequilibration in the starting buffer. (*See* **Subheading 3.** for further details.)

Mono Q (HR5/5) prepacked column. Details specific for this column will be added to the method in italics. A linear gradient will be used to elute bound proteins, although conditions for the use of a step-elution system can readily be substituted for the gradient as will be indicated. An example chromatogram for this application is shown in **Fig. 3**.

3.1. Prepare the Gel/Column

1. Most IEX supports are available as preswollen gels or prepacked columns, but if working with one of the dry powder supports, it must first be preswollen at the pH to be used in the experiment according to the manufacturers instructions (normally 2–3 h at 90°C, or 48 h at room temperature) (*see* **Note 4**).
2. Preswollen gels are supplied in a ready-to-use form. Mix the gel to a slurry, and remove the required amount into a vacuum flask and allow to settle. For both types of gels, swirl-mix the gel approximately five times with an excess of the starting

buffer (~5 volumes) at the correct pH, and after settling discard the supernatant. To test whether the gel is fully equilibrated, monitor the pH and conductivity of the buffer removed from the gel, which should be identical to the starting buffer.

3. If the counter-ion on the IEX support is to be changed from that which the gel was supplied with (which is normally Cl^- for anion exchangers, and Na^+ for cation exchangers), wash the gel first with a 0.5 *M*–1.0 *M* solution of the salt of the new counter-ion, and then equilibrate into the starting buffer as above.
4. Pack the column according to the manufacturer's instructions (*see* also **Subheadings 3.1.–3.3.**, Chapter 5 (Gel filtration) for notes regarding the packing procedure).
5. If a prepacked column is being used, this should be placed in the liquid chromatography system, and the pressure limit of the pump set at the recommended value (5 MPa for Mono Q). The column is then equilibrated with ~5 column volumes (~5 mL, as the column volume of the Mono Q HR5/5 is 1 mL) of the starting buffer (20 m*M* piperazine, pH 9.6) at the recommended flow rate (1 mL/min), and the pH and conductivity checked (as detailed in **Subheading 3.1.2.**) (*see* **Fig. 3**).
6. It is good practice to pre-cycle the column with 5 column volumes each of 1) starting buffer (20 m*M* piperazine, pH 9.6), 2) eluting buffer (20 m*M* piperazine, 300 m*M* NaCl, pH 9.6), and 3) column stripping buffer (1 *M* NaCl), before actual use. Re-equilibrate the column with 5–10 column volumes of starting buffer and ensure that both the pH and conductivity of the eluent are the same as the starting buffer.

3.2. Pilot Study to Choose Optimal pH

1. A pilot test to determine the optimal pH to use for ion-exchange chromatography relies on binding of the molecule(s) of interest to an aliquot of the ion-exchanger, and its subsequent disappearance from solution ***(4)***. The basis of this approach is to set up a series of aliquots (~1 mL) of the ion-exchanger in disposable tubes at pH intervals from 0.5 to 1 pH unit. For an anion-exchanger, start at pH 5.0 using buffers as described in **Table 4**, and prepare aliquots of gel at pH 5.5, 6.0, 6.5, 7.0, 7.5, 8.0, 8.5, and 9.0. Wash the gel aliquots in respective buffers five times to ensure adequate preequilibration (*see* **Subheading 3.1.5.**).
2. Prepare aliquots of the sample molecule in the same buffers. It is important that the buffer pH of the sample is the same as the aliquot of the ion-exchanger to which it is to be added. Similarly, the sample should not contain any salts, as these will interfere with binding. If excess salts are present, these should be removed by either dialysis or gel filtration chromatography.
3. Add the sample aliquot to the ion-exchange gel of the same pH, mix gently on a rotating wheel for 15 min, then allow the gel to settle. Remove the supernatant, and assay it for the presence of the sample of interest. This can be done either as a general test monitoring changes in absorbance, or with a specific assay. The disappearance, or marked decrease of the sample in solution indicates the pH at which the sample binds to the ion-exchanger, and this pH should be used for ion exchange chromatography. Avoid choosing a pH which is more than ~0.5–1.0 pH unit higher

than this optimal pH, as the sample will adsorb more strongly to the ion-exchange column at higher pHs, and hence will require a higher salt concentration to elute it (*see* **Note 5**).

4. This type of pilot test is more appropriate for setting up ion-exchange chromatography using preswollen gels. For prepacked FPLC or HPLC columns operated on chromatography systems with automatic valves controlling buffer selection and sample injection, the effect of different pHs on sample retention can be rapidly evaluated in an automated format.
5. Once the optimal pH for binding has been established, a similar strategy using aliquots of ion-exchanger at this pH can be used to determine the maximum salt concentration needed to cause elution of the sample from the gel. In this case, set up buffers and gel with NaCl increasing in increments of 0.1 *M* from 0 to 1.0 *M* NaCl *(1,4)*. The salt concentration at which all the sample is present in the supernatant indicates the concentration needed in the elution buffer.

3.3. Prepare the Sample for Loading

1. For successful ion-exchange chromatography, the sample should be at the same pH as the starting buffer, and contain either the same (or less) salt. The options for samples containing higher salt concentrations include buffer exchange by dialysis or gel filtration chromatography, or in some cases simple dilution with the starting buffer can be sufficient (*see* **Note 6**).
2. Remove any particulates from the sample by either filtration through a 0.22-μm filter (*see* **Note 7**), or by high-speed centrifugation (15,000*g*, 10 min). Set aside an aliquot of the sample as starting material, which should be assayed in conjunction with the collected fractions in order to determine the yield of sample from the column.

3.4. Prepare the Column for Loading and Injection

1. Establish a flat UV baseline at 280 nm by washing the column with starting buffer (*see* **Fig. 3**). Check the flow rate (1 m/min), chart speed (typically, 0.25 cm/min), and maximum absorbance setting on the detector (0.01–2.0 AUFS).
2. Calculate the gradient time required using the recommended rate of change of counter-ion from the product guidelines. (For the Mono Q column operating at 1 mL/min, this value is 17.5 m*M* Cl^-/min. Hence, with a starting buffer of 20 m*M* piperazine, pH 9.6, and an elution buffer of 20 m*M* piperazine, 300 m*M* NaCl, pH 9.6, the rate of change value to enter into the gradient controller will be 300/17.5 = 17.1 min).
3. If a step elution system is to be used, choose the steps (e.g., increments of 20%, 40%, 60%, and 80% elution buffer), and program the gradient controller accordingly.
4. Inject the protein sample, start the fraction collector (1 min fractions), and continue washing the column until the UV trace returns to baseline (*see* **Fig. 3**).
5. Start the gradient (*see* **Note 8**). At completion of the gradient, continue to wash the column with five column volumes of the elution buffer (5 mL), then apply five

column volumes of stripping buffer (1 *M* NaCl) to elute any strongly bound material.

6. Reequilibrate the column with 5–10 column volumes of starting buffer.
7. Assay collected fractions (and the starting material aliquot) for the presence of the sample of interest. Determine % yield of activity, and pool appropriate fractions ready for the next purification step.

3.5. Column Cleaning and Storage

1. After the final chromatographic run, wash the column extensively and store it in the recommended storage solvent (*see* **Note 9**).
2. If an increased column backpressure has occurred, or if visual fouling of the top of the column is evident, clean the ion-exchanger according to the manufacturers guidelines. A typical cleaning process involves treating the column with 1) ~2 column volumes of 0.1–0.2 *M* NaOH (*not recommended* for silica-based columns, 2) extensive washing with water, 3) ~2 column volumes of 70% ethanol, 4) extensive washing with water. In some cases, it is possible to invert the column prior to these washing steps to assist in removing particulate matter caught in the top of the column, although for pre-packed columns this step should only be used as a last resort.

3.6. Troubleshooting

3.6.1. Sample Elutes With the Starting Buffer

There are several potential reasons for this problem: 1) the pH of the starting buffer is incorrect, 2) the starting buffer has too much salt in it, 3) the sample is not at the correct pH or has too much salt present, 4) the column is not fully equilibrated in the starting buffer. To fix these, firstly check your information about the p*I* of the sample of interest (*see* **Subheading 1.** and **Fig. 1**). If no data is available, do the pilot experiment outlined in **Subheading 3.2.** to choose the optimal pH. To make an anion-exchange column bind more proteins, increase the pH of the starting buffer. Similarly, for a cation exchange column decrease the pH of the starting buffer. For both types of support, remove any salt in the starting buffer.

Ensure that the sample is fully equilibrated in the start buffer before loading (*see* **Subheading 3.1.1.**), and also that the column has been properly preequilibrated (**Subheading 3.1.6.**).

3.6.2. Sample Does Not Elute From the Column

This problem indicates that 1) the salt concentration of the elution buffer is too low, or 2) the pH of the eluent is wrong, 3) the sample has irreversibly bound to the column, or 4) the sample precipitated or was lost prior to loading. To fix these, increase the salt concentration of the elution buffer to a maximum of 1 *M*, and/or change the pH of the elution buffer (decrease the pH for an anion-exchanger, or increase the pH for a cation-exchanger). If the recovered amount of protein is lower than expected after these steps, binding to the column by mechanism(s) other than ion-exchange may be occurring. Hydrophobic interactions can be limited by the addition of 10% ethylene glycol ***(1)*** or acetonitrile to the buffers, but be alert to the fact that some salts precipitate at high organic solvent concentration. Addition of a nonionic detergent (e.g., 0.1% Triton X-100) will limit nonspecific sticking to surfaces. Be sure that the sample was actually loaded onto the column.

3.6.3. Elution of Peaks Requires a Very High Salt Concentration

The pH of the eluent is either too high (anion exchanger) or too low (cation exchanger). Adjust the pH of the buffer system to be closer to the p*I* of the sample of interest.

3.6.4. Sample Elution Pattern Is Not Reproducible

The most obvious cause is that the column has not been fully reequilibrated into the starting buffer (*see* **Subheading 3.1.1.**). Also check that the stripping buffer (normally 1–2 *M* NaCl) contains the same ionic species as the counter-ion on the column (Na^+ for cation exchangers, Cl^- for anion exchangers). If the reproducibility problem occurred between runs carried out with different buffers, check the buffer preparation and calibration of the pH meter.

Ensure that the sample has not changed between the chromatographic runs, by either inappropriate storage, proteolysis, or precipitation. If there is a possibility that the sample has precipitated on the column, clean and regenerate the column (**Subheading 3.5.**) and repeat the chromatography.

3.6.5. Resolution of Sample Is Poor

There are several reasons why resolution can be poor. These are 1) the gradient slope is too steep, 2) there are large precolumn dead volumes in the flow path, 3) the column is poorly packed, 4) the sample has precipitated on the column, 5) the column is contaminated, and 6) the column has been overloaded.

Test the effect of decreasing the gradient slope (by a factor of 2) on sample resolution, or of inserting a step mid-way through the gradient. Make sure that there are no dead volumes prior to or after the column, particularly if using a Superloop. Check the packing of the column by loading a coloured marker protein (myoglobin, haemoglobin, cytochrome-*c*, *see* **Table 1**) and observing the band during chromatography. If it is suspected that components of the sample have precipitated, or that the column is contaminated, clean the column and modify the eluents to prevent reoccurrence (**Subheading 4.2.**). Finally, check the effect that loading less sample has on resolution.

3.6.6. Recovery of Sample Activity Is Poor

If the recovery of sample activity is low compared to the activity present in the starting material aliquot, which was set aside immediately before injection, first check the amount of protein that has eluted from the column. If the peak heights are generally lower than expected, indicating that the recovery of total protein is also low, the problem could be loss of protein on the column (*see* **Note 2**). If total protein recovery appears acceptable, but sample activity is low, check the stability of the sample in the ion-exchange buffers, and make sure that the added salt in the eluted sample is not interfering in the sample assay. Alternatively, the sample may have been separated from an essential cofactor or similar on the column, which can be checked by pooling all the recovered fractions and reassaying for activity.

3.6.7. Backpressure of Column Has Increased

Increasing column backpressure is caused by the introduction of particulate matter (either in the buffers, or in the sample), or precipitation of sample components in the column inlet filter or on top of the column itself. Of these, the main culprit is normally the sample. Ensure that the sample has been properly filtered before loading. If the sample is turbid, either dilute it prior to filtering, or try to improve sample solubility by changing the pH of the buffer, or through the addition of a nonionic detergent or organic solvent (*see* **Note 2**). These additives should also be added to the column eluents.

For laboratory-packed columns, change the filter on top of the column, and if possible make the buffer filtration more stringent. Apply column cleaning procedures as recommended by the manufacturer of the ion-exchanger (*see* **Subheading 3.5.**).

3.6.8. Column Has Run Dry

For laboratory-packed columns this is a terminal problem, requiring unpacking and repacking of the column. For prepacked columns, try introducing a 50:50 solvent/water mix (methanol, acetonitrile, or dimethyl sulfoxide (DMSO), depending on the stability of the exchanger to solvents) at a very slow flow rate overnight to exclude all the air bubbles from the column, after which the column should be washed with 5–10 column volumes of water, and then reequilibrated into the ion-exchange eluents. Test the resolution of the column as per the manufacturers instructions to see whether column performance has been altered.

3.6.9. Peaks Observed in the Blank Chromatogram

The presence of unexplained peaks in the chromatogram indicates 1) impurities in the buffer components, or 2) incomplete elution of a previously loaded sample, which is "ghosting" in each subsequent elution profile. Ghosting can be readily checked, as the peak height should decrease with each subsequent blank run. Clean the column (*see* **Subheading 3.5.**), and then improve the washing conditions at the end of each run to remove all bound material. The presence of buffer impurities indicates that higher quality reagents should be used, or if this is not possible, a guard or precolumn can be placed before the main column.

4. Notes

1. The solubility of many proteins frequently is lowest at (or near) their isoelectric points. Therefore, avoid use of pH conditions near the p*I* of the protein, which can lead to isoelectric precipitation on the column, resulting in blockage of the column and/or poor recoveries. Test the effects of particular pH and salt concentrations on protein stability in solution prior to final selection of ion-exchange support.
2. Ensure that the level of the two buffers remains the same when the mixer is switched on. If the mixer is rotating too quickly, the level of the buffer in the mixing chamber will increase and cause backflow into the second vessel. A simple gradient maker with an overhead stirrer which works very well is available from Amersham-BioSciences.
3. Halide-containing buffers will attack and corrode stainless steel high-pressure liquid chromatography (HPLC) systems, particularly frits, piston seals, and check valves. Leaching of metal ions can also interfere with some metal-sensitive proteins. Therefore, long-term use of these systems for ion-exchange chromatography is not recommended. If a stainless steel HPLC is used for IEX, wash the system with distilled or deionized H_2O as soon as possible after completion of the experiment.
4. Do not use a magnetic stirrer with any ion-exchange supports, as this can break gel particles. Avoid direct application of heat via a hotplate, as this can melt the gel media—use a waterbath instead.
5. The highest pH at which the *least* amount of sample is bound in this pilot test can be used as the pH of the elution buffer if a pH step elution system is to be used.
6. As ion-exchange chromatography is an adsorption technique, it is possible to load large volumes of dilute sample solution onto the column, and then elute the sample with elution buffer. For pre-packed columns on FPLC or HPLC systems, large sample volumes can be loaded using either a large loading loop (e.g., Superloop from Amersham Biosciences), or pumped directly onto the column using a third sample-loading pump.
7. A 0.22-μm filter is suitable for prepacked columns and ion exchangers with particle sizes up to 25 μm. For preparative ion-exchangers with particle sizes >25 μm, filters 0.45–1 μm can be substituted.
8. If a large-volume sample loading loop (Superloop) has been used, it is essential to isolate this from the flow-path before the gradient is initiated. If this is not done, the volume of starting buffer in the loop will significantly delay and dilute the gradient onto the column.

9. Storage solvents can include 20% ethanol, 0.01 *M* NaOH (*not* for silica columns) or equilibration buffer containing recommended bacteriostats (anion exchangers: 0.002% chlorhexidine or 0.001% phenyl mercuric salts, cation exchangers: 0.005% merthiolate) ***(1,4)***. Consult the manufacturers recommendations for each type of ion exchanger.

References

1. *Ion-Exchange Chromatography: Principles and Methods* (1999), Amersham Biosciences, Uppsala, Sweden.
2. Righetti, P. G. and Caravaggio, T. (1976) Isoelectric points and molecular weights of proteins: A table. *J. Chromatogr.* **127,** 1–28.
3. Choudhary, G. and Horvath, C. (1996) Ion-exchange chromatography, in *Methods in Enzymology*, vol. 270 (Karger, B. L. and Hancock, W. S., eds.), Academic, New York, pp. 47–82.
4. Williams, A. and Frasca, V. (1998) Ion-exchange chromatography, in *Current Protocols in Molecular Biology* (Ausubel, F. M., Brent, R., et al., eds.), Wiley, New York, pp. 10.10.1–10.10.30.
5. Henry, M. P. (1989) Ion-exchange chromatography of proteins and peptides, in *HPLC of Macromolecules: A Practical Approach* (Oliver, R. W. A., ed.), IRL Press at Oxford University Press, Oxford, U.K., pp. 91–125.

Web Sites:

Amersham Biosciences-Chromatography: http://www.chromatography.amershambiosciences.com.
Bio-Rad Laboratories: http://www.bio-rad.com/.
Sigma-Aldrich: http://www.sigmaaldrich.com.
Agilent: http://www.chem.agilent.com.
Supelco: http://www.sigmaaldrich.com.
Polymerlabs: http://polymerlabs.com.

4

High-Performance Hydrophobic Interaction Chromatography

Kálmán Benedek

1. Introduction

Hydrophobic interaction chromatography (HIC) is a column chromatographic separation technique frequently used for the purification of macromolecules such as proteins and polynucleotides. Purification schemes often are improved by incorporating HIC along with ion exchange, size exclusion, and affinity chromatography. HIC has found wide use for the purification of membrane proteins ***(1)***, serum proteins ***(2–4)*** nuclear proteins, polynucleotides ***(5)***, receptors ***(6,7)***, cells ***(8,9)*** and recombinant proteins ***(10–12)***.

HIC combines the nondenaturing characteristics of salt precipitation and the precision of chromatography to yield excellent resolution and activity recoveries. HIC is based on the adsorption of biomolecules to a weakly hydrophobic surface at high salt concentrations, followed by elution with a descending salt gradient. HIC was originally developed for protein separations using soft gel-based separation media. During the advance of modern high-performance liquid chromatography (HPLC), the technique was implemented into the new silica format and high-performance mode.

Conditions used in HIC are familiar to biochemists and similar to traditionally used salting out/in protein purification methods. Like in every type of interactive chromatography, the retention mechanism is based on adsorption–desorption equilibrium and preferred solubility conditions. The theoretical foundation of HIC is an extension of the solvophobic theory developed by Imre Molnár, Wayne Melander, and Csaba Horváth ***(13,14)***.

From: *Methods in Molecular Biology, vol. 251, HPLC of Peptides and Proteins: Methods and Protocols*
Edited by: M.-I. Aguilar © Humana Press Inc., Totowa, NJ

2. Materials

2.1. Chemicals

1. Ammonium sulfate $(NH_4)_2SO_4$
2. Milli-Q water.
3. 0.1 *M* potassium phosphate buffer, pH 7.0.

2.2. Equipment and Supplies

1. HPLC solvent delivery system with binary gradient capability and ultraviolet (UV) detector.
2. Hydrophobic Interaction Column (*See* **Note 1**) 4.6 mm ID × 250 mm length, 5 µm particle size, 300Å pore size (*See* **Note 2**).
3. Controlled temperature column jacket is recommended.
4. Guard column made of the same packing material as the analytical column is recommended.
5. Solvent filtration apparatus equipped with a 0.45-µm filter.
6. Sample filters, 0.22 µm porosity.
7. Mobile Phase A: 2 *M* $(NH_4)_2SO_4$ in 50 m*M* buffer (*See* **Note 3**).
8. Mobile Phase B: 50 m*M* buffer (*See* **Note 3**).

3. Method

3.1 Sample Preparation

Dissolve 1 mg of sample in 1 mL of mobile phase B (not in mobile phase A!) and filter the sample through a 0.22-µm filter.

3.2. Solvent Preparation

Prepare all solvents and filter through a 0.45-µm filter before use. This removes any particles remained in the mobile phases after dissolution of the buffer salts. Particles can block solvent lines, inline filters, column and detectors. It is recommended to degas the mobile phases by vacuum or helium purging. $(NH_4)_2SO_4$ is usually dirty at this high concentration. The filtration of the high salt content mobile phase A is very important to extend the instrument and column lifetime.

3.3. Preparations for Chromatography

3.3.1. System Assembly

Connect the guard and the column to the solvent delivery system according to the HPLC system requirements and equilibrate under the following initial conditions:

Solvent:	100% Mobile phase A.
Flow rate:	1.0 mL/min (*see* **Note 3**).
Detection wavelength:	280 nm (*see* **Note 4**).
Temperature:	Ambient (*see* **Note 5**).

3.3.2. System Check

Start the pumps to wash the flow system and equilibrate the column. Once a stable baseline is obtained, inject 10 μL of water or buffer and run a blank gradient. It is recommended to repeat this a couple of times to establish the gradient profile (*see* **Note 6**).

3.4. Separation Optimization

Inject 10 μL of the sample and use a linear gradient from 0 to 100% mobile phase B over 30 min to elute the sample (*see* **Note 6**). Most of the components of a protein mixture will elute from the column under these general conditions. The resolution can be manipulated by the variable separation parameters. The most important parameters are the mobile phase pH, salt type, concentration, and the hydrophobic ligand of the stationary phase. Varying the mobile phase temperature could also lead to changes in the separation. The most critical column parameter is the nature of the immobilized ligand of the stationary phase.

3.5. Sample Analysis

Once the chromatographic conditions are optimized according to the objective of the project, the analysis of the individual samples can be performed. Reproducibility assessment of separation is recommended.

4. Notes

1. Modern HIC can be viewed as an extension of RPLC. The stationary phases suitable for hydrophobic interaction chromatography are much more hydrophilic than the alkyl phases used for RPLC. The stationary phases were developed based on earlier work with soft organic gels, and experience with reversed phase stationary phase syntheses. Varieties of silica-based mildly hydrophobic stationary have been developed and are routinely used for protein separations. The stationary phase could be also highly crosslinked polymer-based bead, which can be modified with appropriate surface chemistry. Ideal HIC stationary phases are nonionic and hydrophilic. The ligand density, charge characteristics and hydrophobicity are variable parameters of the stationary phase. HIC stationary phases usually contain alkyl chains attached to hydrophilic nonionic foundation. The length and density of the alkyl chain has a significant effect on retention and selectivity as was shown with soft gel chromatography *(15)*. A similar result has been shown using silica-based stationary phases prepared by the same bonding chemistry, and attaching different-alkyl homologues to the same base material *(16)*.

   Fig. 1 shows the effect of different ligands on the separation of standard proteins. **Table 1** lists some of the generally used HIC columns.
2. Column geometry has a similar role in HIC as in RPLC and ion exchange chromatography (IEX). Longer columns are recommended for isocratic separations,

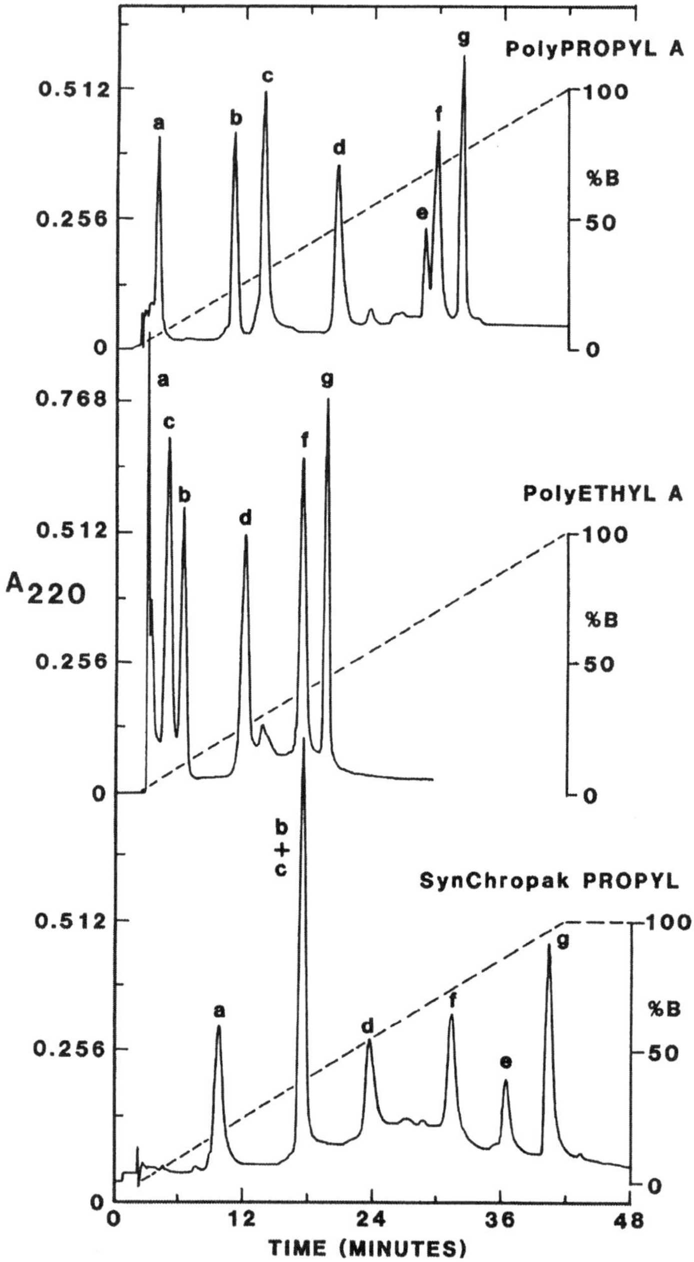

Fig. 1. HIC separations of standard proteins using different stationary phases. Stationary phases applied were PolyPROPYL and PolyETHYL Aspartamide and a PROPYL HIC column from other vendor. The elution condition consists of, mobile phase A: 1.8 *M* ammonium sulfate + 0.1 *M* potassium phosphate, pH 7.0 and mobile phase B: 0.1 *M* potassium phosphate, pH 7.0. A 40-min linear gradient was used from 0 to 100% B at 1 mL/min flow rate. The elution was followed at 220 nm.

Table 1
Stationary Phases[a]

Manufacturer	Product Name	Stationary Phase
Tosoh TSK HIC	Ether-5PW, Butyl-5PW, Phenyl-5PW, Butyl-NPR	Polymer Non-porous
PolyLC	PolyPROPYL Aspartamide, PolyETHYL Aspartamide, PolyMETHYL Aspartamide	Silica
Pharmacia	Butyl-, Phenyl-, Octyl-Sepharose	Polymer
BioChrom Labs	HYDROCELL C3-, C4-, Phenyl C3-NP10, C4-NP10, Phenyl-NP10	Polymer Non-porous

[a] The listed columns are only examples without attempting to be complete.

wider columns for preparative work and precolumns for column protection. More details regarding the role of the column parameters are described in Chapter 2. The pore size of the stationary phase basic material should be at least 300Å to allow the migration of proteins to the pore internal surface. For analytical separation 3–8 µm particles and for preparative purification the larger particles are suggested.

3. Hydrophobic interaction chromatography is fundamentally very close to the generally used salting-out and salting-in purifications method. The fundamental difference is the presence of an adsorptive surface and a flow system. Solvents play a critical role in hydrophobic interaction chromatography (HIC), because the separation of proteins by HIC is based on the hydrophobicity of the proteins presented to the solvents. The high salt content of the starting mobile phase increases the surface tension of the mobile phase, and the solvent-stationary phase interfacial tension. The free energy of adsorption of the protein to the stationary phase is negative. Decreasing the salt concentration thus decreases the interfacial tension and permits the proteins to elute. Various cosolvents have been tested as facilitating binding or elution of the proteins and solvent manipulation can affect retention, resolution, selectivity and peak shape.

   In mobile phase B, which is the elution buffer no sample retention should be observed. The sample adsorption generated by the high salt concentration where the solubility of the protein samples is at minimum. The most frequently used salt is $(NH_4)_2SO_4$, for the preparation of mobile phase A. Other chaotropic salts can also be applied. **Table 2** displays the chaotropic salt series. Strong chaotropic salts disrupt the structure of water and thus tend to decrease the strength of hydropho-

Fig. 1. *(continued)* Peaks: a = cytochrome-*c*, b = ribonuclease A, c = myoglobin, d = conalbumin, e = neochymotrypsin, f = α-chymotrypsin, g = α-chymotrypsinogen A. Reprinted from **ref. *16*** with permission from Elsevier Science, copyright 1986.

Table 2
Anionic and Cationic Ions Used in HIC

Anions	Cations
$(PO_4)^{3-}$	NH_4^+
$(SO_4)^{2-}$	Rb^+
CH_3COO^-	K^+
Cl^-	Na^+
Br^-	Cs^+
NO^-	Li^+
ClO_4^-	Mg^{2+}
I^-	Ca^{2+}
SCN^-	Ba^{2+}

bic interactions; the antichaotropic salts tend to favor them. Organic solvents are also commonly used to alter the polarity of water. Any of those salts can be used for the preparation of the solvophobic agent. The samples elute in the order of their solubility as a function of the salt concentration and their adsorption–desorption equilibrium *(17)*.

Mobile phase A: The adsorption generating solvent is usually $(NH_4)_2SO_4$, dissolved in the mobile phase B buffer.

Mobile phase B: All sample components should be soluble in mobile phase B. The pH and ionic strength of B solvent should be selected accordingly.

Elution mode: During separation condition search, gradient elution mode is recommended. However, ones the elution conditions established than step gradient or isocratic elution can be applied. The elution mode primarily depends on the complexity of the samples.

Flow rate: The usually applied starting flow rate is 1 mL/min. However, higher flow rate can be used in isocratic separation mode.

4. UV or fluorescence detection is usually applied during HIC separations. Because of the high salt concentration, significant background noise can be observed without filtration and use of pure salts.
5. Temperature usually set at a comfortable laboratory temperature (25°C). However, the temperature can have significant effect on the retention, peak shape *(18)*, and activity and mass recovery of proteins. **Figure 2** shows the retention time changes as a function of surface hydrophobicity. Nonlinear changes in retention as a function of temperature are an indication of conformational changes of the sample proteins. Some of the chromatographic changes can be originated from altering the conformational equilibrium of proteins. For sensitive samples, a study of temperature effect is recommended, in order to understand the phenomena involved *(19)*. As an example, **Fig. 3** shows the chromatograms of α-lactalbumin as a func-

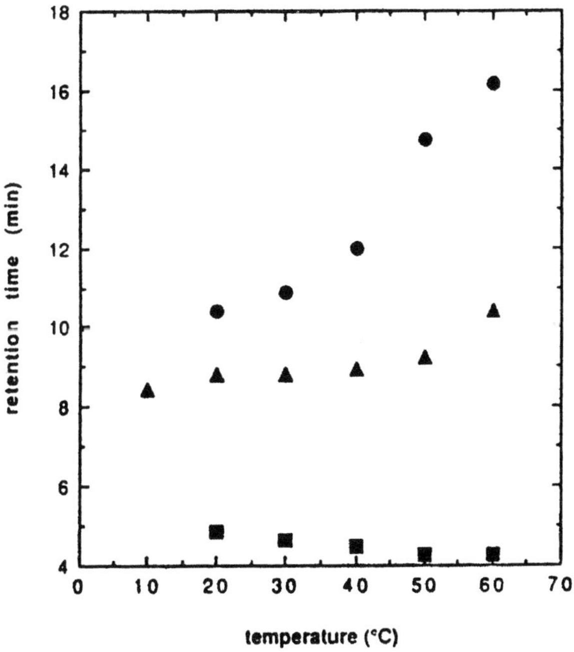

Fig. 2. Effect of surface hydrophobicity and temperature on the retention of α-lactalbumin on the PolyALKYL Aspartamide columns at pH 6.0 using ammonium acetate based mobile phase. Reprinted from **ref.** ***19*** with permission from Elsevier Science, copyright 1988.

tion of temperature. The displayed series of chromatograms clearly illustrate that the retention time and peak width significantly alters with increasing temperature. Normalized retention time-temperature curves are sigmoidal in those cases where denaturation occurs under the given chromatographic conditions. These denaturation curves are very similar to the classical transition curves of protein denaturation and have been analyzed by the same approach. Temperature studies also represent an example, that chromatography is a versatile experimental tool for not only the analysis and separation but physico-chemical studies of protein conformational changes as well ***(19)***.

6. The use of high salt could create erosions and plugging, which are usually not observed with other chromatographic systems. After separations the system should be thoroughly cleaned with the B solvent and water. The high salt mobile phase can precipitate at different part of the chromatographic system. Washing is critically important prior mobile phase and/or chromatographic mode changes. It is important to establish standard conditions to evaluate the separation system with protein standards. Standard conditions can provide information about system reproducibility, separation and system conditions, and column lifetime.

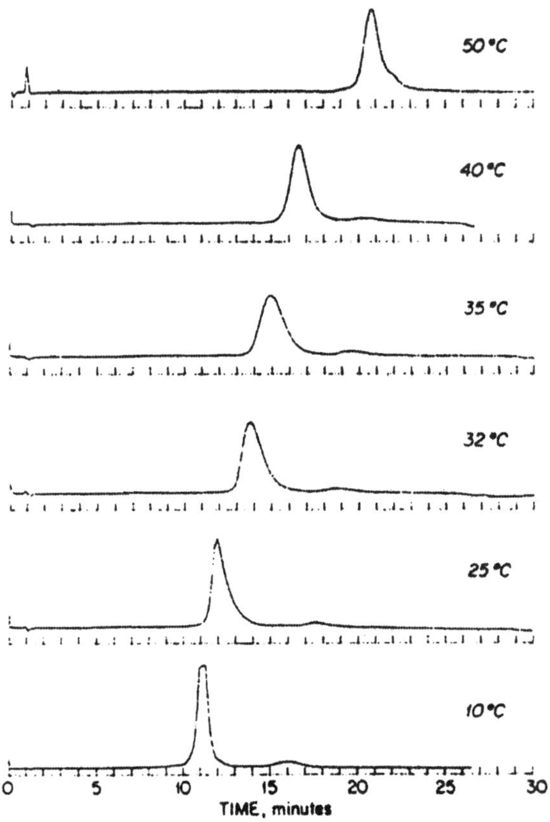

Fig. 3. Effect of temperature on the chromatographic peak width of α-lactalbumin on an HIC stationary phase. Reprinted from **ref.** *18* with permission from Elsevier Science, copyright 1986.

References

1. Welling, G. A. W. K., van der Zee, R., and Welling-Wester, S. (1987) Column liquid chromatography of integral membrane proteins. *J. Chromatogr.* **418,** 223–243.
2. Wilson, M. J., Haggart, C. L., Gallagher, S. P., and Walsh, D. (2001) Removal of tightly bound endotoxin from biological products. *J. Biotechnol.* **88,** 67–75.
3. Karlsson, G. and Winge, S. (2001) Separation of native and latent forms of human antithrombin by hydrophobic interaction high-performance liquid chromatography. *Protein Expr. Purif.* **21,** 49–55.
4. Raynes, J. G. and McAdam, K. P. (1988) Purification of serum amyloid A and other high density apolipoproteins by hydrophobic interaction chromatography. *Anal. Biochem.* **173,** 116–124.

5. Diogo, M. M., Queiroz, J. A., and Prazeres, D. M. F. (2002) Hydrophobic interaction chromatography of homo-oligonucleotides on derivatized Sepharose CL-6B. Application of the solvophobic theory. *J. Chromatogr. A.* **944,** 119–128.
6. Muck, A., Ramm, M., and Hamburger, M. (1999) Efficient method for preparation of highly purified lipopolysaccharides by hydrophobic interaction chromatography. *J. Chromatogr. B. Biomed. Sci. Appl.* **732,** 39–46.
7. Matsumoto, I., Koyama, T., Kitagaki-Ogawa, H., and Seno, N. (1987) Separation of isolectins by high-performance hydrophobic interaction chromatography. *J. Chromatogr.* **400,** 77–81.
8. Peng, J. S., Tsai, W. C., and Chou, C. C. (2001) Surface characteristics of Bacillus cereus and its adhesion to stainless steel. *Int. J. Food Microbiol.* **65,** 105–111.
9. Krishna, M. M., Powell, N. B., and Borriello, S. P. (1996) Cell surface properties of Clostridium difficile: haemagglutination, relative hydrophobicity and charge. *J. Med. Microbiol.* **44,** 115–123.
10. Wu, S. L., Hancock, W. S., Pavlu, B., and Gellerfors, P. (1990) Application of high-performance hydrophobic-interaction chromatography to the characterization of recombinant DNA-derived human growth hormone. *J. Chromatogr.* **500,** 595–606.
11. *Hydrophobic Interaction Chromatography: Principles and Methods*, Pharmacia Biotech, 1993.
12. Lauer, I., Bonnewitz, B., Meunier, A., and Beverini, M. (2000) New approach for separating Bacillus subtilis metalloprotease and alpha-amylase by affinity chromatography and for purifying neutral protease by hydrophobic chromatography. *J. Chromatogr. B. Biomed. Sci. Appl.* **737,** 277–284.
13. Horváth, C., Melander, W., and Molnár, I. (1976) Solvophobic ineractions in liquid chromatography with non-polar stationary phases. *J. Chromatogr.* **125,** 129–156.
14. Melander, W. R., El Rassi, Z., and Horváth, C. (1989) Interplay of hydrophobic and electrostatic interactions in biopolymer chromatography. Effect of salts on the retention of proteins. *J. Chromatogr.* **469,** 3–27.
15. Shaltiel, S. (1984) Hydrophobic chromatography. *Methods Enzymol.* **104,** 69–96.
16. Alpert, A. J. (1986) High performance hydrophobic interaction chromatography of proteins on a series of poly(alkyl aspartamide)-silicas. *J. Chromatogr.* **359,** 85–97.
17. Arakawa, T. and Narhi, L. O. (1991) Solvent modulation in hydrophobic interaction chromatography. *Biotechnol. Appl. Biochem.* **13,** 151–172.
18. Wu, S.-L., Benedek, K. and Karger, B. L. (1986) Thermal behavior of proteins in high-performance hydrophobic-interaction chromatography. On-line spectroscopic and chromatographic characterization. *J. Chromatogr.* **359,** 3–17.
19. Benedek, K. (1988) Thermodynamics of α-lactalbumin in hydrophobic-interaction chromatography and stationary phase comparison. *J. Chromatogr.* **458,** 93–104.

5

Gel Filtration Chromatography

Peter Stanton

1. Introduction

Gel filtration chromatography (sometimes referred to as size exclusion chromatography) separates biomolecules based on differences in their molecular size. The process employs a gel media suspended in an aqueous buffer solution which is commonly packed into a chromatographic column. These columns can vary in size from very small (for example, spin columns of <1 mL bed volume for analytical separations) to very large (for preparative scale applications). The gel media consists of spherical porous particles of carefully controlled pore size through which biomolecules diffuse to different extents based on differences in their molecular sizes. Small molecules diffuse freely into the pores and their movement through the column is retarded, whereas large molecules are unable to enter the pores and are therefore eluted earlier. Hence, molecules are separated in order of decreasing molecular weight, with the largest molecules eluting from the column first.

Applications for gel filtration chromatography typically fall into one of two categories, either: (i) fractionation, or (ii) group separation. In fractionation applications, molecules are separated according to *small* differences in size, as would be required in a purification or characterization protocol. A group separation separates all high-molecular-weight molecules from all low-molecular-weight molecules effectively yielding two groups, as would be required when exchanging buffer components or salts, or removing low-molecular-weight contaminants from valuable higher molecular-weight samples (proteins, oligonucleotides, plasmids, or polysaccharides). Operating procedures and gel selection criteria vary upon the category of separation chosen, as will become evident later.

From: *Methods in Molecular Biology, vol. 251, HPLC of Peptides and Proteins: Methods and Protocols*
Edited by: M.-I. Aguilar © Humana Press Inc., Totowa, NJ

Several other methods are available to separate or characterize biomolecules based on differences in size, including sodium dodecyl sulfate (SDS)-polyacrylamide gel electrophoresis *(1)* and ultracentrifugation *(2)*. The main advantage for gel filtration chromatography over these techniques is the ability to separate analytical to preparative amounts of material under native, non-denaturing conditions with samples recovered in a form suitable for direct downstream processing (such as other chromatographic steps, or analysis in biological assay systems). The instrumentation required for gel filtration chromatography is also relatively inexpensive and readily obtainable. The main disadvantages of gel filtration chromatography are a relative lack of resolution, an inability to analyze more than one sample at any one time (unless multiple columns are utilized), and samples are recovered in a more dilute form than the starting material (frequently ≥ 2–3-fold dilution).

The resolution that can be obtained by gel filtration chromatography is dependent on a number of variables. However, for a fully optimized gel filtration analysis, baseline–baseline resolution can be achieved for proteins that differ in molecular mass by as little as a factor of approx 1.7 fold, although a common rule of thumb is for a two-fold difference in molecular size for separation to occur (*see* **Fig. 1A**). Clearly, if baseline resolution is not critical, then much-lower-fold differences between solutes can be recognized as shoulders on chromatographic profiles. In addition, superior resolution and sensitivity can be achieved with high-performance pre-packed columns *(3–5)*. High-performance gel filtration can also be used for desalting samples through buffer exchange and has also found application in the analysis of peptides *(6)*. High-performance systems can also be used to provide insight into protein folding mechanisms by monitoring changes in protein size as a function of changes in the concentration of chemical denaturants *(7,8)*.

Fig. 1. *(see facing page)* **(A)** Gel filtration chromatography of a series of protein molecular mass standards on a pre-packed Superdex 200 HR 10/30 column (Amersham Biosciences). Standards were 1. thyroglobulin (M_r 669,000), 2. ferritin (M_r 440,000), 3. human IgG (M_r 150,000), 4. human transferrin (M_r 81,000), 5. ovalbumin (M_r 43,000), 6. myoglobin (M_r 17,600), 7. vitamin B12 (M_r 1,355). Conditions were: 50 m*M* sodium phosphate, 150 m*M* NaCl, pH 7.0, flow rate = 0.25 mL/min (19 mL/cm^2/h). Figure reproduced with the kind permission of Amersham Biosciences. **(B)** Diagrammatic representation of the measurement of the void volume (V_0), elution volume (V_e) and total volume (V_t) for a gel filtration column. V_0 is the elution volume of molecules too large to enter the pores of the gel media, whereas V_t is the total volume of the column determined with a small molecule. V_e represents the elution volume of a molecule of intermediate molecular mass. Determination of these parameters is best done with standard

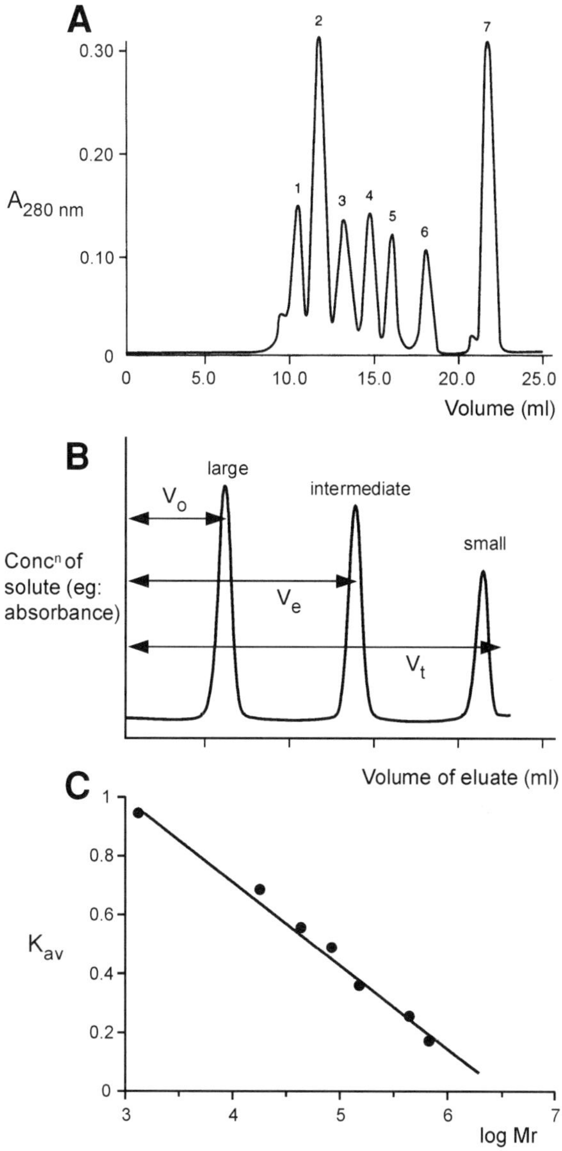

Fig. 1. *(continued)* proteins (*see* panel A) under optimal conditions for flow rate and sample size. From these measurements the coefficient K_{av} can be derived, $K_{av} = (V_e - V_0)/(V_t - V_0)$. **(C)** The selectivity curve for a particular gel filtration media is a plot of K_{av} vs log molecular weight, and the data shown here has been derived from panel A. Selectivity curves as provided by the manufacturer are used to choose the gel filtration media that best suits the application, whereas in the laboratory they are useful for estimation of the molecular mass of an unknown protein.

2. Materials

2.1. Gel Media and Selection Criteria

A variety of gel media are available for gel filtration chromatography, based on either dextrans (Sephadex series, **Table 1**), agaroses (Sepharose, Sepharose-CL, Superose, BioGel A, and Ultrogel A, **Table 1**), polyacrylamide (BioGel P), or mixtures of these components (Sephacryl, Superdex, Ultrogel AcA), all of which are suitable for low- to medium- pressure chromatographic applications. In addition, porous silica-based media (TSK, SynChropak, Bio-Sil, Zorbax) suitable for higher pressure applications are available. A number of these media are prepacked in columns direct from the manufacturer (*see* **Table 1**) whereas others are supplied in bulk as either dry powders requiring swelling in buffer before use, or are supplied preswollen.

Selection of the appropriate media for gel filtration chromatography is heavily dependent on the particular application, hence, some prior knowledge about solute molecular weight(s) is advantageous. Ideally, the solute(s) of interest will elute in the first part of the chromatogram ***(9)***, but should not be eluted in the void volume (V_0), which is the elution volume of molecules too large to enter any of the pores in the gel (*see* **Fig. 1B**). Similarly, solute elution at the total volume of the column (V_t) (*see* **Fig. 1**) yields little information. Most manufacturers provide selectivity curves (molecular weight *v* K_{av}, *see* **Note 1** and **Fig. 1C**) derived from the separation of a mixture of protein standards of varying molecular weight as a guide in media selection. Factors that influence resolution for any particular gel filtration media include column length and width, particle size, and flow rate (see later). If multiple gel filtration media are available in the size range of interest, choose the media with the steepest selectivity curve. In the absence of any corroborating information about the solute of interest, selection of the appropriate media will require an empirical approach.

Although there may appear a large number of media available for gel filtration chromatography (*see* **Table 1**), in reality, the choices can be rapidly minimised based on key variables including molecular-weight range, scale (analytical or preparative), time-scale of the separation (which is influenced by particle size, flow rate, pressure, and expected resolution) and finally, cost and availability.

2.2. Column Length/Width Selection

The ability of a gel filtration column to resolve two components increases in direct proportion to the square root of the column length ***(10)***. Hence, column bed heights tend to be greater in applications requiring maximal resolution

Text continues on page 64

Table 1
Physical Properties of Commonly Used Gel Filtration Media[a]

Mol wt range: Peptides, globular proteins	Mol wt range: Dextrans	Product	Matrix type		Particle size (µm)	Approx bed volume (mL/gm)	Comments	Company[a]
Up to 700	Up to 700	Sephadex G-10	Crosslinked dextran		40–120	2–3	Dry powder	AB
Up to 1,500	Up to1500	Sephadex G-15	Crosslinked dextran		40–120	2.5–3.5	Dry powder	AB
1000–5000	1000–5000	Sephadex G-25	Crosslinked dextran	Coarse	100–300	4–6	Dry powder	AB
				Medium	50–150	"	"	
				Fine	20–80	"	"	
				Superfine	10–40	"	"	
1500–30,000	500–10,000	Sephadex G-50	Crosslinked dextran	Coarse	100–300	9–11	Dry powder	AB
				Medium	50–150	"	"	
				Fine	20–80	"	"	
				Superfine	10–40	"	"	
3000–80,000	1000–50,000	Sephadex G-75	Crosslinked dextran	medium	40–120	12–15	Dry powder	AB
3000–70,000	"			Superfine	10–40	"	"	
4000–150,000	1000–100,000	Sephadex G-100	Crosslinked dextran	Medium	40–120	15–20	Dry powder	AB
4000–100,000	"			Superfine	10–40	"	"	
5000–300,000	1000–150,000	Sephadex G-150	Crosslinked dextran	Medium	40–120	20–30	Dry powder	AB
5000–150,000	"			Superfine	10–40	"	"	
5000–600,000	1000–200,000	Sephadex G-200	Crosslinked dextran	Medium	40–120	30–40	Dry powder	AB
5000–250,000	1000–150,000			Superfine	10–40	20–25		
10,000–4,000,000	10,000–1,000,000	Sepharose 6B " CL-6B[b]	6% Agarose		45–165	—	Preswollen	AB

(Continued)

Table 1 *(continued)*

Mol wt range: Peptides, globular proteins	Mol wt range: Dextrans	DNA exclusion limit (bp)	Product	Matrix type	Particle size (μm)	Approx bed volume (mL/gm)	Comments	Company[a]
60,000–20,000,000	30,000–5,000,000		Sepharose 4B " CL-4B[b]	4% Agarose	45–165	—	Preswollen	AB
70,000–40,000,000	100,000–20,000,000		Sepharose 2B " CL-2B[b]	2% Agarose	60–200	—	Preswollen	AB
1000–300,000	NA			Superose 12	Crosslinked agarose 8–12 prep grade	— 20–40	Preswollen	AB
5000–5,000,000	NA			Superose 6	Crosslinked agarose 11–15 prep grade	— 20–40	Preswollen	AB
1000–100,000	NA	—	Sephacryl S-100 HR	Dextran crosslinked with acrylamide	25–75	—	Preswollen	AB
5000–250,000	1000–80,000	118	Sephacryl S-200 HR	Dextran crosslinked with acrylamide	25–75	—	Preswollen	AB
10,000–1,500,000	2,000–400,000	118	Sephacryl S-300 HR	Dextran crosslinked with acrylamide	25–75	—	Preswollen	AB
20,000–8,000,000	10,000–2,000,000	271	Sephacryl S-400 HR	Dextran crosslinked with acrylamide	25–75	—	Preswollen	AB
NA	40,000–20,000,000	1078	Sephacryl S-500 HR	Dextran crosslinked with acrylamide	25–75	—	Preswollen	AB
NA	500,000–1×10^8	20,000	Sephacryl S-1000SF	Dextran crosslinked with acrylamide	NA	—	Preswollen	AB

100–7000	NA	Superdex peptide	Dextran + crosslinked agarose		11–15	—	Preswollen, and pre-packed column	AB
up to 10,000	NA	Superdex 30	Dextran + crosslinked agarose	prep grade	24–44	—	Preswollen	AB
3000–70,000	500–30,000	Superdex 75	Dextran + crosslinked agarose	 prep grade	11–15 24–44	—	Preswollen, and prepacked column	AB
10,000–600,000	1000–100,000	Superdex 20	Dextran + crosslinked agarose	 prep grade	11–15 24–44	—	Preswollen, and prepacked column	AB
100–1800	NA	Bio-Gel P-2	Polyacrylamide	Extra-fine Fine	<45 45–90	3	Dry powder	Bio-Rad
800–4000	NA	Bio-Gel P-4	Polyacrylamide	Extra-fine Fine Medium	<45 45–90 90–180	4	Dry powder	Bio-Rad
1000–6000	NA	Bio-Gel P-6	Polyacrylamide	Extra-fine Fine Medium	<45 45–90 90–180	6.5	Dry powder	Bio-Rad
1500–20,000	NA	Bio-Gel P-10	Polyacrylamide	Fine Medium	45–90 90–180	7.5	Dry powder	Bio-Rad
2500–40,000	NA	Bio-Gel P-30	Polyacrylamide	Fine Medium	45–90 90–180	9.0	Dry powder	Bio-Rad
3000–60,000	NA	Bio-Gel P-60	Polyacrylamide	Fine Medium	45–90 90–180	11.0	Dry powder	Bio-Rad
5000–100,000	NA	Bio-Gel P-100	Polyacrylamide	Fine	45–90	12.0	Dry powder	Bio-Rad

(Continued)

Table 1 *(continued)*

Mol wt range: Peptides, globular proteins	Mol wt range: Dextrans	DNA exclusion limit (bp)	Product	Matrix type	Particle size (µm)		Approx bed volume (mL/gm)	Comments	Company[a]
					Medium	90–180			
<10,000–500,000	NA	—	BioGel A 0.5 *M*	Agarose	Fine	38–75	—	Preswollen	Bio-Rad
					Medium	75–150			
					Coarse	150–300			
<10,000–1.5×10^6	NA	—	BioGel A 1.5 *M*	Agarose	Fine	38–75	—	Preswollen	Bio-Rad
					Medium	75–150			
					Coarse	150–300			
10,000–5×10^6	NA	—	BioGel A 5 *M*	Agarose	Fine	38–75	—	Preswollen	Bio-Rad
					Medium	75–150			
					Coarse	150–300			
40,000–15×10^6	NA	—	BioGel A 5 *M*	Agarose	Fine	38–75	—	Preswollen	Bio-Rad
					Medium	75–150			
					Coarse 1	50–300			
100,000–50×10^6	NA	350	BioGel A 50 *M*	Agarose	Medium	75–150	—	Preswollen	Bio-Rad
					Coarse	150–300			
25,000–2.4×10^6	NA	Ultrogel A6		Crosslinked agarose	60–140		—	Preswollen	Sigma
55,000–9.0×10^6	NA	Ultrogel A4		Crosslinked agarose	60–140		—	Preswollen	Sigma
120,000–25×10^6	NA	Ultrogel A2		Crosslinked agarose	60–140		—	Preswollen	Sigma

5,000–70,000	NA	Ultrogel AcA 54	Agarose and polyacrylamide	60–140	—	Preswollen	Sigma
10,000–130,000	NA	Ultrogel AcA 44	Agarose and polyacrylamide	60–140	—	Preswollen	Sigma
20,000–350,000	NA	Ultrogel AcA 34	Agarose and polyacrylamide	60–140	—	Preswollen	Sigma
100,000–1.2×10^6	NA	Ultrogel AcA 22	Agarose and polyacrylamide	60–140	—	Preswollen	Sigma
5000–100,000		TSK G2000SW XL	Bonded silica	5	—	Prepacked HPLC column	Supelco, Agilent
5000–150,000		TSK G2000SW	Bonded silica	10,13	—	"	Supelco, Agilent
10,000–500,000		TSK G3000SW XL	Bonded silica	5	—	"	Supelco, Agilent
		TSK G3000SW	Bonded silica	10			
20,000–10×10^6		TSK G4000SW	Bonded silica	13,17	—	"	Supelco, Agilent
5000–100,000		Bio-Sil SEC 125	Bonded silica	NA	—	"	Bio-Rad
10,000–300,000		Bio-Sil SEC 250	Bonded silica	NA	—	"	Bio-Rad
20,000–1×10^6		Bio-Sil SEC 400	Bonded silica	NA	—	"	Bio-Rad
4000–400,000		Zorbax GF-250	Zirconia-stabilised silica	4	—	"	Agilent
10,000–900,000		Zorbax GF-450	Zirconia-stabilised silica	6	—	"	Agilent

[a]Information as provided by the manufacturer. Further details available at: Amersham Biosciences (AB) http://www.chromatography.amershambiosciences.com, Bio-Rad http://www.bio-rad.com/, Sigma-Aldrich http://www.sigmaaldrich.com, Supelco http://www.sigmaaldrich.com, Agilent http://www.chem.agilent.com

[b]CL = cross-linked, which imparts a greater stability (chemical, thermal, physical) on the gel. Noncrosslinked Sepharose supports are also listed.

(for example, for protein purification), with a practical limit in most laboratory situations of approx 1 m. For particularly difficult separations requiring long columns, practical alternatives are to either: (i) link two columns in series, or (ii) recycle selected regions. In contrast, gel bed heights in group separation (desalting) applications are normally shorter (10–30 cm) and wider, to allow for faster flow rates and larger sample volumes.

A key factor in deciding on column dimensions for protein purification purposes is the volume of the sample to be loaded, which should be between 0.5–5% of the bed volume of the column ***(10)***. Amersham Biosciences market a range of columns with internal diameters of 0.9, 1.6, 2.6, and 5.0 cm, with lengths of 15–100 cm, which cover most applications, and comparable hardware is available from the other companies. For desalting purposes, sample volumes may be up to 30% of the column bed volume.

2.3. Eluent

Under native conditions, the composition of the eluent does not contribute directly to the separation, hence, eluent components can be chosen which best suit the particular sample and experiment. For uncharged (electrically neutral) solutes, distilled water is sufficient as an eluent. For charged solutes (proteins, oligonucleotides, and so on), a buffer (for example, 20 m*M* sodium phosphate, 20 m*M* Tris-HCl, 20 m*M* sodium acetate) should be used to control pH, which can be between pH 3–10 for most supports, although the product literature for the gel type chosen should be consulted first (*see* **Note 2**). Buffers should also include up to 100 m*M* NaCl to prevent ionic interactions between the charged solute and the gel media, which can otherwise produce spurious gel filtration behavior (*see* **Subheading 4.3.**). For applications in which limited amounts of sample are available (for example, sub- to low-microgram amounts of proteins, or for trace-labeled proteins) and losses by nonspecific adsorption become important, additives such as nonionic detergents (Tween-20, Brij-35, Triton X-100, at approx 0.01–0.05% v/v) or a protein carrier (bovine serum albumin [BSA], 0.1% w/v) may be included at low concentration (*see* **Note 3**).

Frequently, samples obtained following gel filtration chromatography must be concentrated prior to further processing. If concentration is to be by freeze-drying, buffers containing nonvolatile components can become a problem. Volatile buffers (for example, ammonium bicarbonate, ammonium acetate, or acetic acid) (50–100 m*M*) are a useful alternative. Finally, if the gel filtration experiment is to be carried out under denaturing conditions, the eluent should include 6 *M* guanidine hydrochloride, or 8 *M* urea (*see* **Note 4**). The use of acidic conditions (up to 4 *M* acetic acid) will also dissociate many noncovalent protein–protein complexes.

All buffers for gel filtration chromatography should be prepared using freshly degassed water, to prevent the formation of air bubbles within the tubing lines and column. Buffers should be redegassed (or prepared again) after a 2-d interval for high-resolution applications (longer columns), however, this is not necessary for group separations.

2.4. Flow Rate

Flow rates should be set in conjunction with the manufacturers instructions for the particular gel type and column dimensions, recognising that maximal flow rate for any particular gel type will be dependent on its particle size. In general, improved resolution is obtained with slower flow rates, and an optimum flow rate for protein fractionation of approx 5 $mL/cm^2/h$ is recommended for most soft gels ***(10)***, although acceptable resolution may be achieved with flow rates up to five-fold greater ***(10)***. If maximal resolution is not required, then tradeoffs between flow rate and increased particle size for any particular gel type (*see* **Table 1**) can be considered to shorten the analysis time of the gel filtration experiment.

2.5. Sample

Optimal sample volumes have been mentioned earlier (refer to **Subheading 2.2.**). Samples should be clarified (filtration through 0.22 μ*m* filter, or centrifugation at 10,000*g*) prior to loading to prevent contamination of the support net or the top of the column. High sample viscosities in relation to the eluent should be avoided, as these can lead to skewed peak shapes and apparent lack of resolution (*see* **Note 5**).

In practice, samples are diluted during desalting/buffer exchange by a factor of ≥1.25 fold, whereas dilution factors may be greater during protein fractionation experiments (1.4–3-fold).

2.6. Recommended Hardware Configuration

Simple group separations using prepacked commercially available mini-columns require no more equipment than the column itself, and maybe access to a spectrophotometer to monitor the optical density of eluted fractions. In contrast, fractionation of proteins on larger laboratory-based columns requires the following: (i) a column, flow adaptor, and packing funnel; (ii) a pump (normally peristaltic) to accurately control eluent flow; (iii) an in-line UV detector coupled to a chart recorder, and (iv) a fraction collector. In some cases, the pump and UV detector can be omitted, with flow controlled by gravity and UV monitored on selected fractions on a spectro. Prepacked HPLC gel filtration columns clearly require access to an HPLC preferably with a nonmetallic (glass, teflon, ceramic) flow path to limit potential salt-induced corrosion of metallic components.

A few simple rules should be followed in setting up equipment for gel filtration chromatography. If the column is to be operated at 4°C, choose either a cold-jacketed design that can be hooked up to a recirculating refrigerated cooler in the laboratory, or place the column in the 4°C coldroom. In the latter case, sensitive electronic equipment (control unit for the UV detector, chart recorder) is best located outside the coldroom to avoid condensation problems. If the column eluent is to contain solvents or acids, choose a solvent-resistant column design.

A peristaltic pump can be placed either at the top of the column (to "push" the eluent) or at the base (to "suck" eluent), however, note that placement at the top will prevent potential bubble formation in the lines (*see* **Subheading 4.8.**). In either case, it is essential that the pump tubing be regularly inspected and the flow rate calibrated, as with long use, the tubing cross section can be permanently flattened by the peristaltic action, leading to changes in flow and ultimately to tube failure.

UV detection of proteins in gel filtration chromatography is typically carried out at 280 nm with the detector placed as close to the bottom of the column as possible, and the fraction collector immediately downstream. It is essential to minimize the dead volume of the tubing between column, detector, and collector to avoid post-column mixing of separated components, and also to limit the offset between the UV trace and the collected fraction. Ensure that pathlength of the UV detector cell is compatible with the expected protein load (2–3 mm pathlength for high protein concentrations, 10 mm pathlength for greater sensitivity).

3. Method

The following general instructions apply to both group separations (e.g., desalting, buffer exchange) and fractionation of proteins, although the latter application will require a greater attention to optimisation. For situations in which prepacked columns or spin columns are employed, the manufacturers instructions should be consulted.

3.1. Prepare the Gel

1. If the gel is supplied as a dry powder, calculate the weight of gel needed for the approximate bed volume of the column, and preswell the gel (between 3–72 h at 20°C in separation buffer) according to the manufacturers' recommendation (*see* **Table 1**). Alternatively, dry gels can be preswollen in a water bath at 90°C for 1–5 h (*see* **Note 6**). Swollen gels can be stored indefinitely at 4°C in the presence of a suitable bacteriostat (e.g., 0.02–0.1% w/v NaN_3, 0.005% thimerosal,

0.002% w/v chlorhexidine) (*see* **Note 7**). Avoid the use of a magnetic stirrer with gel filtration media, as this can break beads and generate fine particles.

Preswollen gels are ready for use. Mix the gel to a slurry, remove the required amount into a vacuum flask, and allow to settle.

2. For both gel types, remove the aqueous layer and replace with a volume of the separation buffer to make a final slurry of approx 50–70% gel. Repeat this washing process at least once, or until the gel settles as a defined layer without any suspended 'fines' in the aqueous layer (*see* **Note 8**).
3. Prior to column packing, media and buffers must be degassed on a house vacuum to remove dissolved gas (gels preswollen at 90°C do not need to be degassed). Efficient degassing and packing of gels will not occur if the slurry is too thick.
4. Ensure that the temperature of the gel and buffer is the same as the environment the column is to be packed in, as gel transferred from 4°C to room temperature will absorb gas, potentially resulting in a poorly packed column.

3.2. Prepare the Column

1. Clamp the column in a vertical position, in a position away from drafts or direct sunlight.
2. Ensure that end-filters, nets, and frits are clean and do not have holes.
3. Connect the base of the column to the peristaltic pump (if used), and flush tubing with degassed buffer, removing any bubbles that may be trapped under the net.
4. Set the flow rate of the pump as recommended by the manufacturer of the particular gel media, or to a rate approx 30%–50% higher than will be used for the separation.
5. If the volume of the slurry is greater than the column volume, fit a packing funnel or column extender to the top of the column, to allow all the gel to be added at the one time.

3.3. Pack the Column

1. Reslurry the degassed gel media, and decant it all into the column using a glass rod. Alternatively, the column can be momentarily tipped on its side and the gel poured directly in, after which the column should immediately be placed upright. In both cases, avoid introducing excess bubbles into the gel media.
2. Start the pump, and allow the gel to pack under constant flow until the gel bed height becomes constant.
3. At this stage, stop the pump, remove the packing funnel, and adjust bed height by carefully removing excess gel, leaving the top of the gel bed level and flat. It is important not to disturb the rest of the gel bed at this point. If the column is too short, avoid "topping-up" with extra gel as this can lead to poor chromatography. It is better to unpack the column, add extra gel, and then repack (*see* **Note 9**).
4. At this point, the column should be sealed by adding the top flow adaptor. This is best done in the following steps: (i) gently top the column up with buffer to the top; (ii) use a 10-mL syringe to purge the adaptor and tubing with buffer, making sure that there are no bubbles under the net; (iii) place the buffer vessel above the

top of the column; (iv) connect the adaptor to the buffer vessel and establish buffer flow by syphon; (v) slide the adaptor across the meniscus of the liquid on top of the column, then allow it to sink to the top of the gel bed.

5. Seal the adaptor level with the gel bed, switch the pump on, and equilibrate the column with one column volume of buffer at the packing flow rate. If a small gap appears between the adaptor and the gel top this must be removed prior to sample addition (*see* **Note 10**).

3.4. Test the Packed Column

1. Set the appropriate flow rate for separation, and apply a test sample to the column, comprised of colored molecular-weight markers. These can be blue dextran 2000, cytochrome-*c* (mol. wt. 12,000), myoglobin (mol. wt. 17,300) or other colored reagents of appropriate molecular weights.
2. While loading, visually inspect the colored band which should remain horizontal, and watch for uneven loading caused by "craters" or a gel surface that is not level.
3. The presence of air bubbles or poorly packed gel can also be checked by inspection via transmitted light.
4. Skewing of the colored band(s) during chromatography indicates problems in the packing procedure, and can *only* be fixed by repacking the column.

3.5. Sample Application

1. Columns with flow adaptor. A number of alternatives are available to load the column, ranging from application directly through the buffer line, or via a syringe and valve *(3,4)*, or via a dedicated Superloop *(10)*. In the simplest case, buffer flow is stopped and the column inlet transferred to the sample container, after which the pump is restarted and the sample pumped onto the column (*see* **Note 11**).
2. Columns without flow adaptor. Samples are loaded onto the column by removing the stopper and running the buffer level down to the gel surface, after which the flow is stopped and the sample is carefully pipeted onto the gel. The buffer flow is restarted and when the sample has entered the column, an equivalent volume of buffer is pipeted onto the gel and run in, after which the column is gently refilled and plugged while the flow is running (*see* **Notes 5,11,** and **12**). Alternatively, samples can be layered onto the gel surface under the running buffer (while the column is running) after increasing sample density by the addition of sucrose or glycerol (10% final) (*see* **Note 5**).

3.6. Calibration With Molecular Mass Standards

1. For applications in which the molecular size of an unknown sample is required, the column must first be calibrated with standards of known apparent molecular mass *(Mr)*, which can be done in conjunction with column testing (above). Molecular size standards (*see* **Note 13**) are commercially available from Sigma

Chemical Co., and can also be purchased as kits from Amersham Biosciences and BioRad. These proteins are typically globular in shape, from which linear selectivity curves can be derived (*see* **Fig. 1**). An assumption made in assigning the molecular size of an unknown is that it shares a similar molecular shape with the standard, although obviously this need not be the case (for example, with fibrous proteins). If this is suspected, chromatography of both the standards and the unknown sample can be repeated with eluents containing denaturants (urea, guanidine hydrochloride) to break noncovalent bonds, and/or reducing agents (β-mercaptoethanol, dithiothreitol) to break disulphide bridges.

2. There are several methods by which gel filtration calibration data can be expressed ***(3,4)***, of which the most convenient is the determination of K_{av} **(Fig. 1B)**. A plot of K_{av} vs log molecular mass typically yields a straight line **(Fig. 1C)** from which the molecular mass of an unknown can be calculated. It can be seen that K_{av} values should fall between 0 and 1 if true gel filtration behaviour is occurring. Values of $K_{av}>1$ indicate some level of interaction between the sample and the column (for example, some supports show limited ion-exchange characteristics if the ionic strength of the eluent is too low), which can normally be prevented by modification of the eluent characteristics. K_{av} values <0 indicate formation of channels in the column, and the column must be repacked.

3.7. Trouble-Shooting

3.7.1. Column Runs Dry

The single most common problem with gel filtration columns is running out of eluent, which causes the column to run dry and form cracks and voids. There is no easy fix when this happens, and the column must be repacked. Fortunately, the remedy is simple, which is to always have sufficient eluent prepared for the duration of the chromatographic experiment. It is much easier to prepare more buffer than to repack and restandardize a new column! An electric timer attached a cut-off valve or the peristaltic pump will also suffice. For gravity-fed columns, an alternative solution is to install a loop of tubing alongside the column on the inlet side and place the column outlet above the lowest part of this loop ***(10)***.

3.7.2. Sample Precipitation

Precipitation of the sample on top of the column, or during chromatography, indicates incompatibility between the eluent (in terms of ionic strength, pH, or additives) and the sample. Conditions must be found in which the sample remains soluble, and which are also suitable for gel filtration chromatography. Once formed, a precipitate can significantly interfere with chromatography by either blocking the column, or causing smearing of protein peaks as partial dissolution occurs. Alternatively, precipitates can indicate incomplete clarification of the sample prior to loading on the column (*see* **Subheading 2.5.**).

3.7.3. Sample Elutes After the Included Volume, or Prior to the Void Volume

Peaks that elute after the included volume indicate a departure from ideal gel filtration chromatography to an adsorptive behavior between the sample and the gel media. Frequently this is ionic in character, and can be prevented by increasing the ionic strength of the eluent—typically 50–100 m*M* NaCl is used. Alternatively, hydrophobic interactions between sample and gel media may occur, which can be limited by either decreasing ionic strength, or adding an organic modifier (5–10% methanol, acetonitrile, or isopropanol). If adsorptive behavior continues despite making these changes, choose another type of gel media.

Peaks that elute prior to the void volume generally indicate the formation of channels or voids in the column, and the column must be repacked.

3.7.4. Poor Resolution

Poor resolution can arise through a number of causes. Check that the sample volume is not too large (*see* **Subheading 2.2.**), too viscous (*see* **Subheading 2.5.**), or that the sample has precipitated (*see* **Subheading 4.2.**). Check that the resolution of the column has been maintained using molecular-weight standards. If the resolution of the standards is poor, the problem is more basic and indicates either incorrect choice of gel filtration media (consult selectivity curves provided by the manufacturer) or chromatographic conditions (check column dimensions, flow rate), faulty column packing (repack the column), or a contaminated column (clean and/or repack the column).

3.7.5. Elution Position Not Consistent

A decrease in the column bed volume brought about by increased flow rate or partial column blocking will result in altered elution volumes. Also check the bottom filter or frit of the column and make sure that gel particles are not leaking through. Sample volumes and concentrations should be held constant between runs for good reproducibility.

3.7.6. Poor Recovery of Protein, or Activity

Make sure that an aliquot of the sample was taken for assay directly prior to column load, and assayed under the same conditions as the collected fractions. The mathematical sum of activities of the collected fractions following chromatography should also equal the activity of a combined pool prepared from aliquots of each collected fraction. If these conditions are met and recovery of activity is still low, look for losses occurring through inactivation of the sample via contact with the eluent, or separation of proteins from essential components (for example, enzyme and cofactor), or nonspecific adsorption to the gel media (*see* **Subheading 2.3.**).

3.7.7. Column Blocked/No Flow

If column blockage/slow flow occurs after packing but before samples have been applied, the problem is because of incorrect packing procedure (either flow rate too high, or presence of fines in the gel media), and the column must be repacked. If blockage has occurred due to protein precipitation in the top filter or on the surface of the gel bed, replace both the filter and the top 1–2 cm of contaminated gel.

3.7.8. Air Bubbles in Column, or in Eluent

Air bubbles in the gel bed commonly occur if eluents are not fully degassed, or if the column is moved from the coldroom to the bench. Washing the column extensively with fully degassed buffers of the correct temperature will normally solve these problems, although maximal resolution may require repacking. Large air bubbles under the top filter can be removed by carefully removing the top plunger, and reinstalling it (*see* **Subheading 3.3.4.**).

In situations where a peristaltic pump is installed at the base of the column, air bubbles caused by solvent outgassing can appear in the tubing between the column and the pump. These normally occur due to either partial column blockage (*see* **Subheading 4.7.**), compaction of the gel bed, or use of eluents which have not been fully degassed. Wash the column extensively (as above) in the reverse direction, but if the problem persists consider repacking the column and place the pump in front of the column (*see* **Subheading 2.6.**).

4. Notes

1. K_{av} is a partition coefficient, and is equivalent to $(V_e–V_0)/(V_t–V_0)$. *See* legend to **Fig. 1** for details.
2. In general, HPLC columns with silica-based supports should not be exposed to pHs greater than approx 7.4, otherwise hydrolysis of the silica support can occur. Consult product literature before proceeding.
3. Detergent concentrations should be below their critical micellar concentrations, which is the point at which protein-detergent micelles form. Protein-detergent micelles will chromatograph with larger apparent molecular masses than the protein itself.
4. 6 *M* guanidine hydrochloride and 8 *M* urea readily crystallize on contact with air. Care must be taken that these components do not crystallise inside the pump, detector, or fraction collector, and these items should be washed with water as soon after the experiment is completed as practicable.
5. Do not make the sample too viscous (*see* **Subheading 2.5.**) as peak skewing may result. Sample viscosity should be < two-fold the viscosity of the eluent, corresponding to a maximum protein concentration of approx 70 mg/mL *(4)*.
6. Avoid heating gel directly on a hot plate, as this can melt the gel media.

7. NaN_3 will form explosive salts in lead-based drainage systems, and should be disposed of by appropriate means. NaN_3 also inhibits certain enzymes (for example, peroxidases) and should not be used if gel filtration products are to be subsequently used with live cells or for enzymaticaly catalysed reactions (for example, iodination). Exposure to NaN_3 as a dry powder should also be limited; consult MSDS prior to use.
8. Gel filtration media can stick to glass. Ensure that all glassware is adequately cleaned after use to avoid contamination with other chromatographic media.
9. This description is for larger laboratory-scale columns using reuseable hardware. Where applications require single use or disposable columns (e.g., for radioactive samples) glass tubing, Pasteur pipets, or plastic syringes can be substituted, with a plug prepared from silanized glass wool.
10. For columns without a flow adaptor, carefully fill the space above the gel with buffer, and seal the column with an rubber stopper (or similar) through which the buffer line passes.
11. Make sure that the column inlet tubing is as short as possible with minimal dead volume, and is not looped. In a worst-case scenario, dense protein solutions can sit in the bottom of a tubing loop allowing buffer to flow over the top, increasing the sample application volume and, hence, decreasing resolution. Allowing a very small air bubble (approx 1mm) to enter the line between the sample and the eluent will prevent this problem without any major effects on chromatography.
12. Do not allow the column to run column dry, and avoid any disturbance to the column bed while adding buffer.
13. Commonly used molecular mass standards are bovine thyroglobulin (669,000), horse spleen ferritin (440,000), bovine liver catalase (232,000), γ-globulin (165,000), rabbit muscle aldolase (158,000), phosphorylase b (97,000), BSA (68,000), hen egg ovalbumin (43,000), bovine carbonic anhydrase (29,000), bovine chymotrypsinogen A (25,000), equine myoglobin (17,500), bovine ribonuclease A (13,700), hen egg-white lysozyme (14,300), bovine cytochrome-*c* (12,000), vitamin B_{12} (1,350). Markers for the void volume (V_0) can either be one of the higher molecular-weight proteins, or blue dextran 2000. Markers for the included volume (V_t) can be NaN_3, cytidine, or a low number of counts of a radioactive marker such as ^3H_2O or $Na^{125}I$.

References

1. See, Y. P. and Jackowski, G. (1989) Estimating molecular weights of polypeptides by SDS gel electrophoresis, in *Protein Structure: A Practical Approach* (Creighton, T.E., ed.), IRL Press at Oxford University Press, Oxford, UK, pp. 1–21.
2. Marks, M.S .(1998) Determination of molecular size by zonal sedimentation analysis on sucrose density gradients, in *Current Protocols in Cell Biology* (Bonifacino, J. S., Dasso, M., et al., eds.), Wiley, New York, pp. 5.3.1–5.3.33.
3. Gooding, K. M. and Regnier, F. E. (1990) Size exclusion chromatography in *HPLC of Biological Macromolecules: Methods and Applications* (Gooding, K. M. and Regnier, F. E., eds.), Dekker, New York, pp. 47–75.

4. Gooding, K. M. and Freiser, H. H. (1990) High performance size exclusion chromatography of proteins, in *High Performance Liquid Chromatography of Peptides and Proteins: Separation, Analysis and Conformation* (Mant, C. T. and Hodges, R. S., eds., CRC Press, Boca Raton, FL., pp. 135–144.
5. Anspach, B., Gierlich, H.U., and Unger, K. K. (1998) Comparative study of Zorbax Bio Series GF 250 and GF 450 and TSK-Gel 3000 SW and SWXL columns in size exclusion chromatography of proteins. *J. Chromatogr.* **443,** 45–54.
6. Mant, C. T., Parker, J. M. R., and Hodges, R. S. (1987) Size-exclusion high performance chromatography of peptides. Requirement for peptide standards to monitor column performance and non-ideal behaviour. *J. Chromatogr.* **397,** 99–112.
7. Uversky, V. N. (1993) Use of fast protein size exclusion liquid chromatography to study the unfolding of proteins which denature through the molten globule. *Biochemistry* **32,** 13,288–13,298.
8. Herold, M. and Leistler, B. (1991) Unfolding of truncated and wild type aspartate aminotransferase studied by size exclusion chromatography. *J. Chromatogr.* **539,** 383–391.
9. Hagel, L. (1998) Gel-filtration chromatography, in *Current Protocols in Molecular Biology* (Ausubel, F. M, Brent, R., et al., eds.), Wiley, New York, pp. 10.9.1–10.9.32.
10. *Gel filtration: Principles and Methods*, 8th ed., Amersham Biosciences, Uppsala, Sweden.

Web Sites

Amersham Biosciences-Chromatography: http://www.chromatography.amershambio sciences.com.

Bio-Rad Laboratories: http://www.bio-rad.com/.

Sigma-Aldrich: http://www.sigmaaldrich.com.

Agilent: http://www.chem.agilent.com.

Supelco: http://www.sigmaaldrich.com/Brands/Supelco_Home.html.

6

Hydrophilic Interaction Chromatography

Herbert Lindner and Wilfried Helliger

1. Introduction

In the past, reverse-phase high-performance liquid chromatography (RP-HPLC), which separates solutes based on hydrophobic interactions, has been established as the most commonly used method for separating peptides and proteins. Its advantages include excellent efficiency and resolving power along with reasonable elution times, and the availability of volatile mobile phases that eliminate the need for a desalting step. The application of this method, however, is limited when a low selectivity or the presence of complex protein mixtures does not permit satisfactory separation of individual components. In such cases, methods are successfully applied that are based on separation mechanisms other than hydrophobic interactions, like electrostatic (*see* Chapter 3) or hydrophilic interactions or different molecular size (Chapter 5).

The mode of chromatography utilizing hydrophilic interactions to separate solutes is referred to as normal-phase chromatography. It is characterized by its use of polar stationary phases including bare silica in combination with apolar eluents. In many cases, the selectivity obtained complements that of RP-HPLC in an ideal manner. To emphasize the polar nature of intermolecular forces governing retention with such stationary phases, the term hydrophilic interaction chromatography with the acronym HILIC was suggested by Alpert *(1)*.

Under typical HILIC conditions, retention of solutes increases with increasing solute hydrophilicity and decreasing mobile phase polarity, a behavior, which is obviously the precise opposite of that observed in RP-HPLC.

To illustrate the specific interactions of a hydrophilic solute with a hydrophilic stationary phase which are responsible for this particular chromatographic behavior, **Fig. 1** shows the retention of a basic peptide (expressed

From: *Methods in Molecular Biology, vol. 251, HPLC of Peptides and Proteins: Methods and Protocols*
Edited by: M.-I. Aguilar © Humana Press Inc., Totowa, NJ

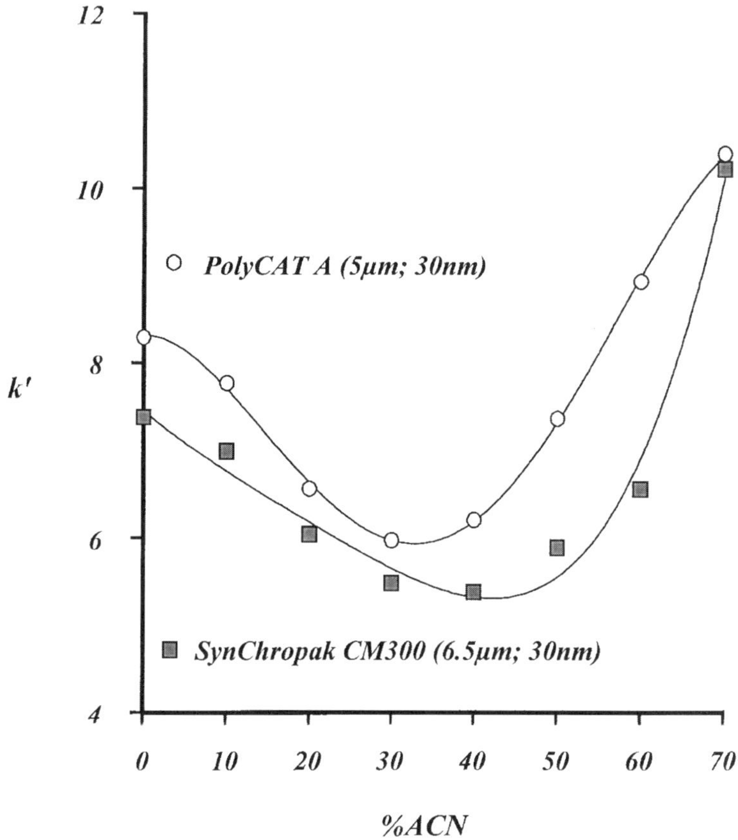

Fig. 1. Effect of organic solvent concentration on protein retention. Columns, SynChropak CM300 (250 × 4.6 mm, 6.5 µm beads, 30 nm) and PolyCAT A (250 × 4.6 mm, 5 µm beads, 30 nm); flow rate, 0.5 mL/min; temperature, 20°C; a linear gradient of sodium perchlorate (0–1 *M*, 40 min); solvent A, 0.010 *M* MPA/TEA (pH 3.0), solvent B 0.010 *M* MPA/TEA (pH 3.0) and 1 *M* $NaClO_4$; solvents A and B with varying acetonitrile concentrations ranging from 0–70% as shown in Fig. 1; $pI_{protein}$=11; detection at 210 nm.

as *k*′) as a function of acetonitrile content, whereby an increase in the acetonitrile concentration corresponds to a decrease in solvent polarity. The polar packing material is a commonly used weak cation exchanger having carboxymethyl groups, and the solute is eluted by an increasing linear salt gradient. It should be noted that the initial conditions with 0% acetonitrile entirely correspond to typical ion-exchange chromatography. However, when acetonitrile is added to the eluent in concentrations of up to about 40%, retention decreases because of a decrease in ionic interactions caused by the presence of an organic solvent in the aqueous phase. Moreover, when present, hydrophobic interactions between solutes and stationary phase are also reduced to a large extent by addition of the organic solvent. A further increase in the acetonitrile level up to 70% produces a significant increase in retention as a result of an increase in hydrophilic interactions, which is promoted by high concentrations of acetonitrile. As can be further seen from **Fig. 1**, the hydrophilic interactions occurring under these special conditions can be even stronger ($k'_{70\%} > k'_{0\%}$) than ionic interactions present in pure aqueous solutions. Therefore, HILIC typically requires the application of polar stationary phases (e.g., weak or strong ion exchangers, uncharged polar stationary phases like diol phases, and so on) *and* high levels of organic solvents (e.g., 70% acetonitrile, 80% methanol, and so on). Solutes are eluted by increasing salt gradients and/or decreasing organic solvent concentration.

The different interactions occurring in IEC and HILIC are best illustrated by the characteristic elution order of phosphorylated proteins. Under cation-exchange conditions, phosphorylated forms of proteins are retained to a lesser degree because of their decreased positive charge in comparison to the unmodified protein. Under HILIC conditions and despite the higher positive charge, the unmodified protein is first eluted followed by the lower positively charged mono-, di-, tri-, and so on phosphorylated forms *(2)*. It is obvious that this elution order is the opposite of that obtained by CEX and corresponds to the order of increasing hydrophilicity caused by the binding of hydrophilic phosphate groups.

When column materials with entirely dissociated functional groups, like strong cation-exchangers, are used in the presence of higher levels of organic solvents, hydrophilic interactions are superimposed by ionic interactions. This is thus referred to as mixed-mode hydrophilic/ion-exchange chromatography (HILIC/IEC) *(3,4)*. The possibility to fine-tune the extent of hydrophilic and ionic interactions between solute and column packing by selecting or adjusting stationary phase polarity, on the one hand, and solvent concentration, on the other, often provides a superior resolution of peptide *(3–8)* and protein mixtures *(2,9–11)*. Hydrophilic interaction chromatography, however, is not limited to the separation of proteins and peptides. On the contrary, this chromatographic method is suitable for fractionation of many other polar compounds such as amino acids *(1)*, oligonucleotides *(1)*, and carbohydrates *(12–15)*. In addition, HILIC is excellently suited for separating posttranslationally modified proteins, e.g., differently acetylated *(9)*, phosphorylated *(2)*, glycosylated *(16)*, as well as deamidated proteins *(10)* can be successfully resolved.

2. Materials

2.1. Stationary Phases

Separation of peptide and protein mixtures by means of hydrophilic interaction chromatography is usually performed with packing materials having average particle diameters of 5–10 µm and a pore size between 30 and 150 nm. Thus, the use of HILIC as high-resolution separation technique reasonably requires an HPLC system with columns, whose lengths and diameters are typical for HPLC. The stationary phases employed are by definition polar and usually include silica modified by chemical bonding of suitable functional groups. These silica-based phases exhibit good mechanical, as well as chemical stability. To achieve a maximum of plate numbers, spherical particles are used, because they provide more homogenous column packings than do the irregularly shaped particles and thus permit highly efficient separation of peptides and proteins. Silica-based packings are supplied by many manufacturers and include weak and strong cation and anion exchangers, amino-, diol-, polyol-, amide- and amino-cyano-phases. They are successfully employed in the pH region between 2.0 and 8.0. Silica materials are normally not suited for use at a pH higher than 8.0, however, because soluble silicate anions are formed under alkaline conditions. It should be noted that the chemical stability of silica is also limited under strong acidic conditions (pH < 2). Thus, despite their relatively low mechanical strength, polymer-based polar sorbents are used in cases requiring such extreme pH values. Moreover, zirconia-based stationary phases have proved especially advantageous for analysing carbohydrates ***(13)***. Compared to the silica-based hydrophilic sorbents, however, all other packing materials play a minor role.

A special kind of silica-based stationary phase having a poly(peptide) coating is particularly promising as a column material for HILIC. Poly (aspartic acid)-silica (PolyCAT A, manufactured by PolyLC, Columbia, MD), for example, is a silica packing with a bonded coating of a hydrophilic aspartic acid anionic polymer. It was originally developed and described as WCX by Alpert in the early 1980s ***(17)***. In addition to its notable cation exchange properties, it shows great potential as a stationary phase for HILIC of peptides and proteins. Peptides and proteins interact with the stationary phase via the β-carboxy group of aspartic acid, which is also responsible for the cation exchange capacity observed in aqueous systems, as well as via the amino and carbonyl groups of peptide bonds of the stationary phase. These functional groups can act as donor/acceptor groups for hydrogen bonds between solutes and stationary phase, which might be one of the factors why these poly(peptide)-coated stationary phases are particularly useful for separating peptides and proteins in the HILIC mode. An alternative to poly(aspartic acid)-silica is the use of poly(2-hydroxyethylaspartamide)-silica known as PolyHydroxyethyl A (PolyLC), which is produced by incorporating of ethanolamine into a coating of poly(succinimide), covalently bonded to silica ***(1)***. This material should be neutral, although minor cation exchange properties are observed. Despite this behavior, solutes are retained mainly through hydrophilic interactions and this would permit HILIC to operate at lower levels of organic solvent and/or proteins to elute at lower salt concentrations.

Another silica material with a poly(peptide) coating is the strong cation exchanger poly(2-sulfoethylaspartamide) (PolySulfoethyl A, PolyLC), which is synthesized by the reaction of taurine with poly(succinimide) covalently bonded to silica ***(18)***. This column material displays mixed-mode effects, i.e., hydrophilic interactions are superimposed by electrostatic effects. The selectivity can be manipulated by varying the organic solvent concentration of the mobile phase. Lower organic solvent levels favor electrostatic interactions, whereas higher levels promote hydrophilic ones. These effects, for example, permit peptides with identical charges (at a certain pH) to be separated using differences in their hydrophilicity caused by different primary sequences or amino acid composition.

2.2. Mobile Phases

Elution of peptides and proteins in HILIC is accomplished by increasing the polarity of the mobile phase. This can usually be done in two ways, namely by increasing the amount of water in the eluent, i.e., using a decreasing organic solvent gradient, or by running an increasing salt gradient. Solvents commonly used in HILIC are acetonitrile, methanol, and 2-propanol at concentrations of up to 85%. Whenever possible, acetonitrile should be used, because of its excellent UV transparency for detecting solutes at wavelengths slightly above 200 nm and because of its lower viscosity. Moreover, as compared to alcohols, solvents containing acetonitrile yield greater retention of solutes than do mobile phases with equivalent amounts of methanol or 2-propanol. It should be noted that these differences in elution behavior can also be utilized in method development. At levels of acetonitrile above 70%, every 1% increase leads to an increase of about 10% in the (isocratic) k' values for small molecules ***(1)***.

To adjust solvent pH and to minimize peak broadening low amounts of buffer salts (5–15 m*M*) should be added to the eluent. In this context, it is important to pay special attention to their UV transparency and solubility in organic solvents. Sodium, ammonium, or triethylammonium (TEA) salts of phosphoric, methylphosphonic acid (MPA), TFA, formic, and propionic acid are suitable. In certain cases, the gradient elution of strong basic peptides or proteins requires high salt concentrations of up to 1 *M* and more. However, the required proper-

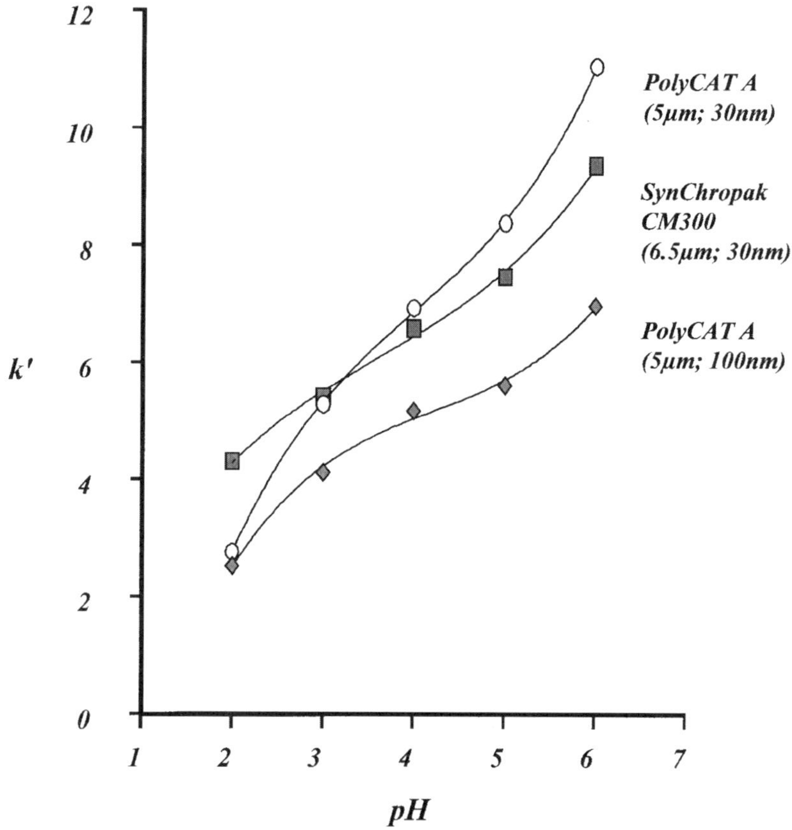

Fig.2. Protein retention on three different HILIC columns as a function of pH. Columns, PolyCAT A (250 × 4.6 mm, 5 μm beads, 30 nm and 100 nm), and SynChropak CM300 (250 × 4.6 mm, 6.5 μm beads, 30 nm); flow rate, 0.5 mL/min; temperature, 20°C; a linear gradient of sodium perchlorate (0–2 *M*, 40 min); solvent A, 70% acetonitrile, 0.010 *M* MPA/TEA, solvent B, 70% acetonitrile, 0.010 *M* MPA/TEA, and 2 *M* $NaClO_4$; in solvents A and B the pH value varies in the range of 2.0–6.0 as shown in Fig. 2; $pI_{protein}$=11; detection at 210 nm.

ties such as good solubility of salts in organic solvents and their UV transparency considerably restrict the number of suitable compounds. Since sodium perchlorate remarkably combines these two properties, this salt has proved to be particularly well suited.

In addition to the nature of the solvent (acetonitrile, alcohols, and so on), the retention of peptides and proteins is substantially influenced by the eluent pH and, to a minor extent, by the temperature, which affect the polarity of both the stationary phase and the solutes. **Figure 2**, for example, illustrates the influence of pH on retention of highly basic protein (p*I* about 11), using two different, weak cation exchangers, a CM, and a PolyCAT A column (latter with a pore size of 30 and 100 nm). As can be seen, retention increases considerably (up to a factor of five) when pH is raised from 2.0 to 6.0. This is mainly caused by an increase in the polarity of the stationary phase. Particularizing, however, the retention behavior of the three stationary phases investigated is quite different, even when they vary in pore size only.

Thus, the type of mobile phase and buffer ions, the pH, and even more importantly, the type of stationary phase are powerful tools for high-resolution separations particularly of proteins. In addition, minor effects can be achieved by altering the temperature, because retention and in most cases resolution generally increase with increasing temperature.

2.2.1. Starting Procedure

1. 20% Nitric acid.
2. Wash buffer containing 0.2 *M* NaH_2PO_4 and 0.3 *M* CH_3COONa.

2.2.2. Sample Application

1. Solvent A: 70% acetonitrile in the presence of 15 m*M* TEA/MPA, pH 3.0.

2.2.3. Sample Elution and Method Development

1. Solvent A: 70% acetonitrile in the presence of 15 m*M* TEA/MPA, pH 3.0.
2. Solvent B: 70% acetonitrile with 15 m*M* TEA/MPA, pH 3.0, containing 1.5 *M* sodium perchlorate.
3. Solvent C: 80% acetonitrile, 15 m*M* TEAP, pH 3.0.
4. Solvent D: 80% acetonitrile, 15 m*M* TEAP, pH 3.0, 0.4 *M* sodium perchlorate.

2.2.4. Column Regeneration and Storage

1. Solvent A: water.
2. Solvent B: filtered salt solution consisting of 0.25 *M* Na_2HPO_4, 0.25 *M* NaH_2PO_4, 0.5 *M* sodium acetate, 0.01 *M* ethylene diamine tetraacetic acid (EDTA), pH 6.5.

3. Methods

3.1. Starting Procedure

1. Especially when cation exchange columns are utilized, the amount of soluble iron (Fe^{3+}) present in the chromatographic system should be minimized, because Fe^{3+} binds very strongly to the stationary phase and decreases column efficiency. The first step, therefore, should be to passivate the HPLC system by treating it with 20% nitric acid (*see* **Note 1**).
2. Most columns are shipped in alcohol (methanol or propanol). To condition a new column, it must be washed thoroughly with at least ten column volumes of HPLC-grade water (20 mL to 40 mL, related to 4.6 mm × 200 mm columns). Extensive

washing with water is important to completely remove the alcohol component. Otherwise, salts that are insoluble in the organic solvent will be precipitated and damage the column.

3. This initial washing is followed by rinsing the column with 50–60 mL of a wash buffer (*see* **Note 2**).
4. The buffer is removed by flushing again with 20–30 mL water and, finally, the column is equilibrated with at least 30 mL of the mobile phase (solvent A) (*see* **Note 3**). Flow rates for 4.6 mm × 200 mm columns typically range from 0.5 to 1.5 mL/min.

3.2. Sample Application

1. To achieve optimum binding of the sample components to the stationary phase, the sample must be dissolved in buffers having a content of organic solvent similar to that of the eluent. When using gradient elution, which in most cases affords a superior resolution, peptide and protein samples, therefore, should be dissolved in eluent A (starting conditions) (*see* **Note 4**). Furthermore, the presence of larger amounts of salt should be avoided. This is especially important when non-charged stationary materials like diol or hydroxyethyl aspartic acid silica phases are employed.
2. Because salts like KCl, NaCl, or $MgCl_2$, and so on, which might be present because of sample preparation, are insoluble in mobile phases containing high levels of organic solvents (typical HILIC conditions), it is important to filtrate the sample before injection in order to prevent clogging of the column. A filter pore size of 0.45 μm is recommended. Moreover, care must be taken to ensure that filter and filter housing are solvent resistant.

3.3. Sample Elution and Method Development

1. For method development the following factors influencing retention of peptides and proteins should be considered:
   a. Retention is proportional to the hydrophilicity of peptides and proteins. Hydrophilic groups are basic amino acids (*His*, *Lys*, *Arg*) and other polar amino acids (*Asn*, *Gln*, *Asp*, *Glu*, *Ser*) as well as phosphate residues.
   b. Retention is influenced by the primary structure. Peptide retention is usually proportional to the number of basic amino acids. Retention behavior is, however, influenced also by neighboring acidic amino acids and, moreover, superimposed by the hydrophilicity/hydrophobicity of other residues present in the molecule.
   c. Retention is influenced by the secondary/tertiary structure.
   d. Solvent pH: Influences the charge (polarity) of peptides and proteins. At pH 2.5–2.8, basic amino acids and phosphate groups, but not acidic amino acids are ionized. At pH values higher than 3.5–4.0 (for *Asp*) and 4.0–4.5 (*Glu*), however, these acidic residues will also be charged.
   e. Temperature: Retention increases (!) with increasing temperature (*see* **Note 4**).

2. The following procedure is recommended as a good starting point for developing or optimizing a HILIC method to separate an unknown peptide or protein sample: The use of a weak cation exchanger (CM phase or poly(aspartic acid) silica) and an eluent system consisting of 70% acetonitrile in the presence of 15 m*M* TEA/MPA, pH 3.0 (solvent A) and 70% acetonitrile with 15 m*M* TEA/MPA, pH 3.0 containing 1.5 *M* sodium perchlorate (solvent B). A proper solvent preparation is very important and described in detail in Note 6.
3. After equilibrating the column with solvent A (*see* **Subheading 2.2.1.**, Starting Procedure), the sample is loaded onto the column and solutes are eluted with a linear gradient increasing from 100% A/0% B to 0% A/100% B in approx 45 min. It should be remembered that salt concentrations higher than 1 *M* are necessary only when eluting highly basic peptides and proteins. Nevertheless, these conditions (salt gradient ranging from 0–1.5 *M* sodium perchlorate) normally ensure the elution of virtually all components in the sample (*see* **Note 7**).
4. Depending on the elution profile obtained, resolution is further optimized by appropriately adjusting the slope of the gradient (and, when necessary, by lowering the salt concentration of solvent B). A shallower gradient will, of course, enhance resolution, but also analysis time. Elution times should preferably be between 10 and 60 min. If higher salt concentrations are required for elution, a two-step gradient may be of some advantage as compared to a one-step linear gradient. Peaks are often sharper and resolution increases when in a first steep step (within only a few minutes) the salt concentration is increased to just below the point where the weakest bound component begins to elute, followed by a longer shallow gradient to complete elution. As usual, optimum gradient elution parameters must be defined in the course of several experiments. Finally, the eluted peptide or protein fractions are desalted most simply by using RP-HPLC, which would even permit further separation of the fractions by using this additional orthogonal chromatographic step.
5. A good alternative to triethylammonium methanephosphonate is triethylammonium phosphate (*see* **Note 8**). In this case, the concentration of sodium perchlorate is limited to approx 0.68 *M*, and solvent preparation is trickier (*see* **Note 9**). Care must be taken, because, if more than 0.68 *M* sodium perchlorate is present, sodium phosphate will precipitate and, consequently, the pump and column inlet frits will clog.
6. As already stated, HILIC separations are influenced by many parameters which can be utilized as powerful tools to modify or optimise a method.

   If solute retention is too strong, for example, the acetonitrile concentration can be decreased or, alternatively, methanol or isopropanol can be used instead of acetonitrile. At comparable concentrations solvents containing methanol or isopropanol retain solutes to a lesser extent than does acetonitrile, however, solvents are more viscous. Also, a decreasing organic solvent gradient and an increasing salt gradient, simultaneously applied, can be used to elute strongly bound solutes.
7. If larger amounts of salt have to be avoided, e.g., certain proteins tend to precipitate in presence of perchlorate, the use of a poly(2-hydroxyethylaspartic

acid)silica column is recommended. Elution is performed with either a decreasing organic solvent gradient (85 to 20% acetonitrile, in the presence of 10 m*M* TEA/MPA, pH 3.0) or an increasing sodium perchlorate gradient, using the same eluents as already described in **Subheading 2.2.3.** Much lower salt concentrations are, however, required for elution with this column material.

8. On the other hand, if peptide or protein retention is too weak, acetonitrile concentration may be increased to up to 85% or even 90%. In addition, changing the buffer pH can emphasize the polarity of solutes prompting a more intensive interaction with the stationary phase. Finally, if retention is inadequate, separation can be performed in a HILIC/CEX mode using strong cation exchangers. A good choice, for example, is poly(2-sulfoethylaspartamide) silica and eluents consisting of 80% acetonitrile, 15 m*M* triethylammonium phosphate (TEAP), pH 3.0 (solvent C), and 80% acetonitrile, 15 m*M* TEAP, pH 3.0, 0.4 *M* sodium perchlorate (solvent D).
9. HILIC coupled with mass spectrometry requires low salt concentrations combined with a completely volatile mobile phase. This can be achieved with poly(2-hydroxyethylaspartic acid) silica and a descending gradient of acetonitrile in the presence of ammonium formate instead of TEAP. Owing to the high absorbance and baseline artefacts in gradient elution at low wavelengths, however, the use of 254 nm or a higher wavelength is recommended.

3.4. Column Regeneration and Storage

1. If peak broadening/loss of resolution or an increasing pressure is observed the column should be cleaned to remove any remaining contaminants. Increased pressure after prolonged use indicate a plugged column inlet frit which should be removed and cleaned with ultrasonication or, even better, replaced with a new one. To restore column efficiency the organic solvent is carefully replaced by extensive washing with at least 10 column volumes of water, followed by a shallow gradient from solvent A to a filtered (!) salt solution consisting of 0.25 *M* Na_2HPO_4, 0.25 *M* NaH_2PO_4, 0.5 *M* sodium acetate, 0.01 *M* EDTA, pH 6.5 (solvent B) (*see* **Note 10**). Washing should proceed for several hours (overnight) at 100% B and at reduced flow rate (0.25 mL/min for 4.6 mm columns). The entire procedure is best performed in reversed flow direction since most of the contaminants are deposited at the top of the column. The column is then thoroughly flushed with water to remove the salts and subsequently equilibrated with HILIC solutions.
2. The columns should be stored at cool temperature and in a solution of 50% acetonitrile in water. To prevent drying the column should be sealed tightly. If column efficiency is inadequate after storage, it must be reconditioned with a salt solution as described in **Subheading 3.1.** (Starting procedure).

4. Notes

1. To passivate the stainless steel components of an HPLC system, which are in contact with the eluent, the column must be removed (to prevent damaging any hardware components the manufacturer's instruction manual should be consulted) and

20% nitric acid pumped through at a flow rate of 1 mL/min for about 1 h. If the solvent present in the pump is not water, it must be removed and replaced with water before the passivating procedure is commenced. Afterward, the nitric acid is thoroughly flushed out with water (flow rate 1 mL/min, 1 h).

2. Buffer solution is prepared by dissolving appropriate amounts of sodium acetate and sodium dihydrogen phosphate in HPLC grade water and filtered (0.45 μm filter pore size). The pH of the buffer solution is not adjusted; it ranges from 5.7 to 5.9 and is used as is. To prevent microbial growth the buffer solution should be freshly prepared and not stored.
3. Ion-exchange columns conditioned in this way and run under HILIC conditions may exhibit a diminished resolution in the ion-exchange mode. For this reason, it is recommended that separate columns be used for each mode to guarantee optimum resolution.
4. Occasionally, however, sharper peaks can be obtained when samples loaded onto the column are dissolved in very small amounts of pure water or in water with lower concentrations of organic solvent.
5. Column temperatures should generally not exceed 50°C, because otherwise the lifetime of the column will be dramatically shortened. Some column materials, however, are even more sensitive to elevated temperatures. Poly(peptide)-coated silica columns (PolyCAT A, and so on), for example, do not tolerate temperatures higher than 30°C for longer periods.
6. The eluents (solvent A: 70% acetonitrile, 15 m*M* TEA/MPA, pH 3.0; solvent B: 70% acetonitrile, 15 m*M* TEA/MPA, pH 3.0, 1.5 *M* sodium perchlorate) are prepared as follows (1000 mL each):

   Solvent A is prepared by mixing 15 mL TEA/MPA stock solution (1 *M*) and 285 mL of water. pH is measured and, if it is too low adjusted to pH 3.0 with TEA. Acetonitrile is then added to give a total volume of 1000 mL. Solvent B is prepared in a slightly different way, namely by mixing 15 mL TEA/MPA stock solution (1 *M*) and 200 mL of water in a graduated cylinder. 212 g Sodium perchlorate monohydrate ($NaClO_4 \cdot H_2O$; FW 140.46) is added to the solution and dissolved by using a magnetic stirrer. The addition of perchlorate causes the pH to drop to about 2.0 or even lower and consequently it must be readjusted by adding TEA (about 100–300 μl) until pH 3.0 is reached. The resulting solution is filtered, water is added to a total volume of 300 mL, and acetonitrile is added until the 1000 mL mark is reached. Finally, solvents A and B are sonicated for 5–10 min to degas the eluents. Water and acetonitrile used for eluent preparation are of HPLC grade quality.

   The TEA/MPA stock solution (100 mL; 1 *M*) is prepared by adding TEA (about 10 mL is needed) to 16 mL methylphosphonic acid (from Merck, a 50% solution in water, which is an inexpensive alternative to the higher concentrated acid from other suppliers) dissolved in about 60 mL of water until pH 2.9 is attained. The solution is filtered and water is added to give a total volume of 100 mL. The pH is measured again and, if necessary, adjusted to 3.0 by adding drop wise several mL of TEA. The stock solution must be refrigerated until use. It should be prepared new after 2 mo except it is frozen and stored under argon. The purity of TEA is critical and, therefore the best quality available should be used for the buffer solution (e.g., from Aldrich, 99.5%). The amine must be absolutely colorless. Otherwise, it must be distilled prior to use to avoid baseline artefacts during gradient elution at low wavelengths.

7. When a column is used for the first time, irreversible binding of sample to both the stationary phase and the column hardware (e.g., sintered metal frits) may occur to a certain extent. This can be a problem if only small sample amounts are available or expensive samples are to be purified. To avoid sample loss, therefore, an inexpensive protein with good binding properties such as protamine sulphate (for neutral and weak cation exchange phases) or BSA (for strong cation exchangers) should be initially injected and eluted from the column.

   Furthermore, ***(1)*** reports from that recovery of phosphorylated peptides from HPLC columns improves significantly when titanium frits are used in place of stainless steel. It is thus recommended that nonferrous frits be used for HPLC of phosphorylated species.

8. The fact that ion-pair reagents influence selectivity is a well-known effect in RP-HPLC. By varying the type of reagent (more or less hydrophobic) and/or its concentration, separations can be affected to a large extent. It should be noted that also in HILIC both cations and anions present in the eluent may influence retention and, even more importantly, resolution, sometimes in a dramatic fashion. Therefore, if resolution is not satisfying it is strongly recommended that different buffer systems and concentrations be tried, although, problems can be caused by the poor solubility of many salts in organic solvents. As a rule of thumb, buffer concentration should be varied from 5–20 m*M*. Ammonium, sodium, or potassium salts can be used instead of TEA salts and formate, acetate, trifluoro acetate, or propionate can be substituted for methane phosphonate, for example. In particular, phosphate is recommended because its notably hydrophilic nature, pronounces the hydrophilic properties of peptides and proteins most strongly.

9. A TEAP stock solution (250 mL; 1 *M*) is prepared according to TEA/MPA, except that 50 mL diluted phosphoric acid (5 *M*) dissolved in about 150 mL water is used instead of methylphosphonic acid. The pH is adjusted with TEA (requiring about 30–31 mL), as already described in detail. Solvent A is prepared by mixing 15 mL TEAP stock solution (1 *M*) and 285 mL of water. The pH is measured and, if it is too low, adjusted to pH 3.0 with TEA. Acetonitrile is then added to give a total volume of 1000 mL. Solvent B is prepared by mixing 15 mL TEAP stock solution (1 *M*) and 200 mL of water in a graduated cylinder. Sodium perchlorate monohydrate ($NaClO_4.H_2O$; FW 140.46) in an amount of 96 g is added to the solution and dissolved by using a magnetic stirrer. Owing to the addition of perchlorate the pH drops to about 2.0 or even less and must be readjusted by adding TEA (about 100–300 µL) until pH 3.0 is reached. The resulting solution is filtered, water is added to give a total volume of 300 mL, and acetonitrile is added until the 1000 mL mark. The resulting solution is not clear because sodium phosphate precipitates. To achieve complete solution of all salts present, water is added slowly

under steady mixing with a magnetic stirrer until the solution becomes clear (about 45 mL of water is required). This produces a saturated sodium perchlorate solution in acetonitrile. Because of the last dilution step, however, the acetonitrile concentration is somewhat below 70%, the precise value being 67%.

10. Generally, the presence of chloride in the regeneration buffer or during storage should be avoided because chloride corrodes steel and Fe^{3+} cations are released.

    Alternatively, the column can be regenerated using a linear gradient starting from 70% acetonitrile (e.g., solvent A, as described), whereby acetonitrile is successively replaced with 50 m*M* formic acid in water. The column is washed for an additional hour with 50 m*M* formic acid and finally flushed with water over a period of 2 h. Before use, it is recommended that the column be reconditioned with the buffer solution described in **Subheading 3.1.**

References

1. Alpert, A. J. (1990) Hydrophilic-interaction chromatography for the separation of peptides, nucleic acids and other polar compounds. *J. Chromatogr.* **499,** 177–196.
2. Lindner, H., Sarg, B., and Helliger, W. (1997) Application of hydrophilic-interaction liquid chromatography to the separation of phosphorylated H1 histones. *J. Chromatogr. A.* **782,** 55–62.
3. Zhu, B.-Y., Mant, C. T., and Hodges, R. S. (1991) Hydrophilic-interaction chromatography of peptides on hydrophilic and strong cation-exchange columns. *J. Chromatogr.* **548,** 13–24.
4. Zhu, B.-Y., Mant, C. T., and Hodges, R. S. (1992) Mixed mode hydrophilic and ionic interaction chromatography rivals reversed phase liquid chromatography for the separation of peptides. *J. Chromatogr.* **594,** 75–86.
5. Oyler, A. R., Armstrong, B. L., Cha, J. Y., et al. (1996) Hydrophilic interaction chromatography on amino-silica phases complements reversed-phase high-performance chromatography and capillary electrophoresis for peptide analysis. *J. Chromatogr. A.* **724,** 378–383.
6. Mant, C. T., Kandejewski, L. H., and Hodges, R. S. (1998) Hydrophilic interaction/cation-exchange chromatography for separation of cyclic peptides. *J. Chromatogr. A.* **816,** 79–88.
7. Mant, C. T., Litowski, J. R., and Hodges, R. S. (1998) Hydrophilic interaction/cation-exchange chromatography for separation of amphipathic alpha-helical peptides. *J. Chromatogr. A.* **816,** 65–78.
8. Litowski, J. R., Semchuk, P. D., Mant, C. T., and Hodges, R. S. (1999) Hydrophilic interaction/cation-exchange chromatography for the purification of synthetic peptides from closely related impurities: serine side-chain acetylated peptides. *J. Peptide Res.* **54,** 1–11.
9. Lindner, H., Sarg, B., Meraner, C., and Helliger, W. (1996) Separation of acetylated core histones by hydrophilic-interaction liquid chromatography. *J. Chromatogr. A.* **743,** 137–144.

10. Lindner, H., Sarg, B., Hoertnagl, B., and Helliger, W. (1998) The microheterogeneity of the mammalian H1(0) histone - Evidence for an age-dependent deamidation. *J. Biol.Chem.* **273,** 13,324–13,330.
11. Mizzen, C. A., Alpert, A. J., Levesque, L., Kruck, T. P. A., and McLachlan, D. R. (2000) Resolution of allelic and non-allelic variants of histone H1 by cation-exchange-hydrophilic-interaction chromatography. *J. Chromatogr. B* **744 (1)**, 33–46.
12. Feste, A.S. and Khan, I. (1993) Separation of glucooligosaccharides and polysaccharide hydrolysates by gradient elution hydrophilic interaction chromatography with pulsed amperometric detection. *J. Chromatogr.* **630,** 129–139.
13. Yu, J. and El Rassi, Z. (1994) Preparation of amino-zirconia bonded phases and their evaluation in hydrophilic interaction chromatography of carbohydrates with pulsed amperometric detection. *J. High Resolut. Chromatogr.* **17,** 773–778.
14. Alpert, A. J., Shukla, M., Shukla, A. K., et al. (1994) Hydrophilic interaction chromatography of complex carbohydrates. *J. Chromatogr. A.* **676,** 191–202.
15. Churms, S. C. (1996) Recent progress in carbohydrate separation by high-performance liquid chromatography based on hydrophilic interaction. *J. Chromatogr. A.* **720,** 75–91.
16. Zhang, J. and Wang, D. I. C. (1998) Quantitative analysis and process monitoring of site-specific glycosylation microheterogeneity in recombinant human interferon-gamma from Chinese hamster ovary cell culture by hydrophilic interaction chromatography. *J. Chromatogr. B.* **712,** 73–82.
17. Alpert, A. A. (1983) Cation-exchange high-performance liquid chromatography of proteins on poly(aspartic acid)-silica. *J. Chromatogr.* **266,** 23–37.
18. Alpert, A. A. (1988) Cation-exchange chromatography of peptides on poly(2-sulfoethyl aspartamide)-silica. *J. Chromatogr.* **443,** 85–96.

7

Immobilized Metal Ion Affinity Chromatography of Proteins

Michael Zachariou

1. Introduction

Immobilized metal ion affinity chromatography (IMAC) of proteins is a high-performance liquid chromatography (HPLC) technique. It has the ability to differentiate a single histidine residue on the surface of a protein *(1)*, it can bind proteins with dissociation constants of 10^{-5} to 10^{-7} *(2)*, and has had wide application in the field of molecular biology for the rapid purification of recombinant proteins.

Since the first set of work was published describing the immobilization of metal ions using a chelating agent covalently attached to a stationary support, to purify proteins *(3,4)*, there have been several modifications and adaptations of this technique over the years. The traditional approach describes the use of immobilized metal ions and in particular borderline Lewis metal ions such as Cu^{2+}, Ni^{2+}, and Zn^{2+}, to purify proteins on the basis of their histidine content *(3)*. This application was further extended to incorporating a hexa-histidine tail onto the protein and using the tail as a purification handle in combination with a highly selective stereochemistry in the form of a Ni-nitrilotriacetic acid (Ni-NTA) complex for binding the purification handle *(5,6)*. In 1985, there were indications that electrostatic interactions were also occurring between proteins and immobilized Fe^{3+}-iminodiacetic acid stationary phases *(7)* and in 1996 it was demonstrated that IMAC adsorbents in general could also be used in pseudocation exchange mode, independently of histidine interaction *(8)*. Yet another mode of interaction involved in the IMAC of proteins was the mixed mode interactions involving aspartate and/or glutamate surface residues on proteins along with electrostatic interactions, again independent of histidine

From: *Methods in Molecular Biology, vol. 251, HPLC of Peptides and Proteins: Methods and Protocols*
Edited by: M.-I. Aguilar © Humana Press Inc., Totowa, NJ

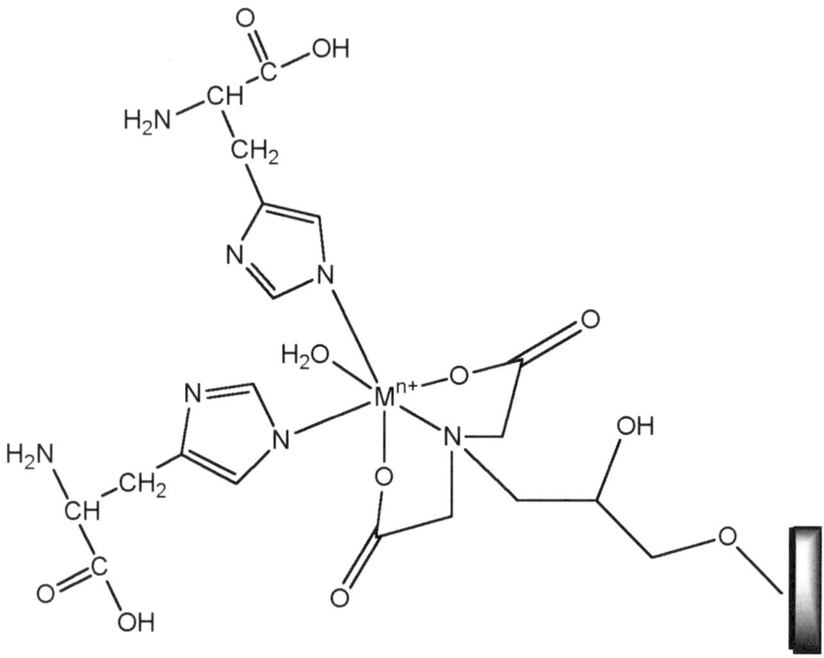

Fig. 1. Histidine residues binding to an immobilized metal ion affinity adsorbent.

interactions *(9)*. It is the purpose of this work to describe the methodologies involved in the traditional histidine-based IMAC interactions (including purification of *his*-tagged proteins), the mixed-mode interactions involving aspartate, glutamate, and electrostatic interactions and then the purely electrostatic interactions. The reader is referred to recent reviews of IMAC of proteins for a more detailed perspective ***(10,11,12,13)***.

The traditional use of IMAC for proteins has been to select proteins on the basis of their histidine content. The approach uses a chelating agent immobilized on a stationary surface to capture a metal ion and form an immobilized metal chelate complex (IMCC) (*see* **Fig. 1**). The chelating agent has usually been the tridentate iminodiacetic acid (IDA), despite a plethora of chelating stationary supports available for such work ***(14)***. Generally, Cu^{2+}, Ni^{2+}, and Zn^{2+}, have been used in this mode but other metal ions such as Co^{2+}, Cd^{2+}, Fe^{2+}, and Mn^{2+} have also been examined as the metal ions of choice. Histidine selection by the IMCC exploits the preference of borderline Lewis metals (*see* **ref. *15*** for a review of the concept of hard and soft acids and bases and their preferred interactions) to accept electrons from borderline Lewis bases such as histidine. With a pKa of 6.0, histidine will be able to donate electrons effectively at

>pH 6.5 and thus bind to the IMCC, although this may vary depending on the microenvironment the histidine finds itself in. Once the protein has bound, a specific elution can be deployed by using imidazole, which is the functional moiety of histidine. Alternatively, the pH may be decreased to <pH 6.5 to prevent histidine from donating electrons thereby inducing elution of the bound protein.

The use of IMAC to purify *his*-tagged proteins employs the same strategy as described for the traditional mode. The difference is that harsher binding, washing, and elution conditions (e.g., higher concentration of imidazole) in combination with a more selective IMCC (Ni-NTA) can be used to obtain a purified protein. This is possible because the target protein has six histidines in sequence to bind to the IMCC. One step purifications that achieve >95% purity of the target protein with >90% yields are routinely achieved from a variety of sources including bacterial and mammalian cells. Other IMCC have also been used for the isolation of *his*-tagged proteins including Ni-IDA (ProBond™) from Invitrogen and Co-carboxy-aspartate (TALON™) from Clontech.

An alternative use of IMAC can be to exploit the interaction between aspartate and glutamic acid residues on the surface of proteins with IMCCs. This was first demonstrated in 1992 when a protein with no surface accessible histidines bound to a group of hard Lewis metal ions immobilized on 8-hydroxyquinoline ***(9,16)***. In this context, at greater than pH 4.0 the carboxyl groups of aspartate and glutamate are fully deprotonated and able to donate electrons. By including imidazole and ≥ 0.5 *M* NaCl in the binding buffer, any histidine or electrostatic interactions will be quenched leaving aspartate and glutamate as the only amino acids able to donate electrons and interact with the IMCC. This type of interaction can be further enhanced by using hard Lewis metal ions as part of the IMCC so as to exploit the preference of hard Lewis metal ions for hard bases such as those found in oxygen rich compounds like the carboxyl groups of aspartate and glutamate. This type of interaction has been observed to occur predominantly in the pH region of 5.5 to 6.5 and may involve some electrostatic component. Above this pH range, electrostatic influence becomes more pronounced and the IMCCs exhibit pseudocation exchange behaviour.

The traditional use of IMAC has involved the inclusion of 0.5–1.0 *M* NaCl in the binding buffer to prevent the protein from interacting with the IMCC on the basis of nonspecific electrostatic interactions. The contribution of such interactions comes from charges presented to the protein by unoccupied chelate sites, a variety of hydrolytic species that exist on the IMCC, as well as the metal ion itself ***(8,17)***. The overall contribution results in a net negative charge on the IMCC, which becomes increasingly negative as the pH becomes more alkaline. This phenomenon occurs with any IMCC and will vary depending on the metal ion and immobilized chelator involved. By encouraging this phenomenon instead of quenching it, IMAC can be used in cation exchange mode. In this mode, the binding buffers are of low ionic strength (<0.1 *M*) and a pH of between 5.0 and 6.5 and elution can be afforded either by increasing the ionic strength and/or increasing pH. Furthermore, by including imidazole in the binding buffer, histidine interactions can be minimized if desired. The use of IMCCs in cation exchange mode has been demonstrated previously using model proteins as well as crude mixtures ***(8,18)***. With the large number of possible combination of metal ion and chelators, use of IMAC in pseudocation exchange mode provides a number of options during purification beyond the traditional use of IMAC or the use of traditional cation exchangers.

2. Materials

2.1. Purification of His-*Tagged Proteins Using Ni-NTA*

1. Stationary support: Ni-NTA-superflow (Qiagen, GmBH, Germany).
2. Charge solution: 0.1 *M* $NiNO_3$.
3. Metal rinsing solution: 0.2 *M* acetic acid.
4. Preequilibration buffer: 0.2 *M* imidazole, 0.5 *M* NaCl, pH 7.0.
5. Equilibration buffer: 0.02 *M* imidazole, 0.05 *M* NaCl, pH 7.0.
6. Elution buffer: 0.2 *M* imidazole, 0.5 *M* NaCl, pH 7.0.
7. Regeneration buffer: 0.2 *M* ethylene diaminetetraacetic acid (EDTA), 0.5 *M* NaCl, pH 8.0.

2.2. Purification of Proteins Using IMAC Based on Histidine But Not His-*Tagged Selection*

1. Stationary Support: chelating-Sepharose FF (Amersham-Pharmacia Biotech, UK).
2. Charge solution: 0.1 *M* $CuNO_3$.
3. Metal rinsing solution: 0.2 *M* acetic acid, pH 4.0.
4. Preequilibration buffer: 0.2 *M* K_2HPO_4/KH_2PO_4, 0.5 *M* NaCl, pH 7.4.
5. Equilibration buffer: 0.02 *M* K_2HPO_4/KH_2PO_4, 0.5 *M* NaCl, pH 7.4.
6. Elution buffer: 0.05 *M* imidazole, 0.5 *M* NaCl, pH 7.0.
7. Regeneration buffer: 0.2 *M* EDTA, 0.5 *M* NaCl, pH 8.0.

2.3. Purification of Proteins Using IMAC Based on Aspartate and Glutamic Acid Selection and High Ionic Strength (see Note 11)

1. Stationary support: Chelating-Sepharose FF (Amersham-Pharmacia Biotech, UK).
2. Charge solution: 0.05 *M* metal salts.
3. Metal rinsing solution: 0.05 *M* acetic acid, 0.1 *M* KNO_3.
4. Preequilibration buffer: None.
5. Equilibration buffer: 0.03 *M* morpholinoethane sulphonic acid (MES), 0.03 *M* imidazole, 0.5 *M* NaCl, pH 5.5/pH 6.0.

6. Elution buffer: 0.03 *M* MES, 0.03 *M* imidazole, 0.1 *M* K_2HPO_4, 0.14 *M* NaCl, pH 5.5/pH 6.0.
7. Regeneration buffer: 0.2 *M* EDTA, 0.5 *M* NaCl, pH 8.0.

2.4. Purification of Proteins Using IMAC in Pseudocation Exchange Mode (see Note 15)

1. Stationary support: chelating-Sepharose FF (Amersham-Pharmacia Biotech, UK).
2. Charge solution: 0.05 *M* metal salts.
3. Metal rinsing solution: 0.05 *M* acetic acid, 0.1 *M* KNO_3.
4. Preequilibration buffer: None.
5. Equilibration buffer: 0.03 *M* MES, 0.03 *M* imidazole, 0.05 *M* NaCl, pH 5.5/pH 6.0.
6. Elution buffer: 0.03 *M* HEPES, 0.03 *M* imidazole, 0.5 *M* NaCl, pH 8.0.
7. Regeneration buffer: 0.2 *M* EDTA, 0.5*M* NaCl, pH 8.0.

3. Methods

3.1 Purification of His-Tagged Proteins Using Ni-NTA

1. Wash packed Ni-NTA column with two column volumes (CVs) of metal rinsing solution, 0.2 *M* acetic acid (*see* **Note 1**).
2. Wash column with 5 CV of Milli-Q water.
3. Prewash packed Ni-NTA column with 10 CVs of 0.2 *M* imidazole + 0.5 *M* NaCl, pH 7.0 (*see* **Note 2**).
4. Equilibrate the column with 10 CVs of 20 m*M* Imidazole and 50 m*M* NaCl pH 7.0 (*see* **Note 3**). Confirm equilibration by measuring pH and conductivity. Continue equilibration until pH and conductivity of effluent matches equilibration buffer.
5. Load sample containing target protein ensuring pH is between pH 7.0 and 7.2. As a general rule, loading linear velocities should be between 10 and 33% the maximum operating linear velocity allowed by the stationary support (*see* **Note 4**). Assume a loading of no more than 1 mg target protein per ml of stationary support (*see* **Note 5**). Target proteins in ratio volumes of 300:1 cell culture per Ni-NTA have been successfully loaded by the author (*see* **Note 6**).
6. Wash stationary support with 10 CVs of equilibration buffer at the loading linear velocity or until the A280nm reading is at baseline (*see* **Note 7**).
7. Subsequent wash steps can be carried out if deemed necessary (*see* **Note 8**). If a wash step is required follow **step 6** with the appropriate wash buffer.
8. Elute protein with up to 5 CVs of 0.2 *M* imidazole, 0.5 *M* NaCl pH 7.0 at 33% of the recommended maximum linear velocity of the stationary support. If this is insufficient, imidazole should be taken up to 0.5 *M*. If the target protein is still bound then elution with 0.5 *M* imidazole, 0.5 *M* NaCl at pH 5.5 should be tried (*see* **Note 9**). Samples should be examined on sodium dodecylsulfate-polyacrylamide gel electrophoresis (SDS-PAGE) for purity ***(19)***.

9. After elution of the target protein the column should be regenerated using 3 CVs of 0.2 *M* EDTA, 0.5 *M* NaCl, pH 8.0. Washing linear velocity is not critical as long as it does not exceed the maximum linear velocity of the stationary support (*see* **Note 10**).
10. Wash with 10 CVs of Milli-Q water.
11. Load column with 2 CVs of 0.1 *M* $NiNO_3$ (*see* **Note 11**). Silica based supports should be stored free of metal ion and only charged when required.
12. Wash with 10 CVs of Milli-Q water.
13. Store column at 4°C.

3.2. Purification of Proteins Using IMAC Based on Histidine But Not His-Tagged Selection

1. Wash packed Cu-IDA column with 2 CVs of metal rinsing solution, 0.2 *M* acetic acid, pH 4.0 (*see* **Note 1**).
2. Wash column with 5 CV of Milli-Q water.
3. Prewash packed Cu-IDA column with 10 CVs of 0.2 *M* K_2HPO_4/KH_2PO_4, 0.5 *M* NaCl, pH 7.4.
4. Equilibrate the column with 10 CVs of 20 m*M* K_2HPO_4/KH_2PO_4, 0.5 *M* NaCl, pH 7.4.
5. Confirm equilibration by measuring pH and conductivity. Continue equilibration until pH and conductivity of effluent matches equilibration buffer.
6. Load sample containing target protein ensuring the sample pH is between pH 7.0 and 7.2. As a general rule, loading linear velocities should be between 10 and 33% the maximum operating linear velocity allowed by the stationary support (*see* **Note 4**). Assume a loading of no more than 1 mg target protein per mL of stationary support (*see* **Note 5**). Target proteins in ratio volumes of 300 : 1 cell culture per Ni-NTA have been successfully loaded by the author (*see* **Note 6**).
7. Wash stationary support with 10 CVs of equilibration buffer at the loading linear velocity or until the A_{280nm} reading is at baseline (*see* **Note 7**).
8. Subsequent wash steps can be carried out if deemed necessary (*see* **Note 8**). If a wash step is required follow **step 7** with the appropriate wash buffer.
9. Elute protein with up to 5 CVs of 50 m*M* imidazole, 0.5 *M* NaCl pH 7.0 at 33% of the recommended maximum linear velocity of the stationary support. If this is insufficient, imidazole should be taken up to 0.5 *M*. If the target protein is still bound then elution with 0.5 *M* Imidazole, 0.5 *M* NaCl at pH 5.5 should be tried (*see* **Note 9**). Samples should be examined on SDS-PAGE for purity ***(19)***.
10. After elution of the target protein the column should be regenerated using 3 CVs of 0.2 *M* EDTA, 0.5 *M* NaCl, pH 8.0. Washing linear velocity is not critical as long as it does not exceed the maximum linear velocity of the stationary support (*see* **Note 10**).
11. Wash with 10 CVs of Milli-Q water.
12. Load column with 2 CVs of 0.1 *M* $CuNO_3$ (*see* **Note 11**). Silica-based supports should be stored free of metal ion and only charged when required.

13. Wash with 10 CVs of Milli-Q water.
14. Store column at 4°C.

3.3. Purification of Proteins Using IMAC Based on Aspartate and Glutamic Acid Selection and High Ionic Strength (see Note 11)

1. Load column with 2 CVs of 50 m*M* metal salt (*see* **Note 12**).
2. Wash packed M^{n+}-IDA column with 2 CVs of metal rinsing solution 50 m*M* Acetic acid + 0.1 *M* NaCl pH 4.0 (*see* **Note 1**).
3. Wash column with 5 CV of Milli-Q water.
4. Equilibrate packed M^{n+}-IDA column with 10 CVs of 30 m*M* MES + 30 m*M* Imidazole + 0.5 *M* NaCl pH 5.5 or pH 6.0 (*see* **Note 13**). Confirm equilibration by measuring pH and conductivity. Continue equilibration until pH and conductivity of effluent matches equilibration buffer.
5. Load sample containing target protein that has been preequilibrated in equilibration buffer. As a general rule, loading linear velocities should be between 10 and 33% the maximum operating linear velocity allowed by the stationary support (*see* **Note 4**). Assume a loading of no more than 1 mg target protein per ml of stationary support (*see* **Note 5**). Target proteins in ratio volumes of 300:1 cell culture per M^{n+}-IDA have been successfully loaded by the author (*see* **Note 6**).
6. Wash stationary support with 10 CVs of equilibration buffer at the loading linear velocity or until the A_{280nm} reading is at baseline (*see* **Note 7**).
7. Subsequent wash steps can be carried out if deemed necessary (*see* **Note 14**). If a wash step is required follow **step 6** with the appropriate wash buffer.
8. Elute protein with up to 5 CVs of 30 MES, 30 m*M* imidazole, 0.1 *M* K_2HPO_4, 0.14 *M* NaCl pH 5.5 or 6.0 at 33% of the recommended maximum linear velocity of the stationary support. If this is insufficient, phosphate should be taken up to 0.2 *M*. If the target protein still remains bound then elute with 30 n*M* HEPES, 30 m*M* imidazole, 0.1 *M* K_2HPO_4, 0.14 *M* NaCl pH 8.0, Samples should be examined on SDS-PAGE for purity ***(19)***.
9. After elution of the target protein the column should be regenerated using 3 CVs of 0.2 *M* EDTA, 0.5 *M* NaCl pH 8.0. Washing linear velocity is not critical as long as it does not exceed the maximum linear velocity of the stationary support (*see* **Note 10**).
10. Wash with 10 CVs of Milli-Q water.
11. Store column at 4°C.

3.4. Purification of Proteins Using IMAC in Pseudocation Exchange Mode (see Note 15)

1. Carry out **steps 1, 2,** and **3** of **Subheading 3.2.**
2. Equilibrate packed M^{n+}-IDA column with 10 CVs of 30 m*M* MES, 30 m*M* Imidazole, 0.05 *M* NaCl, pH 5.5 or pH 6.0 (*see* **Note 13**). Confirm equilibration by measuring pH and conductivity. Continue equilibration until pH and conductivity of effluent matches equilibration buffer.

3. Carry out **steps 5, 6,** and **7** of **Subheading 3.2.**
4. Subsequent wash steps can be carried out if deemed necessary (*see* **Note 16**). If a wash step is required follow **step 7** of **Subheading 3.2.** with the appropriate wash buffer.
5. Elute protein with up to 5 CVs of 30 MES, 30 m*M* imidazole, 0.5 *M* NaCl, pH 5.5 or 6.0 at 33% of the recommended maximum linear velocity of the stationary support. If this is insufficient, NaCl should be taken up to 1.0 *M*. If the target protein still remains bound then elute with 30 m*M* HEPES, 30 m*M* imidazole, 1 *M* NaCl, pH 8.0. Samples should be examined on SDS-PAGE for purity *(19)*.
6. Follow **steps 10** and **11** of **Subheading 3.2.**

4. Notes

1. All columns precharged with metal should be washed with acid to release any loosely bound metal ions.
2. This step serves to totally quench the immobilized metal ion with imidazole, improving selectivity of the IMCC for proteins. Furthermore, it creates a uniform surface by eluting weakly bound hydroxide species bound to the IMCC surface. Such species have been observed previously and if not controlled can significantly contribute to nonspecific electrostatic interactions during IMAC *(8)*. Lower imidazole concentrations are not as effective. In addition, the precharge buffer approximates the elution buffer and so can reduce metal ion leakage attributable to such a high imidazole concentration even before elution occurs.
3. The pH of equilibration is varied throughout the literature and can range from 7.0 to 8.0. By operating closer to pH 7.0 than to pH 8.0 during protein binding, a greater selectivity may be achieved which would ultimately yield greater purity of the final product. Improved capacity may also result because less nonspecific interactions will occur. Most *his*-tagged proteins will bind within pH 7.0 to 8.0 range and should be determined empirically. Other buffers such as 100 m*M* phosphate are commonly used at pH 7.0 to 7.5. In these instances, the Ni-NTA becomes less selective and proteins containing histidine regions are more likely to bind, than if imidazole was used, leading to potential problems downstream of the process.
4. A slow loading velocity improves the diffusion of proteins (particularly large proteins) through pores and onto the IMCC and hence improves yields. The stated linear velocities have been derived from the author's personal experience and will vary depending on the stationary support. For example, Poros and Hyper D supports can have linear dynamic capacities, in some cases up to 7000 cm/h, before decreases in capacities are observed. Care must also be taken to ensure that if prolonged loading times are chosen, the target protein is not subject to destabilizing factors such as proteolysis or any intrinsic instability such as deamidation or oxidation and should be monitored during the process. In these instances, the molecules stability needs to take precedence over slow loading velocities.

5. This amount is conservative relative to the manufacturer's claims of 5–10 mg of protein per mL Ni-NTA resin ***(20)***. However, capacities of <5 mg/mL have been observed often particularly when cell culture is used where proteins with a higher *his*-content are present as well as the presence of a large amount of free amino acids used as media components which can bind through their α-amino groups ***(21)***. Hence, allowing for excess stationary support will reduce the possibility of your target protein not binding because of capacity issues. Furthermore, any nonspecific interactions that may occur because of excess stationary support not interacting with the target protein is addressed through the proposed stringent preequilibration, equilibration and washing regimens.
6. In these instances, significant metal leaching may occur during loading, reducing the capacity of the Ni-NTA but not below 1 mg of protein per mL of Ni-NTA.
7. If monitoring A_{280nm} note that imidazole absorbs at this wavelength and so achieving baseline should only be relative to the absorbance of the equilibration buffer at A_{280nm}.
8. Washing options that can be used when IMAC is used in the amine (histidine) binding mode to maximize purity are indicated in **Table 1**.
9. If this does not elute your protein then it is highly probable that harsher specific elution conditions are required or that it is nonspecifically bound to the backbone matrix. In the former case, elution with regeneration buffer can be tried so as to strip the metal of the chelator. In the nonspecific case, chaotropes (e.g., 4 *M* Urea) or detergents (e.g., 1% Triton X-100) can be tried.
10. This step should also be considered as a last resort for eluting the target protein in the event that all other elution conditions failed.
11. Metal ions that could be used for this work are preferably the hard-Lewis metal ions such as Fe^{3+} and any of the lanthanides. Hard-Lewis metal ions such as Ca^{2+} could also be used however, a good chelating stationary phase to use this metal ion in IMAC for the purification of proteins does not exist commercially. Al^{3+} is also another example however, the commercially available 8-hydroxyquinoline support would be more useful over IDA stationary phases for this metal ion. Borderline Lewis metal ions like Cu^{2+} can also be used in this mode ***(9,18)***.
12. This section is not applicable to purification of *his*-tagged proteins.
13. Under these conditions, histidine interaction with the IMCC should be quenched ***(9)***. Furthermore, the use of oxygen rich buffers such as phosphate, acetate, carbonate, and so on should be avoided whilst equilibrating hard Lewis IMCCs. Sulphonic acid-based buffers such as MES and other Good's buffers used at ≤ 20 m*M* have minimal interference and can be used.
14. Washing options that can be used when IMAC is used in the aspartate/glutamate binding mode to maximize purity are indicated in **Table 2**.
15. Any metal ion that can be hydrolysed can be employed with any commercially available chelating stationary support for this section of work.

Table 1
Washing Options on IMAC During Amine(Histidine) Binding Mode

Wash type	Effect	Comment
Glycine, Arginine, NH_4Cl, approx 0.5 *M* and pH 7.0	Mild eluents that compete for Ni with histidine	These are mild eluents that will not elute the His-tag protein but may displace weaker bound proteins
Nonamine salts, e.g., NaCl; approx 0.5–1.0 *M* in 20 m*M* imidazole + 50 m*M* NaCl, pH 7.0	Will disrupt any nonspecific electrostatic interactions	Such interactions are common in IMAC particularly if the equilibration and wash steps had <0.2 *M* salt present
Non-ionic detergents, e.g., Triton, Tween-20. Not more than 1% v/v	Disrupts hydrophobic interactions	In particular, will disrupt any interactions between the spacer arm and proteins as well as any protein-protein hydrophobic interactions that may be occurring with the target protein. This is more effective when applied as part of the equilibration conditions so as to prevent such interactions from taking place. Inclusion of detergent will also assist in removing lipids or DNA ***(20)***.
Chaotropic agents, e.g., 4 *M* urea or 4 *M* GuHCl		
Decreasing pH (<7.0) and/or increasing imidazole concentration (>20 m*M*)	Disrupts the histidine bond to the IMCC	This step can also be used to elute the target protein so care must be taken to select a condition that ensures good differentiation between contaminants and target protein.

Table 2
Washing Options on IMAC During Aspartate and Glutamate Binding Mode

Wash type	Effect	Comment
Oxygen-rich buffers such as phosphate, glutamate, aspartate, acetate; at 0.1 *M* strength	Eluents competing with metal ion for aspartate and glutamate surface residues	This step can also be used to elute the target protein so care must be taken to select a condition that ensures good differentiation between contaminants and target protein. Acetate is the mildest and phosphate the strongest eluent from this set.
Non-ionic detergents, e.g., Triton, Tween-20. Not more than 1% v/v	Disrupts hydrophobic interactions	In particular, will disrupt any interactions between the spacer arm and proteins as well as protein–protein hydrophobic interactions that may be occurring with the target protein. This is more effective when applied as part of the equilibration conditions so as to prevent such interactions from taking place. Inclusion of detergent will also assist in removing lipids or DNA ***(20)***.
Chaotropic agents, e.g., 4 *M* urea or 4 *M* GuHCl		
Increasing pH (>6.0) and/or increasing phosphate concentration (>0.1 *M*)	Disrupts the aspartate and glutamate bonds to the IMCC as well as disrupting electrostatic interactions if protein has bound in mixed mode.	This step can also be used to elute the target protein so care must be taken to select a condition that ensures good differentiation between contaminants and target protein.

Table 3
Washing Options on IMAC During Pseudocation Exchange Mode

Wash type	Effect	Comment
Nonionic detergents, e.g., Triton, Tween-20. Not more than 1% v/v	Disrupts hydrophobic interactions	In particular, will disrupt any interactions between the spacer arm and proteins as well as protein–protein hydrophobic interactions that may be occurring with the target protein. This is more effective when applied as part of the equilibration conditions so as to prevent such interactions from taking place. Inclusion of detergent will also assist in removing lipids or DNA ***(20)***.
Increasing pH (>6.0)	Adjusting the pH to beyond the isoelectric point of the protein will make it more negative and interefere with the interactions on the adsorbent.	This step can also be used to elute the target protein so care must be taken to select a condition that ensures good differentiation between contaminants and target protein.
Increasing ionic strength to between 0.5–1.0 *M*	Disrupts electrostatic interactions	NaCl is used traditionally as an eluent however, other similar salts could also be used.

16. Washing options that can be used when IMAC is used in the pseudocation exchange binding mode to maximize purity are indicated in **Table 3**.

References

1. Hemdan, E. S., Zhao, Y. J., Sulkowski, E., and Porath, J. (1989) Surface topography of histidine residues: a facile probe by immobilized metal ion affinity chromatography. *Proc. Natl. Acad. Sci. USA* **86(6),** 1811–1815.

2. Wirth, H-J., Unger, K. K., and Hearn, M. T. W. (1993) Influence of ligand density on the proteins of metal-chelate affinity supports. *Anal. Biochem.* **208,** 16–25.
3. Porath, J., Carlsson, J., Olsson, I., and Belfrage, G. (1975). Metal chelate affinity chromatography a new approach to protein fractionation. *Nature* **258,** 598–599.
4. Everson, J. R. and Parker, H. E. (1974) Zinc binding and synthesis of 8-hydroxy-quinoline-agarose. *Bioinorg.Chem.* **4,** 15–20.
5. Hochuli, E., Dobeli, H., and Struber, A. (1987) New metal chelate adsorbents selective for proteins and peptides containing neighbouring histidine residues. *J. Chromatogr.* **411,** 177–184.
6. Hochuli, E., Banworth, W., Dobeli, H., Gentz, R., and Struber, A. (1988) Genetic approach to facilitate purification of recombinant proteins with a novel metal chelate adsorbent. *Bio\Technol.* **6,** 1321–1325.
7. Ramadan, N. and Porath, J. (1985) Fe(III)hydroxamate as immobilized metal affinity-adsorbent for protein chromatography. *J. Chromatogr.* **321,** 93–104.
8. Zachariou, M. and Hearn, M. T. W. (1996) Application of immobilized metal ion-chelate complexes as pseudocation exchange adsorbents for protein separation. *Biochemistry* **35,** 202–211.
9. Zachariou, M. and Hearn, M. T. W. (1995) Protein selectivity in immobilized metal affinity chromatography based on the surface accessibility of aspartic and glutamic acid residues. *J. Protein. Chem.* **14,** 419–430.
10. Beitle, R. R. and Ataali, M. M. (1992) Immobilized metal affinity chromatography and related techniques. *AIChE Symp.Series.* **88,** 34–44.
11. Wong, J. W., Albright, R. L., and Wang, N-H., L. (1991). Immobilized metal ion affinity chromatography (IMAC) chemistry and bioseparation applications. *Sep. Purif. Meth.* **20,** 49–106.
12. Arnold, F. H. (1991). Metal-affinity separations: A new dimension in protein processing. *Bio\Technol.* **9,** 151–156.
13. Porath, J. (1992) Immobilized metal ion affinity chromatography. *Protein Expr. Purif.* **3(4),** 263–81.
14. Sahni, S. K. and Reedijk, J. (1984) Coordination chemistry of chelating resins and ion-exchangers. *Coord. Chem. Rev.* **59,** 1–139.
15. Pearson, R. G. (1990) Hard and soft acids and bases. The evolution of a chemical concept. *Coordin. Chem. Rev.* **100,** 403–425.
16. Zachariou, M. and Hearn, M. T. W. (1992) High performance liquid chromatography of amino acids, peptides and proteins. CXXI. 8-hydroxyquinoline-metal chelate chromatographic support:an additional mode of selectivity in immobilized metal affinity chromatography. *J. Chromatogr.* **599,** 171–177.
17. Zachariou, M. and Hearn, M. T. W. (1997) Characterization by potentiometric procedures of the acid-base and metal binding properties of two new classes of immobilized metal ion affinity adsorbents developed for protein purification. *Anal. Chem.* **69,** 813–822.
18. Zachariou, M. and Hearn, M. T. W. (2000) Adsorption and selectivity characteristics of several human serum proteins with immobilised hard Lewis metal ion-chelate adsorbents. *J Chromatogr.* **890(1),** 95–116.

19. Laemmli, U.K. (1970) Cleavage of structural proteins during the assembly of the head of bateriophage T4. *Nature* **227,** 680–685.
20. Qiagen (1998) The QIA*expressionist* A handbook for high-level expression and purification of 6xHis-tagged proteins, p. 66.
21. Hansen, P., Lindeberg, G., and Andersson, L. (1992) Immobilized metal ion affinity chromatography of synthetic peptides. Binding via the alpha-amino group. *J. Chromatogr.* **215,** 333–339.

8

Immunoaffinity Chromatography of Proteins

Stuart R. Gallant

1. Introduction

Immunoaffinity chromatography offers to the chromatographer the exquisite specificity of the antibody's complementarity determining regions (CDRs), or hypervariable regions ***(1,2)***. These highly selective loops on the antibody surface capture an antigen with high affinity, while having little interaction with impurities that may also be present. When appropriately immobilized, an antibody retains its affinity for its antigen while being held covalently on the chromatographic support. With appropriate wash conditions, 1000-fold or higher impurity clearances are common. With appropriate elution conditions, 90% or more of the activity present in the chromatographic injection may be recovered in the elution peak. And, quite frequently, it is possible to regenerate the immunoaffinity column to allow its use for 100 or more cycles depending on the impurity levels in the injection.

Choice of the activated matrix has a significant effect on the ultimate purity of the product because nonspecific adsorption to the support is one of the important ways that trace impurities are retained during the wash step. Fortunately, there is a wide variety of activated supports available. A number of manufacturers (e.g., Perceptive Biosystems and JT Baker) now offer rigid supports for conducting immunoaffinity chromatography at relatively high flow rates and pressures. For high-performance liquid chromatograpyhy (HPLC) purposes, rigid supports are clearly superior, although soft gels may be used for low-pressure chromatography applications.

The surface of activated affinity resins is coated with reactive groups capable of forming covalent bonds with ligands such as antibodies. A wide variety of reaction chemistries are available ***(3–11)***. Aldehyde groups represent one particularly

From: *Methods in Molecular Biology, vol. 251, HPLC of Peptides and Proteins: Methods and Protocols*
Edited by: M.-I. Aguilar © Humana Press Inc., Totowa, NJ

useful chemistry which is widely available, typically results in high coupling efficiencies, and utilizes relatively mild reaction conditions. Aldehyde supports couple through the amine groups of the lysine side-chains of the antibody.

During immobilization, key variables to be monitored are the coupling efficiency, surface concentration, and activity of the antibody. Couple efficiency can be calculated by collecting the solution used to wash the resin after coupling:

$$\text{Efficiency} = 100\% \ \frac{A_{280} \text{ of wash solution} \times \text{volume of wash solution}}{A_{280} \text{ of antibody solution} \times \text{volume of antibody solution}}$$

A high coupling efficiency is desirable so as not to waste antibody. Information from the support manufacturer regarding surface area can be utilized to calculate the surface concentration (moles of antibody per unit surface area) and the surface area occupied on average by a single antibody. Care should be exercised in such calculations because micropores may contribute a significant proportion of the total surface area, but may not be available to the antibody. A representative length for an antibody is 250 angstrom; pore size should be chosen to allow diffusion of the antibody into the pores during coupling *and* diffusion of the antigen into the pores of the resulting immunoaffinity column. The activity of the immobilized antibody can be calculated based on the static binding capacity of the support for its antigen. Coupling efficiency and antibody activity can be increased by appropriate selection of the immobilization conditions (*see* **Table 1**).

In this chapter, a protocol for immobilization of an antibody on an aldehyde coated support is provided. Often the protocol provided by the manufacturer of the support works quite well. If acceptable results are not found quickly, selection of a different activation chemistry is advised (rather than spending time to optimize what may be an inferior chemistry for the application at hand).

Two chromatography protocols are given. The first protocol elutes the antigen in a single step (*see* **Fig. 1**). This method is suitable for quantitation of the antigen and could be substituted for an enzyme-linked immunosorbent assay (ELISA). (This method would be particularly useful if an HPLC system was available, but the number of assays required did not justify the investment in an ELISA plate reader.) The second chromatography method utilizes gradient elution. Further consideration to the method of elution is given in the notes section.

2. Materials

2.1. Immobilization

1. 20% Ethanol.
2. PBS: 0.14 *M* NaCl, 2.7 m*M* KCl, 1.5 m*M* KH_2PO_4, 8.1 m*M* Na_2HPO_4.
3. 1 *M* Ethanolamine, titrated to pH 8.0 with HCl.
4. Storage buffer: 0.2 *M* Imidazole, 0.5 *M* NaCl, titrated to pH 7.0 with HCl.

Table 1
Antibody Immobilization Conditions

Condition	Recommendation
pH	Coupling efficiency increases at higher pH for many immobilization chemistries; pH 8.0 is typically acceptable; pH 9.0 and above should be avoided if possible.
Antibody concentration	Higher antibody concentration increases coupling efficiency; the liquid volume should not be substantially larger than required make a loose slurry of the resin for batch coupling. The antibody stock solution may be concentrated using an ultrafiltration membrane, but precipitation due to overconcentration should be avoided.
Coupling mode	Batch coupling is typically employed, but it is possible to immobilize the antibody in-column with recirculation through the pump. To maximize coupling efficiency, reaction volume should be just large enough to permit recirculation.
Time and temperature	Higher reaction time and temperature increase coupling efficiency, but may adversely impact antibody activity.
Salt addition	If a stationary phase with a somewhat hydrophobic surface is employed, sodium chloride addition to the reaction solution may increase the hydrophobic interaction between the stationary phase and the antibody; this interaction may increase coupling efficiency by temporarily holding the antibody on the stationary phase until a covalent bond forms. Quite high NaCl concentrations are typically required.
Blocking protocol	Blocking with a reagent (such as ethanolamine for aldehyde activated supports) will decrease nonspecific interactions during use of the resulting immunoaffinity resin.

5. Corning 50-mL tube top filters (430320) or similar filtration device
6. Centrifuge tubes.

2.2. Immunoaffinity Chromatography (see *Note 1*)

1. Equilibration buffer: 10 m*M* HEPES, 700 m*M* NaCl, 0.01% Tween, pH 7.0.
2. Elution buffer: 10 m*M* HEPES, 500 m*M* NaCl, 1.0 *M* $CaCl_2$, 0.01% Tween-20, pH 7.0.

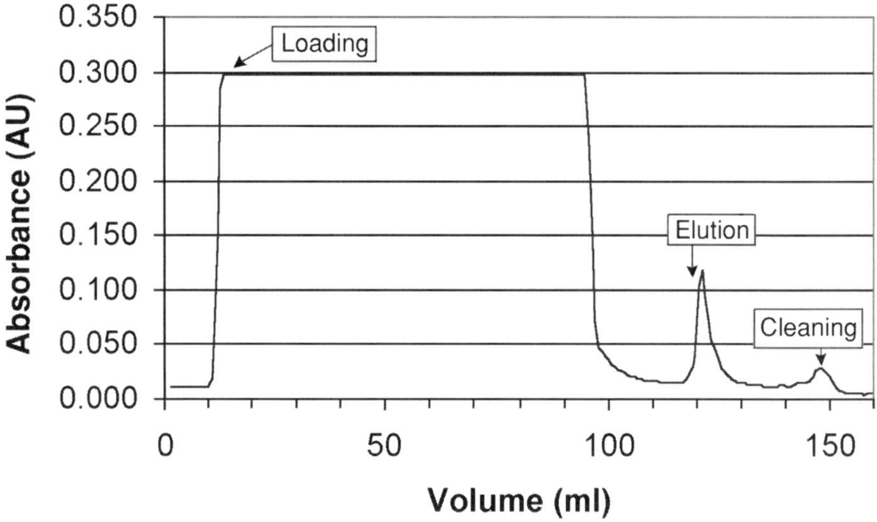

Fig. 1. Step elution in immunoaffinity chromatography.

3. Buffer A: 10 m*M* HEPES, 100 m*M* NaCl, 0.01% Tween, pH 7.0.
4. Buffer B: 10 m*M* HEPES, 500 m*M* NaCl, 1.5 *M* $CaCl_2$, 0.01% Tween, pH 7.0.
5. Cleaning: 24% ethanol/2% acetic acid, pH 2.5–3.0.
6. Storage buffer: 0.2 *M* Imidazole, 0.5 *M* NaCl, titrated to pH 7.0 with HCl.

3. Method

3.1. Immobilization on Aldehyde Support

1. Wash aldehyde support (using a tube top filter) with 5 volumes 20% ethanol; wash support with 10 volumes of PBS. Vacuum dry to remove excess PBS; residual PBS will dilute the reaction solution.
2. Place support in a centrifuge tube and add antibody in PBS (approx 5 mg antibody per mL of support). If support is not wet enough to flow during gentle rotation, add a small amount of PBS. Add 0.05 mL of 22 mg/ml Na-cyanoborohydride per ml of support. Rotate overnight at 5°C. (Increase temperature if efficiency is not adequate.)
3. Wash with PBS; save wash for coupling efficiency calculation.
4. Place support in a centrifuge tube and add 1 *M* ethanolamine, pH 8.0, enough to allow the support to flow under gentle rotation. Add 0.05 mL of 22 mg/mL Na-cyanoborohydride per ml of support. Rotate overnight at 5°C.
5. Wash with storage buffer. Place in test tube with storage buffer.

3.2. Step Elution of Antigen (see Note 2)

1. Prior to a column's first use pass 3 column volumes of cleaning solution to remove any antibody which is not covalently bound.
2. Pass 3 column volumes of storage buffer.

3. Pass 6 column volumes of equilibration buffer.
4. Load the solution containing the antigen. If the antigen solution contains phosphate or other components which precipitate in the presence of calcium; a dialysis membrane or a desalting column should be used to remove these components prior to immunoaffinity chromatography.
5. Pass 6–8 column volumes of equilibration buffer to remove unbound impurities.
6. Pass 6 column volumes of elution buffer. During elution, a single elution peak should be observed although shoulders are not uncommon, owing to the mobile phase disturbance of the elution buffer. The peak may be collected; integration of the peak may be used to quantify the amount of the antigen in the original sample. If this technique is employed, care should be taken to maximize the loading volume (to increase the signal) without saturating the column (to insure that no antigen flow through the column during the loading/washing).
7. Pass 3 column volumes of cleaning solution and 3 column volumes of storage solution. Storage solution should be passed regardless of whether the assay is to be repeated in order to bring the column back to the correct pH.

3.3. Gradient Elution of Antigen (see Note 3)

1. Prior to a column's first use pass 3 column volumes of cleaning solution to remove any antibody which is not covalently bound.
2. Pass 3 column volumes of storage buffer.
3. Pass 6 column volumes of Buffer A.
4. Load the solution containing the antigen. If the antigen solution contains phosphate or other components which precipitate in the presence of calcium; a dialysis membrane or a desalting column should be used to remove these components prior to immunoaffinity chromatography.
5. Pass 3 column volumes of Buffer A to remove unbound impurities.
6. Run a linear gradient from 100% Buffer A to 100% Buffer B in 10–30 column volumes. Collect fractions and analyze for activity. The elution peak will be sharper with a steeper gradient; however, it may not be possible to observe a single peak.
7. Pass 3 column volumes of cleaning solution and 3 column volumes of storage solution. Storage solution should be passed regardless of whether the assay is to be repeated in order to bring the column back to the correct pH.

4. Notes

1. The elution strategy of these methods involves increasing the ionic strength of the mobile phase. It is possible that higher affinity antibody antigen interactions will not be disrupted by this change in the mobile phase composition. Stronger elution protocols may be employed; however, some of these strategies may denature the antibody. In such a case, the column will not be able to be reused. See alternate elution strategies listed in **Table 2**.
2. The wash volume maybe significantly decreased for cleaner samples. The wash volume need only be large enough that the UV signal returns to baseline prior to the emergence of the elution peak from the column.

Table 2
Alternate Elution Strategies

Property	Example
Organic content	Without additional of other mobile phase modifiers, ethylene glycol typically cannot elute an antigen alone; however ethylene glycol can significantly increase the elution efficiency of sodium chloride and other salts. In many cases, up to 50% by volume may be used without denaturing the antibody.
Salts	Water disordering salts may be substantially more effective than sodium chloride. Urea, guanidine HCl, and thiocyanate may all be used up to 2 *M* concentration with little danger to the antibody; higher concentrations carry some risk.
Detergents	SLS, SDS (CMC = 8 m*M*), and other detergents can significantly improve the elution efficiency of salts ***(12)***; however, detergents cannot typically elute an antigen alone. Use the detergent at a concentration higher than its critical micelle concentration for greatest effectiveness.
pH	Antibodies may be exposed to pH values as low as 2.5 and as high as 9.0 without significant denaturation. Utilized formic and acetic acid to buffer near pH 2.0 ethanolamine buffers near pH 9.0 ***(13)***.

3. Gradient elution offers the advantage of providing separation of the antigen based on differences in interaction energy with the antibody. A homogeneous antigen may elute in a small volume while a heterogeneous antigen (such as a protein with significant variation in its folding or glycosylation) may elute over a larger gradient volume. Collection of fractions during the gradient for subsequent analysis of activity and electrophoretic mobility can provide useful information even if a more compact step gradient separation is adopted for routine analytical use.

References

1. Voet, D. and Voet, J. G. (1995) *Biochemistry*. Wiley, New York.
2. Nelson, D. L. and Cox, M. M. (2000) *Lehninger: Principles of Biochemistry*. Worth, New York.
3. Scouten, W. H. (1987) A survey of enzyme coupling techniques. *Methods in Enzymology* **135,** 30–65.
4. Narayanan, S. R. (1994) Preparative affinity chromatography of proteins. *J. Chromatogr.* **658,** 237–258.

5. Narayanan, S. R., Kakodkar, S. V., and Crane L. J. (1990) "Glutaraldehyde-P," a stable, Reactive aldehyde matrix for affinitychromaography. *Anal. Biochem.* **188,** 278–284.
6. Narayanan, S. R., Knochs, S., and Crane L. J. (1990) Preparative affinity chromatography of proteins: influence of the physical properties of the base matrix. *J. Chromatogr.* **503,** 93–102.
7. Highsmith, F., Regan, T., Clark, D., Drohan, W., and Tharakan, J. (1992) Evaluation of CNBr, FMP, and hydrazide resins for immunoaffinity purification of Factor IX. *BioTech.* **12,** 418–423.
8. Nakamura, K., Hashimoto, T., Kato, Y., Shimura, K., and Kasai K. (1990) Tresyl-activated support for high-performance affinity chromatography. *J. Chromatogr.* **510,** 101–113.
9. Cress, M. C. and Ngo, T. T. (1989) Site specific immobilization of immunoglobins. *Am. Biotechnol. Lab.* **7,** 16–19.
10. Matson, R. S. and Siebert, C. J. (1988) Evaluation of a new N-hydroxysuccinimide activated support for fast flow immunoaffinity chromatography. *Prep. Chromatogr.* **1,** 67–91.
11. van Sommeren, A. P. G., Machielsen, P. A. G. M., and Gribnau, T. C. J. (1993) Comparison of three activated agaroses for use in affinity chromatography: effects on coupling performance and ligand leakage. *J. Chromatogr.* **639,** 23–31.
12. Neugebauer, J. M. (1990) Detergents: an overview. *Methods in Enzymology* **182,** 239–253.
13. Stoll, V. S. and Blanchard, J. S. (1990) Buffers: principles and practice. *Methods in Enzymology* **182,** 239–253.

9

Liquid Chromatography–Mass Spectrometry and Tandem Mass Spectrometry of Peptides and Proteins

Tun-Li Shen and Kathleen R. Noon

1. Introduction

Mass spectrometry has become one of the preferred methods of detection for high-performance liquid chromatography (HPLC) analyses of biopolymers for a broad range of applications. The primary reason for its widespread use is the value of the information obtained from these types of measurements. When liquid chromatography is interfaced directly to mass spectrometry (LC–MS), molecular weight information can be retrieved from the mass spectra acquired continuously as various components elute from the chromatographic column. In some cases, the identity of the protein or peptide can be assigned immediately if the mass measurements are determined with sufficient accuracy. However, it is more likely that additional information regarding molecular structure is required for identification. In tandem mass spectrometry experiments (LC–MS/MS), structural data are readily generated by fragmentation of peptides (or protein-derived peptides) in the mass spectrometer using the technique of collision-induced dissociation (CID). The fragments generated with CID all originate from the precursor; thus, supplementary information relating to the primary sequence and post-translational modifications of the protein or peptide is obtained.

Over the last two decades, improvements in the direct coupling of HPLC with mass spectrometry has advanced this technique from the level of a challenging problem in analytical instrumentation to a routine laboratory tool. One of the key advancements in successfully interfacing HPLC with mass spectrometry was the development of ionization methods that are compatible with non-

From: *Methods in Molecular Biology, vol. 251, HPLC of Peptides and Proteins: Methods and Protocols*
Edited by: M.-I. Aguilar © Humana Press Inc., Totowa, NJ

volatile, thermally labile biomolecules. Several different approaches to ionization were introduced to address these issues. They include thermospray ***(1)***, particle beam ***(2)***, continuous-flow FAB ***(3)***, and atmospheric pressure ionization (API) ***(4,5)***. The more recent advent of API techniques, primarily electrospray (ESI) and atmospheric pressure chemical ionization (APCI), combined sensitivity and robustness with relatively mild ionization conditions which made them universal interfaces between HPLC and mass spectrometry. Both techniques can be performed directly on HPLC effluent without modification, and can be coupled directly to most common mass analyzers. Other technical issues that must be considered are the compatibility of the flow rate from the LC with the vacuum system of the mass spectrometer, the compatibility of the mobile phase of the LC system and the operation of the mass spectrometer, and interferences with ionization of analyte molecules. Much of the progress made in source designs for atmospheric pressure ionization focused on optimizing the nebulization process. **Figure 1** illustrates the essential features of three common electrospray ionization sources (curtain gas, orthogonal heated capillary, and "Z-spray") that have been successfully adapted by commercial mass spectrometer manufacturers. The curtain gas in the API/MS interface was first introduced by SCIEX (Toronto, Canada). It generates a counter gas flow toward the electrospray that helps sweep away large, undesolvated droplets. Perpendicular and other off-axis source designs such as heated capillary/orthogonal spray and "Z-spray" (Micromass) are different from on-axis configurations in that they offer the advantage of minimizing contamination and enhancing signal-to-noise (*see* **Fig. 1**).

The most common types of mass analyzers combined with HPLC systems are time-of-flight, quadrupoles, ion traps, and various combinations of the three. In recent years, there have been significant improvements in several areas of mass spectrometric instrumentation, most notably in sensitivity, resolution, automation of data collection, and postacquisition processing. As a result, it is now possible to analyze large numbers of samples with very little operator intervention, but it is important to understand how the data are collected.

The time-of-flight mass spectrometer is the simplest mass analyzer in theory. Mass-to-charge (*m/z*) ratios are determined from the flight times of analyte ions traveling through a field free region or a flight tube. At a constant accelerating voltage, the flight time for an ion is proportional to the square of the *m/z* ratio. The lighter ions (with lower *m/z* values) have a higher velocity and are recorded on a detector earlier than the heavier ones (higher *m/z* values), thus producing the TOF spectrum as a function of time. In a quadrupole instrument, mass separation is achieved by creating an electric field in which ions of certain *m/z* value will travel without interference. The quadrupole mass filter consists of four rods to which a DC voltage and an AC voltage with RF frequencies are applied. By

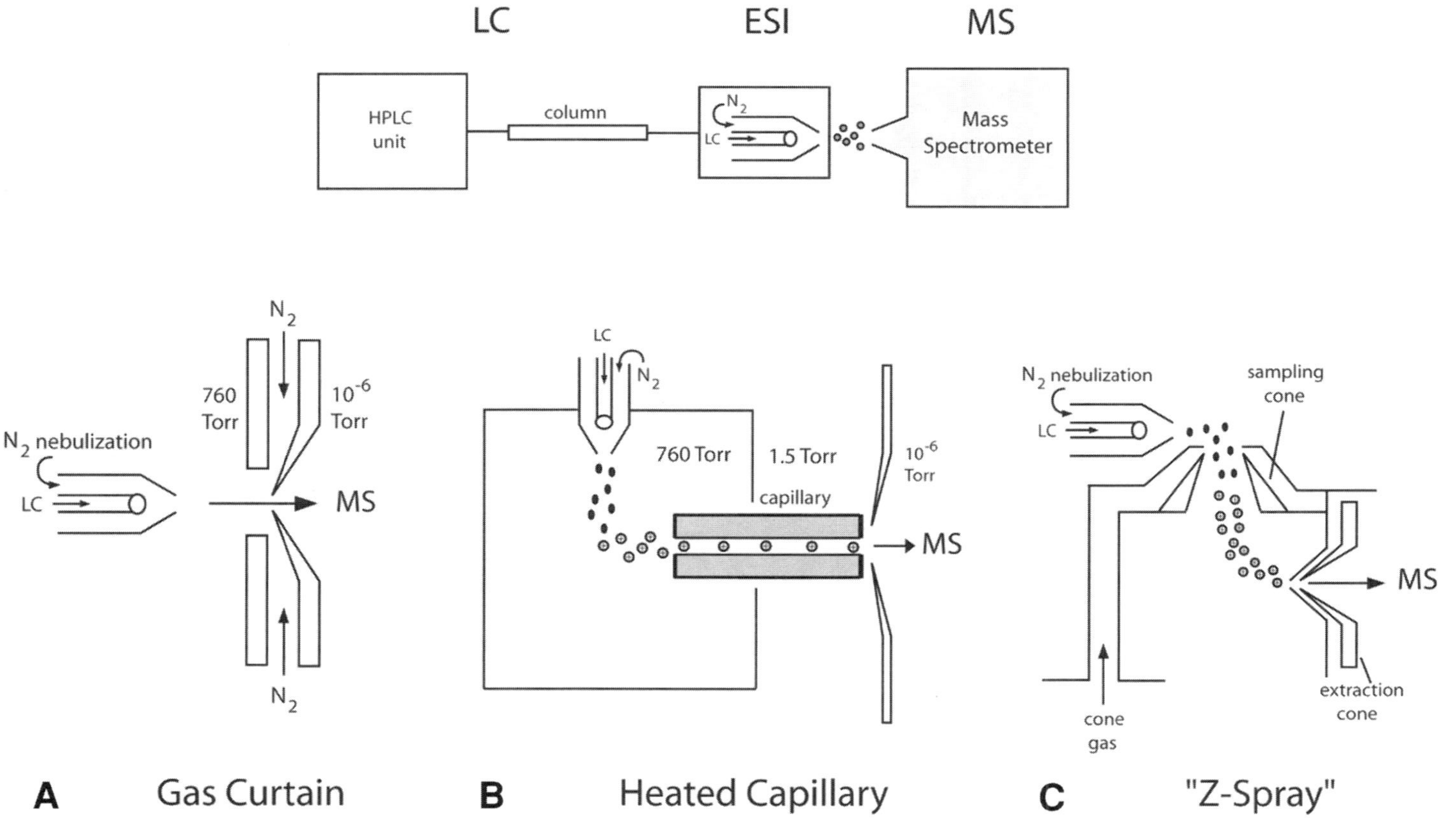

Fig. 1. The interface of liquid chromatography and mass spectrometry. **(A)** Introduction of ions into vacuum by passing through a gas curtain. **(B)** passing though a heated capillary tube. **(C)** a two stage orthogonal "Z" sampling technique.

scanning the amplitudes of the DC and RF voltages while keeping the DC to RF ratio constant, stable trajectories are created for ions of different *m/z* to pass through the quadrupole and exit to the detector. Both single and triple quadrupole mass spectrometers are used in LC–MS systems. **Figure 2A** shows a typical triple quadrupole mass spectrometer consisting of three regions. The first quadrupole is mass analyzer 1, the middle quadrupole is operated with only RF voltage and acts as an ion guide or a collision cell, and the third quadrupole is used to analyze product ions (mass analyzer 2). The triple quadrupole can be operated in different scan modes (*see* **Fig. 2A**) enabling a variety of tandem mass spectrometry experiments. With the introduction of the hybrid quadrupole time-of-flight instrument (Q-TOF) (*see* **Fig. 2B**), the third quadrupole is replaced by a time-of-flight analyzer. Tandem mass spectrometry is performed in a manner similar to the triple quadrupole, but it provides much greater mass accuracy and sensitivity ***(6)***. The quadrupole ion trap mass spectrometer (*see* **Fig. 2C**) is based on a device that trap ions in a three-dimensional electric field. Externally generated ions are injected into the ion trap where their energies are stabilized by collisions with a buffer gas. The trapped ions are then subjected to changes in the electric fields, which results in ejection of one ion after another from the trap. A detector records the ejected ions as a function of time. MS/MS experiments performed with an ion trap are "tandem-in-time" as opposed to the "tandem-in-space" experiments performed using a triple quad or a Q-TOF mass spectrometer. All ions except those with the selected *m/z* value are first ejected. The ions remaining in the trap are fragmented by vigorous collisions with the buffer gas and the product ions are analyzed as described earlier.

Fragmentation of peptides by collision-induced processes in tandem mass spectrometry has become one of the most useful applications of LC–MS in peptide analyses. Under conditions where ions undergo multiple collisions with an inert gas, peptide fragmentation takes place mainly at the amide bonds along the backbone, producing various ions series that are helpful for structural determination *(7)* (*see* **Subheading 3.4.**). Although single mass analyzers (e.g., single quadrupole mass spectrometer) have been used to induce fragmentation (*see* **Note 1**), most conventional tandem mass spectrometry employed in protein and peptide studies involves two separate stages of mass analysis. **Figure 2** illustrates the implementation of tandem mass spectrometry in **(A)** triple quadrupole, **(B)** quadrupole time-of-flight, and **(C)** ion trap instruments.

One of the most important advancements in the application of LC–MS and LC–MS/MS to biological problems that require protein identification, particularly proteomics, is the development software tools to identify proteins by matching mass spectrometric data to protein sequence databases ***(8,9)***. This

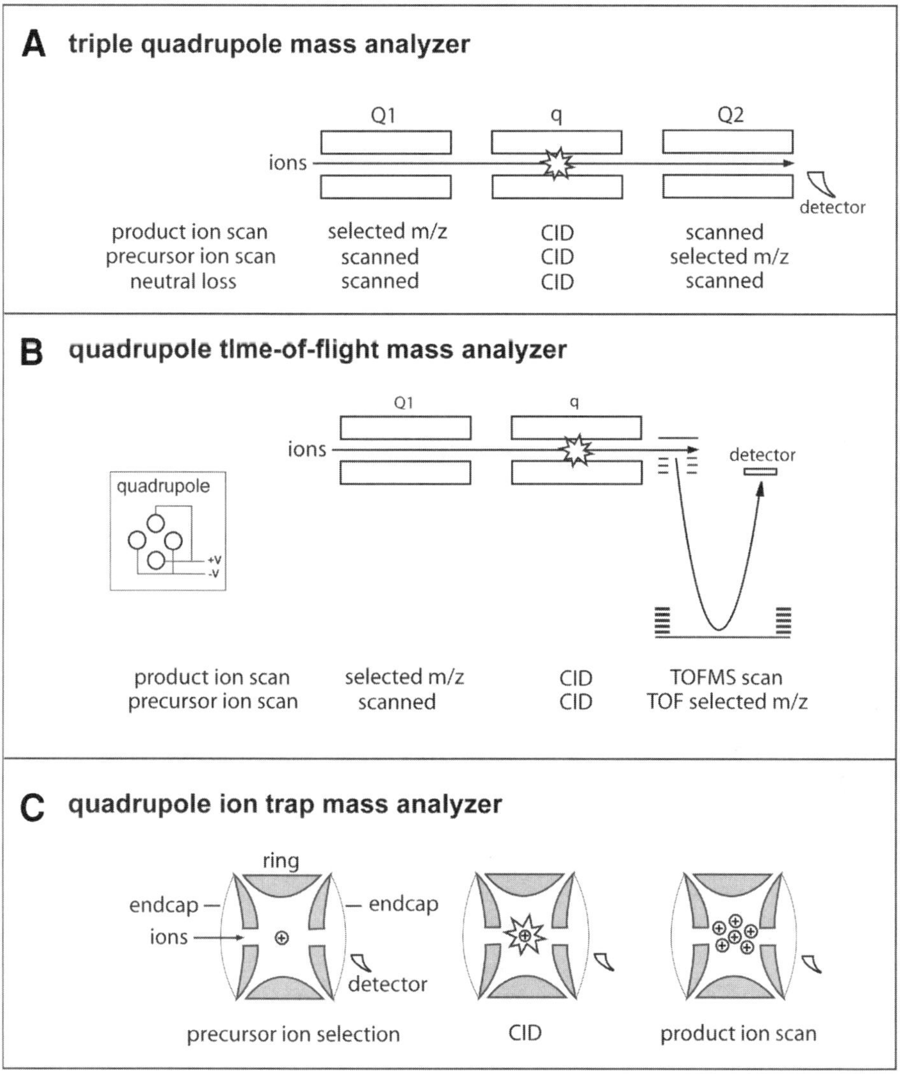

Fig. 2. Schematic diagram of the common mass analyzers used to perform tandem mass spectrometry. **(A)** Triple quadrupole mass analyzer and the types of experiments performed. **(B)** Quadrupole time-of-flight mass analyzer. (Inset) A cross-section view of the quadrupole. **(C)** Quadrupole ion trap mass analyzer. In the MS/MS experiment, the precursor ion is isolated by ejecting all other ions then increased its motions causing CID. The resulting product ions are scanned and spectrum recorded.

computational approach involves performing a proteolytic digest on the protein of interest, analyzing the peptides with mass spectrometry or tandem mass spectrometry, and submitting the mass spectral data to a computer program that searches a database and predicts the identity of the protein. Some searches are based simply on the accurate mass values of proteolytic peptides (peptide mass fingerprinting). Although LC–MS can provide high quality data for peptide mass fingerprinting searches, it is far more efficient to obtain such information with a technique called matrix-assisted laser desorption ionization (MALDI) mass spectrometry. Alternatively, a second approach utilizes the masses of individual proteolytic peptides and their fragment ions (peptide sequence tag) to generate a series of amino acid sequences that are matched to theoretical MS/MS spectra from proteins in the database. Identification is assigned from the consensus sequence.

2. Materials

2.1. Chemicals and Reagents

1. Calibration standards: Poly(propylene glycol) (PPG) and cesium iodide were purchased from Aldrich (Milwaukee, WI). The octapeptide ALILTLVS was provided by BACHEM (King of Prussia, PA).
2. Cytochrome-*c*, Gly-Tyr, Val-Tyr-Val, methionine-enkephalin, angiotensin II, and leucine-enkephalin were purchased from Sigma (St. Louis, MO) and used without purification.
3. Ammonium bicarbonate and formic acid were obtained from either Aldrich (Milwaukee, WI) or Sigma (St. Louis, MO). High-purity buffer salts and acids are readily available from a number of laboratory chemical suppliers.
4. Proteolytic digests were performed with sequencing grade modified trypsin from Promega (Madison, WI).
5. Acetonitrile and water were HPLC grade obtained from Burdick and Jackson (Muskegon, MI), J.T. Baker (Phillipsburg, NJ), or Fisher Scientific (Optima grade; Pittsburgh, PA). Water was also purified in house with a Milli-Q system (Millipore, Bedford, MA).

2.2. Analytical Liquid Chromatography

1. Shimadzu (Columbia, MD) HPLC system consisting of LC-10AD liquid chromatography pumps, SPD-10A UV detector, SIL-10AXL autosampler, and SCL-10A system controller.
2. HPLC Column: Supelco reversed phase C_{18} column (Bellefonte, PA) with dimensions 2.1 mm id x 10cm length or 4.6 mm id × 10 cm length, 5-μm particles, and 300 Å pores.
3. Solvents: mobile phase A: 0.1% formic acid in water; mobile phase B: 0.1% formic acid in ACN.

2.3. Nanoflow Liquid Chromatography

1. Pumps: MicroTech Ultra Plus II (Sunnyvale, CA) with binary gradient capability, set to deliver 50 µL/min.
2. Autosampler: LC Packings FAMOS (San Francisco, CA) configured with a 5-µL sample loop.
3. HPLC Column: LC Packings PepMap C_{18} column (San Francisco, CA) with dimensions 100 µm id × 15 cm length, 3 µm particles, and 100 Å pores.
4. Solvents: mobile phase A: 2% acetonitrile, 98% water, 0.3% formic acid; mobile phase B: 90% acetonitrile, 10% water, 0.3% formic acid.

2.4. Mass Spectrometry

1. Q-TOF: The hybrid mass spectrometer used for this study is a PE-Sciex QSTAR (Concord, ON, Canada) with Analyst QS software for instrument control and data analysis. Ionization is accomplished with a pneumatically assisted electrospray source. The pre-TOF ion path consists of three quadrupoles—Q0, Q1, and Q2. Q1 is the mass filtering device, whereas Q0 and Q2 are operated in the RF-only mode and function as ion transport devices with essentially no mass-discriminating capability. Ions are transferred from the source through Q0 into the high vacuum region of the first mass analyzer (Q1). Q2 is housed inside the collision cell and transmits ions from Q1 into the TOF region. In tandem mass spectrometry, Q1 can be operated with a transmission window of either 1 or 3 Daltons. Generally, the collision energy is kept under 10 eV (laboratory frame) in full scan mode, and typically set to 25–35 eV (laboratory frame) for product ion scans. Collision gas is present in both MS and MS/MS modes but it is increased for MS/MS experiments. The gas is required for collisional cooling of the ions to minimize their kinetic and spatial energy distributions, thus improving ion transmission and resolution. The TOF mass analyzer has an effective flight path of 2.5 m, a four anode detector with a time-to-digital converter (TDC) (625 ps resolution) that extends the dynamic range of the detector to 40,000 counts/s/peak (the range of a single anode detector is 10,000 counts/s/peak). The time-of-flight analyzer was operated with a pulsing frequency of 10 kHz and the resolution was measured to be 8000 full-width-half-maximum at *m/z* 829. The quadrupole mass analyzer was set to unit resolution.
2. Quadrupole Ion Trap: Ion trap experiments were performed on a ThermoFinnigan LCQ DECA (San Jose, CA) with Xcalibur software for instrument control and data analysis. Sequest, a computer program developed at the University of Washington ***(10)***, was used for database searching. The ion trap is interfaced to a New Objective PicoView nanospray ionization source (Cambridge, MA) configured with the uncoated tip module and fitted with a New Objective fused silica PicoTip emitter of dimensions 75 µm tubing id, 30 µm tip id, and 360 µm tubing od (FS360-75-30-D-5). Ions formed in the source are drawn into a heated capillary where desolvation occurs, then pass through a tube lens which focuses them into the

skimmer. The tube lens also functions as a gate to halt the injection of ions into the mass analyzer during the detection phase. The skimmer physically reduces the number of ions passing into the ion optics region containing a quadrupole, an octapole, and an interoctapole lens. Both the quadrupole and the octapole are operated in RF-only mode; hence, they function as ion transport devices. The interoctapole lens focuses the ion beam and acts in concert with the tube lens to deflect the beam away from the mass analyzer during ion detection. Mass analysis occurs in four stages: ion injection, storage, ejection, and detection. Helium is used as a buffer gas inside the trap to reduce the energy spread of the trapped ions and to improve resolution. Upon ejection from the trap, ions strike a conversion dynode, and the secondary particles emitted enter the electron multiplier where the signal is amplified by a factor of approximately 10^5. In MS/MS mode, precursor ions are isolated in the trap then subjected to collision induced dissociation using helium as the collision gas. The normalized collision energy is generally set to 35% of the RF amplitude required to isolate the ion of interest. Although the resolution of the ion trap is fairly low (unit resolution up to 2000 Daltons), higher resolution can be achieved over a narrow mass range using the zoom scan function.

3. Methods

3.1. Instrument Preparation

QSTAR: The mass analyzer must be calibrated before beginning experiments. The mass accuracy of the quadrupole (Q1) is ensured by calibrating with a standard mixture of PPGs that produce known mass peaks at *m/z* 59.0, 175.1, 906.0, and 2010.5. A 10^{-4} *M* PPG solution is prepared by mixing volumes of PPG 2000, PPG 1000, and PPG 425. Calibration of the TOF analyzer was carried out with a standard solution containing 2×10^{-5} *M* cesium iodide (CsI, *m/z* 132.9054) and 2×10^{-6} *M* ALILTLVS, a synthetic peptide (*m/z* 829.5398). A two point external quadratic calibration file was generated using the exact masses of the calibrants to solve for the slope a and the offset t_0 in the equation $(m/z)^{1/2} = a(t - t_0)$.

LCQ: Calibration of the ThermoFinnigan LCQ DECA is extremely stable and need only be performed once every 1–2 mo, unless drift is observed in mass assignments. Complete tuning of all instrument voltages is usually required to ensure successful calibration. This procedure can be accomplished manually or using the auto tune function available through the Xcalibur software. Comparable results are obtained with both methods. The same set of compounds is used for tuning and calibration. It consists of caffeine, a small peptide (MRFA), and Ultramark 1621. The mass spectrum of this mixture is well characterized and produces abundant singly charged ions over the entire mass range of the instrument. The calibration routine only accepts data input from these standard compounds; therefore, it is not necessary to create different calibration files for different types of experiments (positive ion vs negative ion, MS vs MS/MS). Unfortunately, Ultramark 1621 is not compatible with biological samples so the mass spectrometer source and focusing elements up to the ion optics must be washed extensively with solvents, particularly acetone, immediately after calibration to prevent background contamination for long periods of time.

3.2. Optimizing Ion Source Parameters and Lens Voltages

QSTAR: All source and instrument parameters are optimized by direct infusion of a standard solution at 5–10 μL/min. Critical ion source parameters are the gas curtain, nebulizing gas, source position, and ion spray voltage. The curtain gas should be set to a flow rate as high as possible without compromising sensitivity. The nebulizing gas affects the stability of the spray (ion signal) and the sensitivity. It should be adjusted until a stable ion signal and the best sensitivity are obtained. The spray voltage and source position also affect the quality of the ion signal. Higher flow rates require the source position to be backed away from the MS orifice. Again, it is best to use the lowest voltage possible to achieve maximum sensitivity. Ion transmission is fine tuned by tweaking the voltages that focus ions through the ion path chamber such as the orifice plate, the focusing ring, the interquad lens, and the dual element exit lens.

LCQ: Several instrumental parameters should be optimized on a daily basis to ensure consistent instrument performance. The spray voltage, capillary voltage, and tube lens offset voltage all affect the sensitivity and stability of the ion beam. Because these parameters are interdependent, it is usually necessary to arrive at a compromise tune to avoid frequent adjustments. Fine-tuning of the nanospray source position often improves the quality of the signal, particularly when the sample tip has been replaced. If the instrument is not equipped with a nanospray source, then the source position is fixed. At higher flow rates, the sheath and auxiliary gas pressures should also be optimized.

3.3. Coupling of HPLC Flow to the Mass Spectrometer

For LC–MS flow rates of 200 μL/min or less (column id < 1 mm), the entire effluent can be directed into the electrospray source. At flow rate higher than 200 μL/min, postcolumn splitting is employed. The split ratio is obtained by devising a flow restriction between the column outlet, the line going into the source, and the fraction collector or waste line. The simplest flow splitter consists of a tee and two pieces of fused silica or PEEK tubing. The dimensions of the tubing are used to control the back pressure used create the proper split ratio. The desired flow is set at the splitter by adjusting the lengths of the capillaries between the splitter and the ion source, and between the splitter and the waste line. Although the exact split is usually not critical, for these experiments it was maintained at a ratio of 20:1 (*see* **Note 2**).

The buffers or solvents used for chromatographic separation and electrospray ionization must be compatible (*see* **Note 3**). We used a solvent system consisting of 0.1% formic acid in water for Solvent A and 0.1% formic acid in acetonitrile for Solvent B or similar solutions. Ammonium acetate and ammonium bicarbonate also work well because these buffers are volatile and do not interfere with ionization. Compared to the more commonly employed TFA and phosphate buffer systems, the solvents mentioned earlier will produce slightly broader chromatographic peaks, but ion intensities in the mass spectra are higher. In addition, the mass spectra are relatively free from background ions originating from the solvents.

3.4. LC–MS of Peptide Mixtures

LC–MS experiments produce complex data sets that may contain several hundred mass spectra. In a typical experiment, the total ion current chromatogram (TIC) consists of the summed intensities of all ions in each scan plotted as a function of time. It is a convenient way to view an entire data set. Simultaneously, the data can be presented in an extracted ion chromatogram (XIC). These functions are available through the Analyst QS software. Like most commercial software used for LC–MS analysis, Analyst QS enables the user to display, manipulate, and process data. It also contains programs for performing qualitative as well as quantitative analyses (*see* **Fig. 3**). In this example, Analyst QS was used to focus on the chromatographic peaks of interest and examine the underlying mass spectra. The XIC plots the intensity of a selected mass as a function of time.

As shown in **Fig. 3**, the Analyst QS window is divided into separate panes, each displaying a different view of the same data. **Figure 3A** is a chromatogram of the full scan TIC from an LC–MS analysis of a five-component peptide mixture (1. GY; 2. VYV; 3. met-enkephalin YGGFM; 4. angiotensin II, DRVYIHPF; and 5. *leu*-enkephalin, YGGFL) separated on a 4.6 mm id × 10 cm length C_{18} column. The conditions for separation were as follows: 5–65% B in 8 min at a flow rate of 0.8 μL/min using the acetonitrile/water/formic acid solvent system described in **Subheading 3.3.** This experiment can be used as a benchmark to examine the performance of the HPLC, electrospray ionization source, and the mass spectrometer. The quality of the chromatographic separation and the stability of the ion current throughout the gradient can be evaluated by examining the TIC . It should be noted that a wide range of solvent gradients may be used to fine tune the separation as long as there are no adverse effects on the performance of the electrospray interface. The major advantage to coupling HPLC with MS detection is the added dimension of information gained, namely molecular mass data. Because electrospray is a "soft ionization"

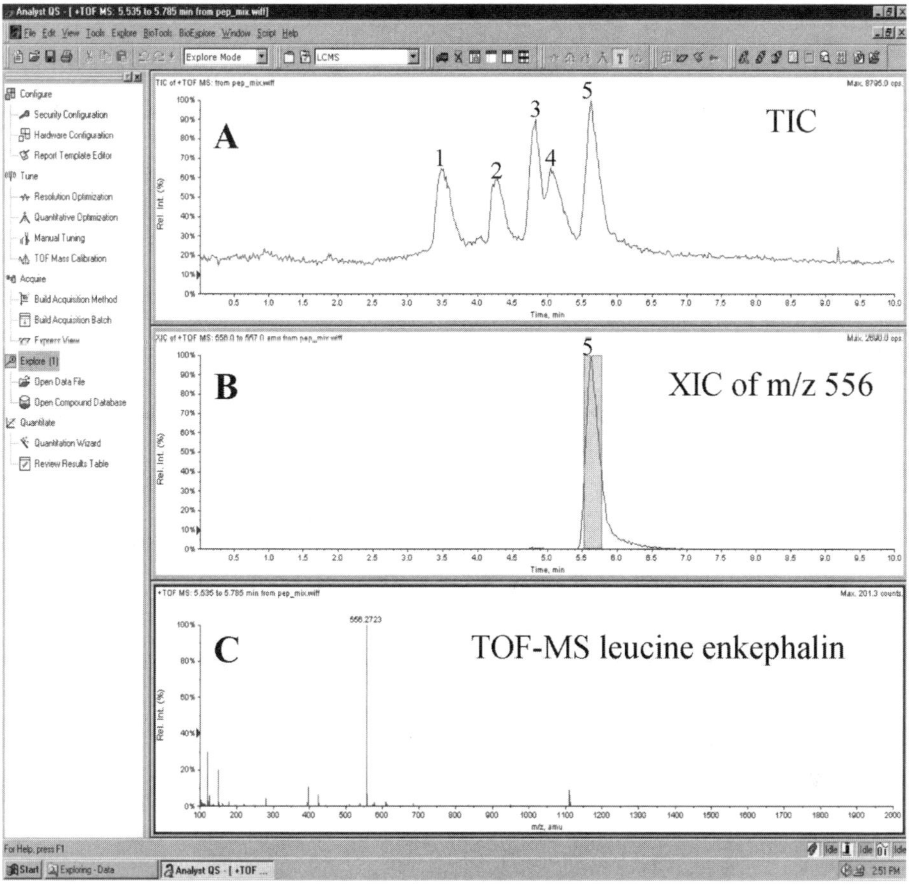

Fig. 3. Screen capture of the LC–MS analysis of a peptide mixture processed by Analyst QS. **(A)** The total-ion current chromatogram from the LC–MS experiment is shown in the top pane. **(B)** The middle pane illustrated the extracted total ion chromatogram for *m/z* 556.2732. **(C)** The averaged full scan mass spectrum from (gray area indicated the scans being averaged) the XIC of *m/z* 556.2732, the mass peak represent [M+H]+ ion of leucine-enkaphalin.

technique, the molecular weight of a compound is readily obtained from the mass spectrum. **Figure 3B** shows the extracted ion current of the ion observed at *m/z* 556. The complete mass spectrum corresponding to this component is shown in **Fig. 3C**. The most abundant ion was observed at *m/z* 556.2723, which correspond to the singly charged protonated molecular ion $[M+H]^+$ of leucine-enkephalin.

3.5. Tandem Mass Spectrometry of Peptides

In addition to molecular weight information, some structural information can be obtained from CID of peptide molecular ions *(11)*. **Figure 4** depicts the process of acquiring tandem mass spectra from peptides and the nomenclature used to describe the products of dissociation. The a_n, b_n, and c_n ions are generated when peptide bonds break with the charge retained on the N-terminal portion of the molecule. Conversely, when x_n, y_n, and z_n ions are formed, the charge remains on the C-terminal portion.

In MS/MS experiments, the exact mass of the precursor ion must first be determined using optimized instrument parameters for full scan mode as previously described in **Subheading 3.2.** The CAD gas should initially be set to a minimum value while optimizing the collision energy (RO2). Once the collision energy is set, the collision gas pressure should be adjusted until a stable signal is obtained. For some methods it may be necessary to lower the resolution of the first mass analyzer (Q1) in order to achieve a reasonable level of sensitivity. Collision energy (CE) is a term used to describe the velocity of the ions entering the collision cell. In the QSTAR, it is estimated by calculating the potential difference between Q0 and RO2 (CE = Q0 – RO2). The value can range from 0 to 90 V.

Figure 5 shows tandem mass spectra taken of the protonated molecular ion $[M+H]^+$ of Leu-enkephalin (Tyr-Gly-Gly-Phe-Leu) at three different collision energies and a fixed collision gas pressure. The spectra were acquired with the collision cell offset voltage held at 15, 26, and 34 V, respectively. The choice of applied CE affects the abundances and types of ions observed in the MS/MS spectra. The range of suitable CID energies is defined by conditions where almost no fragmentation of the precursor ion occurs to those where the precursor ions are too weak to be observed. The kinetic energy of ions (in the laboratory frame) is $E_{lab} = zeV_c$, where z is the number of charges on the ion; e is the charge of an electron, and V_c is the collision cell offset voltage. For the experiment where E_{lab} = 15 eV, only a few product ion peaks are observed. Spectra from the other two collision energies (26 and 34 eV) exhibit the same fragmentation

---

Fig. 4. *(see facing page)* The principle of sequencing peptides by tandem mass spectrometry. The fragmentation of peptides is accomplished when a peptide is ionized to form an $[M+H]^+$ ion and selected in first mass analyzer (MS 1) so that only this ion is passed through the collision cell to produce fragment peptide ions upon collision-induced dissociation (CID). MS 2 is used to record the product ion spectrum. The nomenclature of fragment ions produced by CID for a tetrapeptide ions by Roepstorff and Biemann *(10)* is shown (middle of **Fig. 3**). Also shown is the schematic presentation of the mechanism of formation of y_2, b_2, and a_2 fragment ions.

Sequencing of Peptides by Tandem Mass Spectrometry

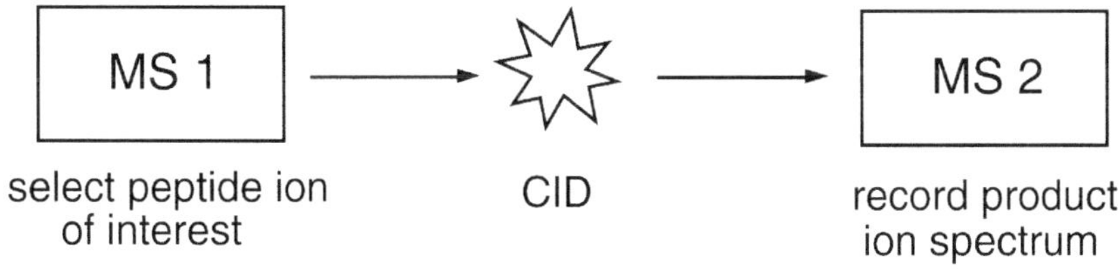

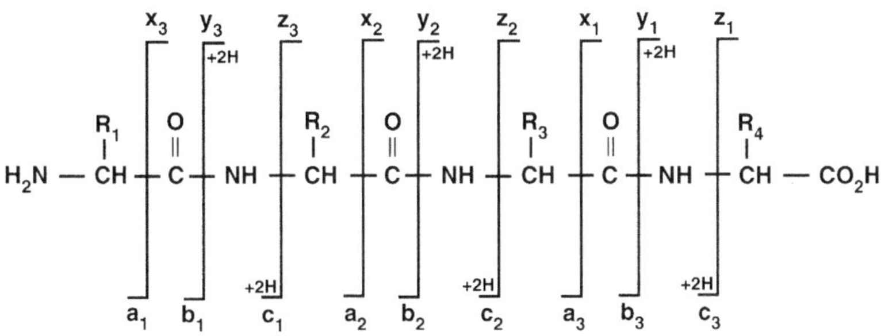

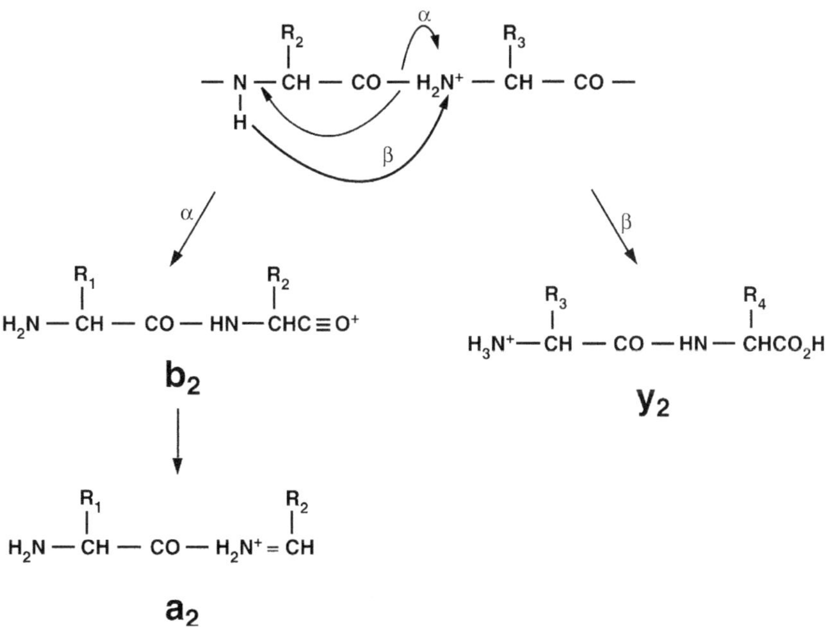

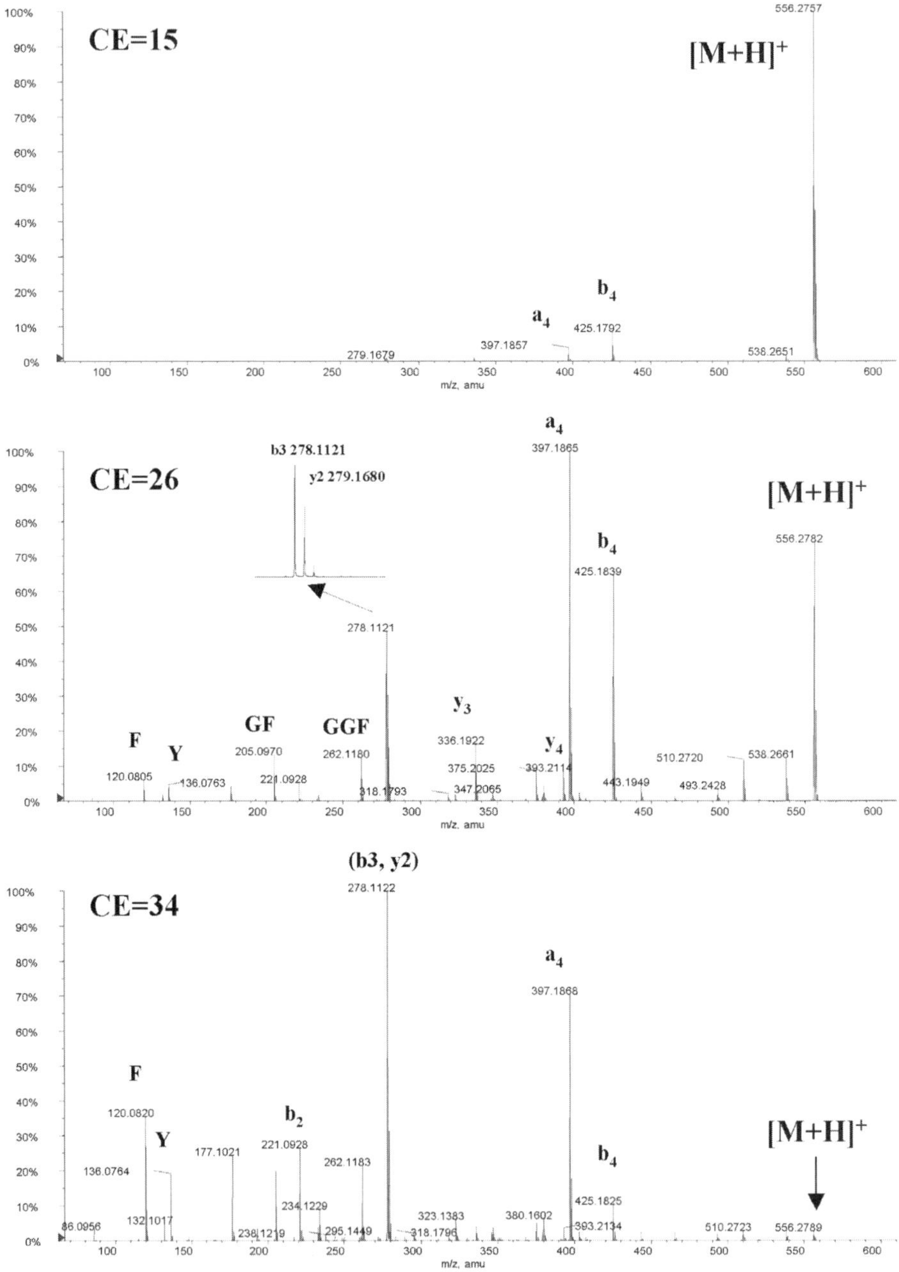

Fig. 5. CID spectra of Leu-enkaphalin (Tyr-Gly-Gly-Phe-Leu) at three different collision energies: CE = 15 V; CE = 26 V; CE = 34 V. Note the decrease of precursor ion intensity and the concurrent increase in the smaller fragment ions at higher collision energies.

pattern but the fragment ions have different intensities. Most members of the y-ion series (y_4, *m/z* 393.2114; y_3, *m/z* 336.1992; y_2, *m/z* 279.1680) and the b-ion series (b_4, *m/z* 425.1825; b_3, *m/z* 278.1122; b_2, *m/z* 221.0928) are present. The inset in the spectrum collected with a CE = 26V shows details of the ions at *m/z* 278.1122 and 279.1680, the b_3 and y_2 fragment ions, respectively. The doublet can easily be resolved with the high resolution capability of a Q-TOF instrument. It is also common to find ions that are not sequence specific, such as internal fragment ions resulting from y- and b-type cleavages at two peptide bonds, and immonium ions in the low mass region of the spectrum. Immonium ions have the general structure $^{+}NH_2{=}CH(R)$ so they provide diagnostic information on the amino acid content of the peptide. The peaks representing ions at *m/z* 86.0956, 120.0820, and 136.0763 are assigned to the immonium ions from Leu (L), Phe (F), and Tyr (Y). Other immonium ions often present in all types of MS/MS spectra are His (H), Trp (W), and Met (M).

3.6. Enzymatic Digestions

Standard procedures were followed for the tryptic digests performed prior LC–MS analysis. Generally, nonionic denaturants should be avoided, and volatile buffers are substituted for Tris-HCl. Typically, 400 pmole of cytochrome-c in 50 µL of 50 m*M* ammonium bicarbonate buffer, pH 8.5, was mixed with sequencing grade trypsin to give a final enzyme-to-substrate ratio of 1:50 (w/w). The digestion was allowed to proceed at 37°C for 18 hours then quenched by freezing.

3.7. LC–MS of Protein Digests

At present, online LC–MS analysis of protein digests has become one of the most popular techniques for characterizing proteins. There are two basic types of MS experiments that can be performed on digests: accurate mass measurements of the peptides formed from the digest, and amino acid sequencing of the peptides by tandem mass spectrometry. Most often the digestion mixture is fractionated by reversed-phase HPLC before it is introduced into the mass spectrometer. Electrospray ionization coupled with tandem mass spectrometry is ideally suited to the analysis of tryptic peptides. Because there are two sites for protonation located on every tryptic peptide (the basic N-terminus and the C-terminus lysine or arginine), they generally form a doubly charged ions. MS/MS of doubly charged tryptic peptide ions yields primarily singly charged fragment ions, and the spectra are dominated by b- and y-series ions.

Guidelines for tuning the mass spectrometer in MS/MS mode are outlined in **Subheading 3.5.** Most importantly, appropriate conditions for CID are determined by adjusting collision energy and/or collision gas while observing both the precursor and the fragment ions. The intensity ratio of precursor to product is not critical as long as the fragment ion intensities are sufficiently high for structural assignment. To produce a product ion scan, the second mass analyzer (TOF in quadrupole time-of-flight or quadrupole in a triple quadrupole mass spectrometer) scans repetitively to cover the mass range of all possible fragment peptide ions that could be produced (e.g., *m/z* 50–1800). It should be noted that the presence of product ions at *m/z* values higher than the precursor ion is a unique feature of the fragmentation of multiply charged ions.

With manual data acquisition, analysis of complex protein digests by LC–MS/MS would require large amounts of sample (*see* **Note 4**). However, when the data are collected in a fully automated manner, very small sample sizes are consumed. Information dependent acquisition (IDA) is a term used by the Analyst QS software to describe automatic collection of MS/MS data. In IDA, peptide ions that meet preestablished selection criteria are isolated in real time and subjected to tandem mass spectrometry analysis. Experimentally, the IDA function will acquire a TOF-MS survey scan and search for the most intense ions in the spectrum. Ions with intensities that exceed the threshold level are selected by the first mass analyzer (Q1) and subjected to collision-induced dissociation. A high-resolution mass spectrum of the fragment ions is collected with the TOF analyzer. In Analyst QS, up to eight of the most intense ions can be selected as precursors for MS/MS experiments and three different collision energies can be used to collect product ion scans for each precursor ion. Chromatographic peaks are usually 10–30 s wide with the LC system described in **Subheading 3.4.**; consequently, for typical LC–MS/MS IDA experiments on peptide mixtures, accumulation times of 0.5–1 s for TOF-MS survey scans and 2–3 s for product ion scans are suitable. It is important that the complete cycle time (survey scan + product ion scan) is not too long so as to allow for the collection of 3–10 cycles across the LC peak (*see* **Note 4**).

Figure 6A shows the TIC chromatogram from a tryptic digest of cytochrome-*c* analyzed by LC–MS. The separation was achieved with a 2.1 mm × 10 cm long C_{18} column by applying a linear solvent gradient from 5% B to 60% B in 18 min using a flow rate of 0.3 mL/min. **Figure 6B** shows the mass spectrum from the chromatographic peak eluting at 12.8 min. The ion observed at *m/z* 748.3496 corresponds to the molecular mass of the component since electrospray ionization rarely results in fragmentation. The charge state of the ion is determined from the spacing of the isotopic peaks. A separation of 0.5 and 1.0 Da indicates doubly and singly charge ions, respectively. Although molecular weights of the separated peptides are easily determined by LC–MS, this information alone is often insufficient for protein identification. Tandem mass spectrometry provides the amino acid sequence data required for structural characterization. The MS/MS spectrum of the doubly charged ion at *m/z* 748.3496 is shown in **Fig. 6C**. Inspection of the product ion spectrum reveals an abundance of frag-

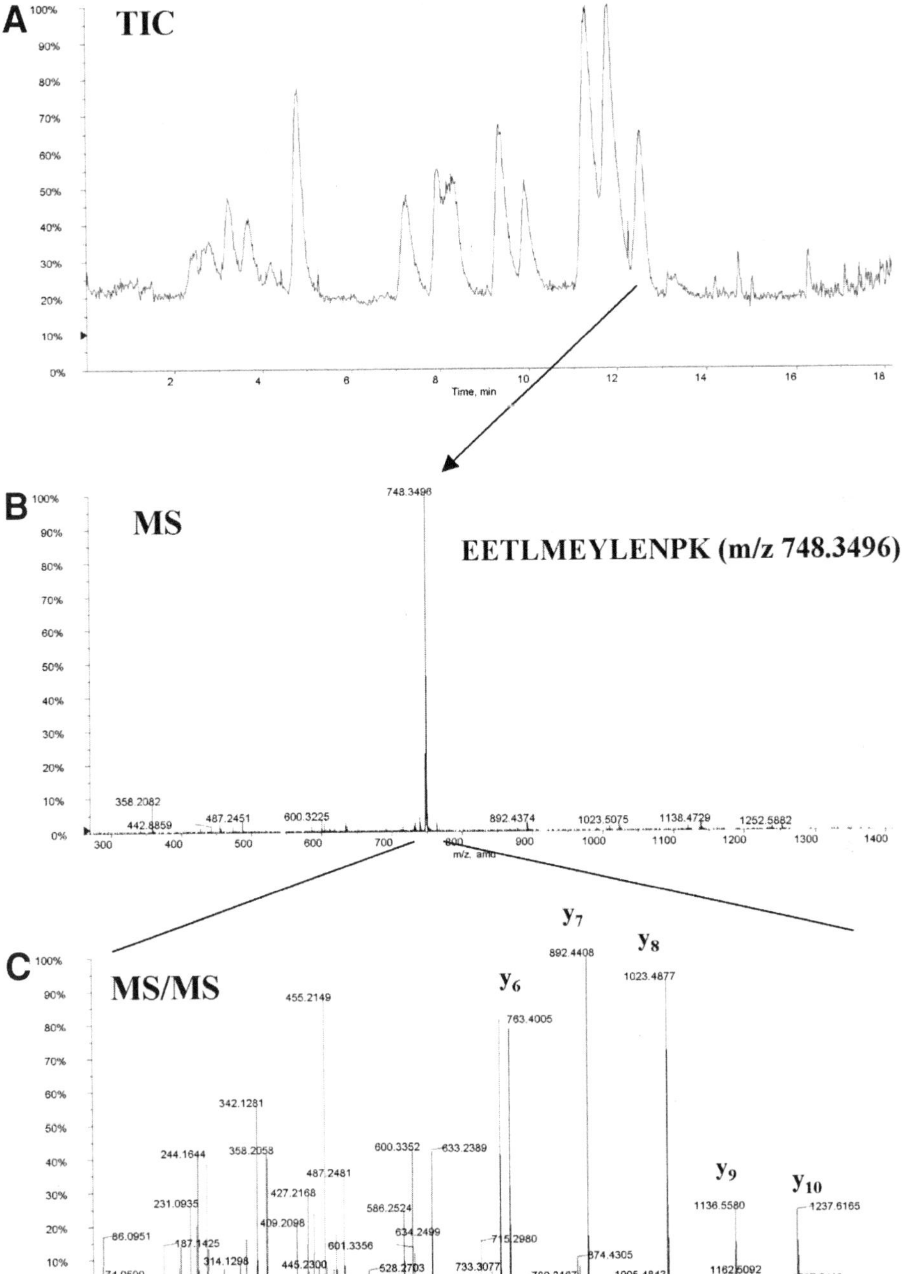

Fig. 6. (**A**) The total-ion current chromatogram of the LC–MS analysis of a tryptic digest of cytochrome-*c*. (**B**) Mass spectrum of the doubly charged peptide ion *m/z* 748.3496 at retention time 12.6 min. (**C**) The MS/MS spectrum of the doubly charged ion $[M+2H]^+$.

Table 1
Mass Spectrometric Strategies for Protein Identification Based on Protein Sequence Database (*see* Note 6)

MS data	MS/MS data
Peptide mass mapping	1. Uninterpreted peptide fragment mass data 2. Peptide sequence tag *(13)* or peptide fragment ion masses

ment ions including the clearly distinguishable y-series ions. The interpretation is straightforward because the spectrum displays a simple fragmentation pattern that is typical of tryptic peptides (*see* **Note 5**). In the next section, protein identification using database searching with MS/MS data from tryptic peptides will be demonstrated.

3.8. Manual Database Searching With Tandem Mass Spectrometry Data

There are two approaches to searching protein sequence databases using mass spectrometric data *(**12**; see* **Table 1***)*.

The first approach is called peptide mass mapping. This method uses accurate mass measurements of proteolytic peptides and searches for a correspondence between the theoretical peptide masses of each protein in the database and the experimental data (*see* **Note 7**). As mentioned earlier, peptide fragmentation mass spectra obtained from tandem mass spectrometry can also be used for sequencing and identification of proteins . The search can be performed either by automated correlation of uninterpreted MS/MS spectra (*see* **Subheading 3.9.**) or by using MS/MS spectra that have the peptide fragment ions series assigned. In this example, data obtained from **Fig. 6C** on the fragmentation of *m/z* 748.3496 was used to conduct a protein sequence database search with the MS-Tag tool from Protein Prospector, available free through internet (http://prospector.ucsf.edu). Five y-type fragment ions (y_6, *m/z* 763.4005; y_7, *m/z* 802.4408; y_8, *m/z* 1023.4877; y_9, *m/z* 1136.5580; y_{10}, *m/z* 1237.6165) characteristic of tryptic peptides were first identified. The database was searched for a peptide with fragments that would match the masses of these five product ions within a specified mass tolerance of 100 ppm. Other information used in this search include the charge state (+2) and mass of the precursor ion (*m/z* 748.3496), the mass range of protein (10–20 kDa), the enzyme used for digestion (trypsin), and the database (SwissProt.9.30.2001). The results of the search are summarized in **Fig. 7**. Identical scores are obtained

MS-Tag Search Results

Press stop on your browser if you wish to abort this MS-Tag search prematurely.
Sample ID (comment): **tryptic peptide**
Database searched: **SwissProt.9.30.2001**
Full Molecular Weight range: **100225** entries.
Full pI range: **100225** entries.

Pre searches select **100225** entries.
Ion Types Considered: **a a-NH3 a-H2O b b-NH3 b-H2O b+H2O y y-NH3 y-H2O I i m**

Search Mode	Peptide Masses are	Digest Used	Max. # Missed Cleavages	Cysteines Modified by	Peptide N terminus	Peptide C terminus
identity	**monoisotopic**	**Trypsin**	**1**	**unmodified**	**Hydrogen (H)**	**Free Acid (O H)**

Number of sequences passing through parent mass filter: **1833**
MS-Tag search selects **2** entries.
Parent mass: **748.3496**$^{+2}$ **(+/- 100.0000 ppm)**
M+H equivalent: **1495.6914 (+/- 0.1497 Da)**
5 Fragment Ions used in search: **763.4000, 892.4400, 1023.4900, 1136.5600, 1237.6100 (+/- 100.00 ppm)**
Max # Unmatched Ions = **0**

Result Summary

Rank	# Unmatched Ions	Sequence	MH$^+$ Calculated (Da)	MH$^+$ Error (ppm)	Protein MW (Da)/pI	Species	SwissProt.9.30.2001 Accession #	Protein Name
1	0/5	(K)EETLMEYLENPK(K)	1495.6990	-5.1	11687.6 / 9.59	EQUAS	P00005	CYTOCHROME C
1	0/5	(K)EETLMEYLENPK(K)	1495.6990	-5.1	11701.7 / 9.59	HORSE	P00004	CYTOCHROME C

Detailed Results

Rank	# Unmatched Ions	Sequence	MH$^+$ Calculated (Da)	MH$^+$ Error (ppm)	Protein MW (Da)/pI	Species	SwissProt.9.30.2001 Accession #	MS-Digest Index #	Protein Name
1	0/5	(K)EETLMEYLENPK(K)	1495.6990	-5.1	11687.6 / 9.59	EQUAS	P00005	4122	CYTOCHROME C
1	0/5	(K)EETLMEYLENPK(K)	1495.6990	-5.1	11701.7 / 9.59	HORSE	P00004	4134	CYTOCHROME C

Fragment-ion (m/z)	763.4000	892.4400	1023.4900	1136.5600	1237.6100
Ion-type	y_6	y_7	y_8	y_9	y_{10}
Delta ppm	1.3	-1.8	7.7	-5.4	-3.1

Fig. 7. Database search result based on the **Fig. 6C** data, entering the masses of y-series fragment ions (763.4000, 892.4400, 1023.4900, 1136.5600, 1237.6100), the doubly charged peptide ion mass (748.3496), and other search constraints. Under the results summary, all matched database entries and detailed resulted are listed, the peptide sequence that matched the input information is underlined.

for the cytochrome-*c* from equas (donkey) (accession # P0005) and horse (accession # P0004), two closely related proteins. The number of top ranked protein matches generally decreases and the reliability of the results increases when the search is carried out with high resolution mass spectral data such as that obtained from a Q-TOF instrument. For example, if the mass uncertainty is 1–2 Da, then Ile/Leu (113.084) residues could not be differentiated from Asn (14.043) and possibly Asp (115.027), Gln (128.059) and Lys (128.095), would be indistinguishable from Glu (129.042), and the difference between oxidized Met (147.035) and Phe (147.068) could not be measured. In the present study, five fragment ions correctly identify the peptide sequence EETLMEYLENPK which matches the two cytochrome-*c* proteins including the correct species of cytochrome-*c* from horse. Note that the mass accuracies of the fragment ions were all less than a few ppm.

3.9. Automated Database Searching With Tandem Mass Spectrometry Data

An example of the fully automated approach to database searching with MS/MS data is shown in **Figs. 8** and **9**. A small sample of a cytochrome-*c* tryptic digest (90 femtomoles injected) was chromatographically separated on a capillary column then analyzed in an LCQ ion trap using data dependent acquisition. Full scan mass spectra were acquired every 0.15 s (*see* **Fig. 8**, Step **A**) until the MS to MS/MS switching criteria were met (*see* **Fig. 8**, Step **B**). Once the requirements for switching were satisfied, MS/MS spectra were collected every 1.05 s (*see* **Fig. 8**, Step **C**) until the number of repeat cycles was completed or the ion was placed on a dynamic exclusion list for a length of time corresponding to the width of a typical chromatographic peak. By using a dynamic exclusion process instead of permanent removal, peptides with similar or even identical masses that elute at later times in the chromatogram will still trigger MS/MS scans. Fragmentation data were collected sequentially on the three most intense ions in each full scan MS spectrum. Consecutive MS/MS analyses of more than one ion offers the advantage of generating primary sequence data on multiple co-eluting peaks so that very complex mixtures may be analyzed without optimizing the chromatographic separation. Switching criteria also include provisions for the precursor ion charge state and collision energy profiles used in fragmentation. Tryptic peptides most often produce doubly charged ions in electrospray; therefore, the default charge state was set to 2, but exclusion of singly charged species was not enabled for MS/MS scans. Because it was not necessary to determine charge state for each ion, the zoom scan function was eliminated. Fragmentation can be carried out at multiple col-

Fig. 8. *(see facing page)* A data dependent experiment performed on a tryptic digest of cytochrome-*c*. (Step **A**) The peptide mixture is separated on a capillary column using a short solvent gradient. The results show incomplete separation of the peptides. (Step **B**) Full scan mass spectra are collected until all switching criteria are satisfied. At this point, the mass spectrometer shifts into MS/MS mode. (Step **C**) MS/MS spectra of the most intense ion are collected until the switching criteria are no longer met.

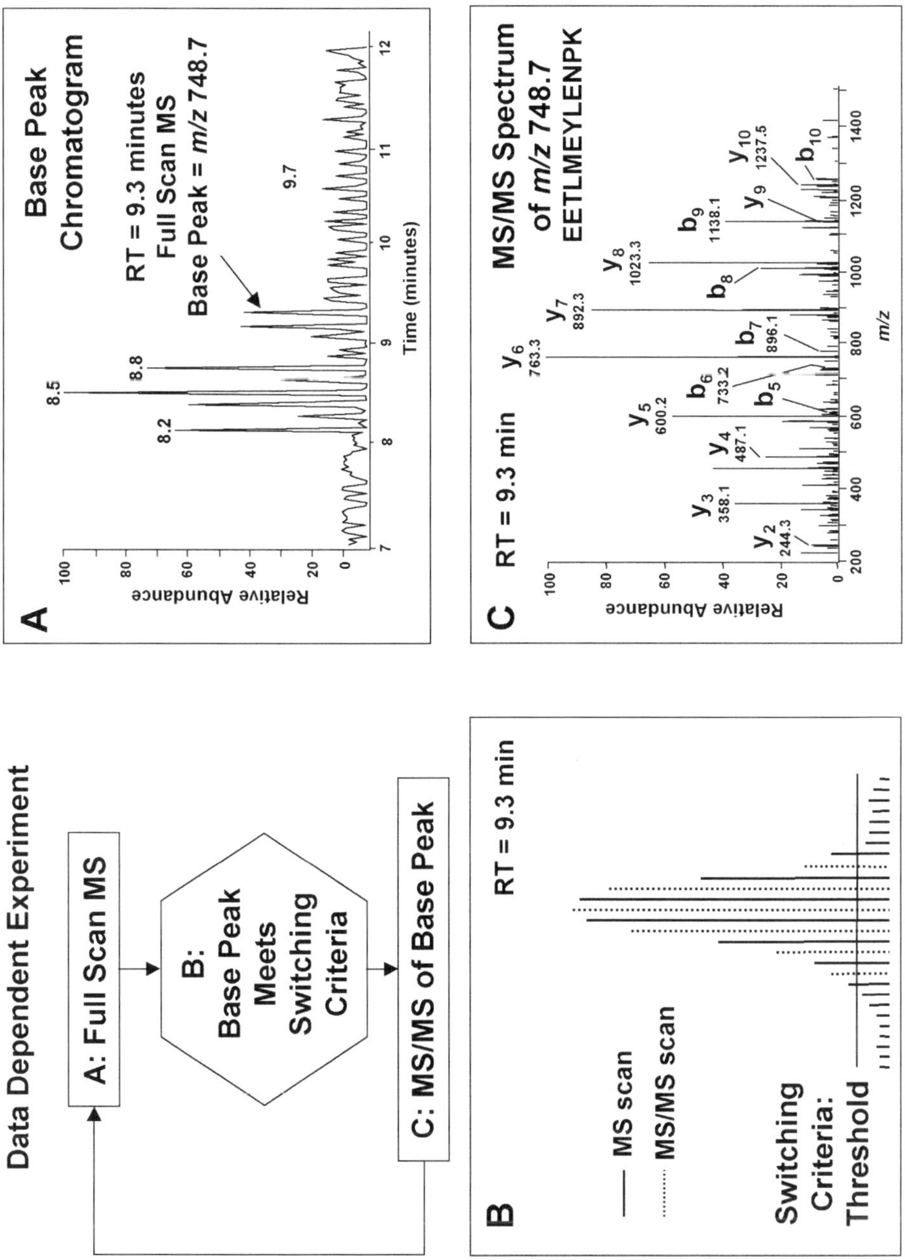

Data Dependent Experiment
A: Full Scan MS
B: Base Peak Meets Switching Criteria
C: MS/MS of Base Peak
A
Base Peak Chromatogram
RT = 9.3 minutes
Full Scan MS
Base Peak = m/z 748.7
Relative Abundance
Time (minutes)
B
RT = 9.3 min
MS scan
MS/MS scan
Switching Criteria: Threshold
C
RT = 9.3 min
MS/MS Spectrum of m/z 748.7
EETLMEYLENPK
Relative Abundance
m/z

lision energies by setting the repeat cycle to a value greater than one. In this experiment, the normalized collision energy was held constant at 35% because the majority of peptides fragment readily at this value. Known contaminants such as peptides from keratin and trypsin autodigestion can be placed on a separate list of exclusion masses that are ignored throughout the duration of the chromatographic run. The MS/MS spectrum of *m/z* 748.7 corresponding to the doubly charged peptide EETLMEYLENPK from cytochrome-*c* shown in **Fig. 8** was collected under the conditions for data dependent acquisition listed above. The spectrum displays a near complete series of y-ions and several prominent b-ions; thus, the amino acid sequence can be deduced with a high degree of certainty, even when manual interpretation is employed.

Once the LC–MS with data dependent MS/MS acquisition was completed, the uninterpreted results were submitted for automatic identification of protein components using SEQUEST to search the SwissProt 9.30.2001 database. A portion of the results is shown in **Fig. 9**. Briefly, the program first calculates theoretical tryptic peptides for every protein in the specified database and generates a predicted MS/MS spectrum for each peptide. The data are indexed on the basis of peptide *m/z* values to enable rapid yet thorough searching of the database. In the second step, all MS/MS scans are separated from the raw data then sorted according to the mass value of the precursor ion. Spectra from the same precursor are grouped together into one SEQUEST input data file to avoid redundant searches. Each theoretically generated MS/MS spectrum in the database within a specified mass tolerance is compared to the MS/MS experiment input files in a reverse search using a formula that evaluates several parameters such as number of ions matched and their intensities. The top 500 peptides from the database are then evaluated using a cross-correlation algorithm that determines the consistency of the predicted MS/MS spectrum to the observed fragmentation, and calculates a score (Xcorr value) representing the goodness of fit. Peptides from the database with the highest Xcorr values are placed on the list of top sequence candidates. In the last step, the database is searched with the entire set of SEQUEST input data files and the identity of the protein is determined based on the best match. A summary of the search is saved in a results file (*see* **Fig. 9**).

Fig. 9. *(see facing page)* A portion of the summary output from a SEQUEST search against the SwissProt database for all the MS/MS spectra collected in a data dependent experiment of the tryptic digest of horse cytochrome-*c* shown in **Fig. 8**. The SEQUEST search parameters are listed in the header information. Results from the first 10 MS/MS experiment input files and the top five candidate proteins are displayed.

SUMMARY2HTML v.8 (rev. 1.2), Copyright 1996

Molecular Biotechnology, Univ. of Washington, J.Eng/J.Yates

Licensed to Finnigan Corp., A Division of ThermoQuest Corp.

10/29/01, 10:51 AM, C:\Xcalibur\database\swissprot.fasta, AVG/AVG

Database type = Peptide

Start_scan=283; End_scan=351; Precursor_mass_tol =1.40; Group_scan_tol =1; Min_group_count =1;

Min_ion_count =35; Low_mass_limit =300.00; High_mass_limit =4000.00; Use_charge_state =0; Subsequence =[];

peptide_tol=1.0000; Fragment_Ion_Tol=1.0000; B_ion=Yes; Y_ion=Yes;

Enzyme:Trypsin (1)

#	File*	MassA*	Xcorr*	DelCn*	Sp	RSp	Ions	Reference*	MW*	Sequence**
1	101001_05.0288.0288.1	589.7 (-0.2)	1.4838	0.231*	607.9	2	8/8	gi|3023673|sp|Q59110|DSRB_ARCFU	41551	K.VDVEK.C
2	101001_05.0290.0290.2	1746.9 (-0.8)	0.8086	0.196	149.4	50	9/32	gi|3914056|sp|O24617|MSH2_ARATH	105485	R.PEITSSDAGDIVLEGSR.H
3	101001_05.0290.0290.3	2617.9 (+0.6)	0.9294	0.079	467.3	10	26/96	gi|2499495|sp|Q27684|PGKB_LEIME	44950	K.AVDSPLVTEDVDVPDGYMALDIGPK.T
4	101001_05.0291.0291.1	538.6 (-0.2)	0.4387	0.053	105.9	4	4/8	gi|119236|sp|P08016|EGGS_SCHMA	18820	K.NGKGY.-
5	101001_05.0292.0292.2	1464.6 (+0.0)	1.0615	0.194	230.7	20	10/28	gi|2497834|sp|P55652|Y4SI_RHISN	79960	R.TTVSNIAATATSQAK.H
6	101001_05.0292.0292.3	2196.5 (-0.1)	0.9955	0.003	354.3	66	22/92	gi|3121735|sp|O25283|ACCA_HELPY	34863	K.VPTISVIIGEGGSGGALAIAVADK.L
7	101001_05.0294.0294.2	1268.4 (-0.4)	1.3515	0.043	369.7	9	12/20	gi|399137|sp|P30996|BXF_CLOBO	146691	R.GVTYEETIEVK.Q
8	101001_05.0294.0294.3	1901.1 (+0.4)	1.3610	0.108	287.8	104	20/64	gi|466085|sp|P34616|YOG7_CAEEL	375734	K.IQAISYDGERVGFCDVK.I
9	101001_05.0295.0295.2	1355.5 (+0.5)	0.8871	0.013	204.3	12	12/24	gi|548613|sp|P35820|PSC_DROME	169981	K.LVNGGQPQSAQQK.T
10	101001_05.0295.0295.3	2034.4 (-0.9)	1.2009	0.021	175.3	324	21/72	gi|133717|sp|P28079|RS10_THEAC	11929	K.EIASRTGVEIHGPMPLPTK.R

1 gi|117995|sp|P00004|CYC_HORSE 77.8 (6,1,1,0,0){ 13 15 25 46 48 56 : 31 : 32 } _ BEST FIT

2 gi|117982|sp|P00005|CYC_EQUAS 54 (4,1,1,0,0){ 25 46 48 56 : 31 : 32 }

3 gi|118006|sp|P00013|CYC_MINSC 28.4 (2,1,0,0,0){ 25 32 : 31 }

4 gi|118001|sp|P00014|CYC_MACGI 28 (2,1,0,0,0){ 13 15 : 31 }

5 gi|117963|sp|P00006|CYC_BOVIN 28 (2,1,0,0,0){ 25 48 : 31 }

The protein in this experiment was correctly identified as cytochrome-c from horse. In the top portion of the report, every MS/MS input data file is listed along with the results of the reverse search. The summary at the bottom of the page indicates that this protein has an overall score of 77.8 with data from eight MS/MS experiment input files mapped to the cytochrome-*c* sequence. Out of the eight input files, six were ranked the top candidate, one was ranked second, and the last ranked third. The numbers corresponding to these MS/MS input files are listed in parentheses at the end of the entry. The second most likely protein has an overall score of only 54.0 because the number of peptides and the quality of the matches is not as good as cytochrome-*c*. The differences become even larger as the remaining candidates are considered. This is a good indication that the top choice is the correct identification.

The complete amino acid sequence of a candidate protein can be accessed by clicking on the accession number. In addition, information about the database, protein accession number, molecular mass, a list of amino acid sequences of the peptides mapped to the protein, and percent protein coverage are also displayed. The entry for cytochrome-*c* shows the location of five out of the six top-ranked peptide fragments. The sixth peptide could not be located in cytochrome-*c* because the proposed sequence contains one isoleucine residue while the corresponding amino acid in cytochrome-*c* is a leucine. The protein coverage calculated from these five peptides is 34.6%; however, if you take into account that leucine and isoleucine are isobaric, then the sixth peptide can be mapped to positions 28–38 and the coverage would be 45.2%.

4. Notes

1. The "in-source fragmentation" technique has been used in single quadrupole mass spectrometers to produce fragmentation similar to the CID process in a tandem mass spectrometer. In-source fragmentation is based on the production of fragment ions by increasing the energy of the ions formed in electrospray as they pass through the atmospheric pressure/vacuum interface. The energy acquired by the ions increases as the electric field in the sampling region increases. Depending on the design of the interface, collisions occur between the capillary and the skimmer or sampling cone, and the degree of fragmentation can be controlled by varying the potential of the capillary and the counter electrode.
2. The following equation can be used to determine the proper tubing lengths and inner diameters for a particular split ratio (S):

$$S = F_W/F_E = (l_E d_W^4)/(l_W d_E^4)$$

where F_W is the flow rate exiting the waste line; F_E is the flow rate entering the ion source; l_E is the length, and d_E is the diameter of tubing used from the splitter to the ion source; l_W is the length, and d_W is the diameter of tubing used for the waste line. To minimize dead volume and peak broadening, used 50 μm id tubing for the ESI transfer line.

3. Commonly used solvents for reverse phase LC such as water, acetonitrile, and methanol are ideal electrospray ionization; although it is sometime necessary to increase desolvation temperature and gas flows when using a high percentage of water. If the pH of the mobile phase must be reduced to enhance separation efficiency, then acetic acid and formic acid can be used to adjust the pH in the 2.5–3.5 range. Normally the amount is kept below 1%. For separations that require a pH in the range of 3.0–6.0, ammonium and triethlyamine salts of the above acids are suitable. Ammonium acetate is a volatile buffer which can be used in the mobile phase without interfering with the performance of the mass spectrometer provided it is present in a concentration less than 100 m*M*. Ammonium bicarbonate and carbonate can be used as buffers for pH 6.0–8.0 and pH 9.0–11.0, respectively. Note that trifluoroacetic acid (TFA) is widely used in the isolation and purification of proteins and peptides. The role of TFA appears to be one of ion pairing with positively charged and polar groups on peptides and proteins. This effectively prevents polar interaction between the analytes and active sites on the column resulting in narrow chromatographic peaks. However, TFA should be avoided in LC–MS analyses. Owing to the low pKa and abundant production of the trifluoroacetate ion, it can form very strong ion pairs with the analyte, thus decreasing sensitivity. Nonvolatile buffers used in reversed-phase and ion exchange chromatography such as phosphate, citrate, Tris-HCl, and borate are also incompatible with mass spectrometry because they can crystallize in the ion source block sample introduction, although recent changes in source design (orthogonal ion source geometry) have minimized this problem.
4. Automation of LC–MS/MS (data dependent experiment): Recent instrument developments have greatly improved the efficiency of LC–MS/MS data acquisition ***(14,15)***. Many commercially available instrument control programs are capable of switching between MS and MS/MS experiments based on the following criteria: (1) ion intensities (e.g., the most abundant ion, the second most abundant ion, and so on); (2) *m/z* value (present or absent on a predetermined list); and (3) charge state (e.g., doubly charged for tryptic peptides). The MS/MS instrumental conditions can be placed under data system control including collision energy and scan range of the second mass analyzer.
5. Tandem mass spectrometry of tryptic peptides: Protein digestion with trypsin is widely used for protein characterization in conjunction with tandem mass spectrometry. Tryptic digests produce peptides with molecular weights around 2–3 kDa and lower that provide useful fragmentation data. Also, trypsin cuts on the C-terminal side of arginine (R) and lysine (K) residues. With basic termini, peptide fragmentation is more predictable. Typical MS/MS spectra of tryptic peptides obtained from a triple quadrupole or a Q-TOF contain y ions and b ions only in the low mass region (below the *m/z* value of the precursor ion). Ion trap MS/MS spectra are slightly different in that they usually contain both y- and b-type ions throughout the spectrum.

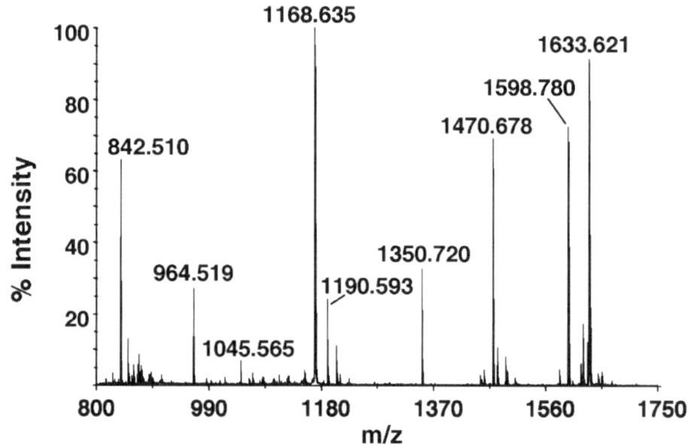

Fig. 10. MALDI-TOF mass spectrum of a tryptic digest of cytochrome-*c*. The data were collected in reflection mode using α-cyano-4-hyroxycinnamic acid as the matrix. External calibration of the *m/z* axis were performed.

6. Peptide mass mapping is a very popular method for the identification of proteins. Accurate mass measurements of the peptide fingerprint formed from proteolytic enzyme digestions is the key to this technique. The experimentally derived peptide masses are compared to the protein database of choice using one of the utilities listed in **Note 7**. An example of a MALDI-TOF mass spectrum of a cytochrome-*c* trypsin digest is shown in **Fig. 10**. The spectrum was acquired on a PerSeptive Voyager Elite DESTR mass spectrometer (Framingham, MA) in the reflectron mode.

   The masses of 21 peptides were submitted to Protein Prospector for a database search using the MS Fit tool. A portion of the results are shown in **Fig. 11**. The SwissProt database was searched for proteins with molecular weights in the range of 1–25 kDa, with the mass tolerence set to 50 ppm. The other search parameters are listed in the results summary. The two best matches are cytochrome-*c* variants, one from horse and one from donkey (Equas). Horse cytochrome-*c* is a better candidate as judged by the higher MOWSE score and the slightly larger number of peptides that were matched (8 vs 7). Whereas the cytochrome-*c* used in this experiment was from horse, it is interesting to note that the donkey and horse proteins differ by a single amino acid.
7. There are a number of software tools for searching protein sequence database. They are available either integrated into the data management software of com-

Fig. 11. *(see facing page)* Results from a database search using the MS Fit tool of Protein Prospector. The masses of 21 tryptic peptides from a cytochrome-*c* digest were measured by MALDI-TOF mass spectrometry and submitted for matching against the SwissProt database.

Parameters Used in Search

Considered modifications: I Peptide N-terminal Gln to pyroGlu I Oxidation of M I Frotein N-terminus Acetylated I

Min. # Peptide to Match	Peptide Mass Tolerance (+/-)	Peptide Masses are	Digest Used	Max. # Missed Cleavages	Cysteines Modified by	Peptide N terminus	Peptide C terminus
4	50.00 ppm	Monoisotipic	Trypsin	1	Unmodified	Hydrogen (H)	Free Acid (OH)

MS-Fit Search Results

Sample ID (comment): K Noon Cyto C
Database searched: SwissProt.9.30.2001
Molecular weight search (1000 - 25000 Da) selects 38010 entries.
Full pI range: 100225 entries.

Combined molecular weight and pI searches select 38010 entries.
Pre searches select 38010 entries.
MS-Fit search selects 4 entries.

Result Summary

Rank	MOWSE Score	# (%) Masses Matched	Protein MW / pI	Species	Accession #	Protein Name
1	7.16e+004	8/21 (38%)	11701.7 / 9.59	HORSE	P00004	CYTOCHROME C
2	1.33e+004	7/21 (33%)	11687.6 / 9.59	EQUAS	P00005	CYTOCHROME C
3	1.85e+003	4/21 (19%)	15398.0 / 5.64	ONCMA	P48252	GONADOTROPIN BETA-1 CHAIN PRECURSOR (GTH-I-BETA)
4	468	4/21 (19%)	11572.4 / 9.52	BOVIN	P00006	CYTOCHROME C

mercial mass spectrometers or as stand alone search engines available on the internet. For peptide mass mapping:

PepSea (http://pepsea.protona.com/PA_PepSeaForm.html), PeptIdent/MultiIdent (http://www.expasy.ch/tools/peptident.html),

MS-Fit (http://prospector.ucsf.edu/ucsfhtml3.2/msfit.htm). For searches using mass spectral information obtained from peptide fragmentation: PepSea (http://pepsea.protona.com/PA_Peptide/PatternForm.html), SEQUEST (http://fields.scripps.edu/sequest/index.html),

PepFrag (http://www.proteometric.com/prowl/pepfragch.html),

MS-Tag ((http://prospector.ucsf.edu/ucsfhtml3.2/mstagfd.htm),

Mascot (http://www.matrixscience.com/cgi/search_form.pl?SEARCH=MIS)

Acknowledgments

The authors express their sincere thanks to Dr. R. D. Edmondson and K. J. Biederman of Proteomic Research Services, Inc., Ann Arbor, MI, for their assistance in collection of the mass spectral data for peptide mass mapping and data dependent experiments.

References

1. Vestal, M. L. (1983) High performance liquid chromatography—mass spectrometry. *Science* **226,** 275–281.
2. Cappiello, A. (1996) Is particle beam an up-to-date LC–MS interface? State of the art and perspectives. *Mass Spectrom. Rev.* **15,** 283–296.
3. Caprioli, R. M. and Suter, M. J.-F. (1992) Continuous-flow fast atom bombardment-recent advances and applications. *Int. J. Mass Spectrom. Ion Processes.* **118,** 449–476.
4. Huang, E. C., Wachs, T., Conboy, J. J., and Henion, J. D. (1990) Atmospheric pressure ionization mass spectrometry: Detection for the separation sciences. *Anal. Chem.* **62,** 713A–725A.
5. Fenn, J. B., Mann, M., Meng, C. K., Wong, S. F., and Whitehouse, C. M. (1989) Electrospray ionization for mass spectrometry of large biomolecules. *Science* **246,** 64–71.
6. Chernushevich, I. V., Loboda, A. V., and Thomson, B. A. (2001) An introduction to quadrupole-time-of-flight mass spectrometry. *J. Mass Spectrom.* **36,** 849–865.
7. Papayannopoulos, I. A. (1995) The interpretation of collision-induced dissociation tandem mass spectrometry spectra of peptides. *Mass Spectrom. Rev.* **14,** 49–73.
8. Mann, M., Hendrickson, R. C., and Pandey, A. (2001) Analysis of proteins and proteomes by mass spectrometry. *Annu. Rev. Biochem.* **70,** 437–473.
9. Yates, J. R. (1998) Database searching using mass spectrometry data. *Electrophoresis* **19,** 893–900.
10. Eng, J., Ashly, L., McCormack, A. L., and Yates, J. R. III (1994) An approach to correlate tandem mass spectral data of peptides with amino acid sequences in a protein database. *J. Am. Soc. Mass Spectrom.* **5,** 976–989.

11. Biemann, K. (1991) Nomenclature for peptide fragment ions (positive ions). *Methods Enzymol.* **193,** 886–887.
12. Fenyo, D. (2000) Identifying the proteome: Software tools. *Curr. Opinion in Biotechnol.* **11,** 391–395.
13. Mann, M. and Wilm, M. (1996) Error-tolerant identification of peptides in sequence database by peptides sequence tags. *Anal. Chem.* **66,** 4390–4399.
14. Stahl, D. C., Swiderek, K. M., Davis, M. T., and Lee, T. D. (1995). Data-controlled automation of liquid chromatography/tandem mass spectrometry analysis of peptide mixture. *J. Am. Soc. Mass Spectrom.* **7,** 532–540.
15. Yates, J. R., McCormack, A. L., Link, A. J., Schieltz, D., Eng, J., and Hays, L. (1996) Future prospects for analysis of complex biological systems using micro-column liquid chromatography-electrospray tandem mass spectrometry. *Analyst* **121,** 65R–76R.

II

Preparative Methodologies

10

Capillary Separations

Montserrat Carrascal and Joaquin Abian

1. Introduction

Capillary liquid chromatography (capLC) utilizes columns with inner diameters below 0.5 mm and flow rates from some few µL/min down to the µL/min range. Miniaturization of chromatographic procedures using capLC offers substantial advantages over conventional LC separation methods. Capillary columns show increased separation efficiency, minimal solvent consumption, higher peak concentration at the detector, and higher peptide recovery. Several reviews on the state of the art can be found in the literature **refs.** ***1–4***.

CapLC is probably to LC the same as some decades ago open tubular capillary chromatography was to gas chromatography. Horvarth first reported capLC in 1967 ***(5)***. The preparation, analytical characteristics, and some applications of capLC columns were already described in the late 1970s by Scott and Kucera ***(6)***, Novotny et al. ***(7)***, and Ishii et al. ***(8,9)***. Two main classes of capLC columns were developed: wall coated open tubular (OTLC) columns and packed columns. So far, several practical problems (column preparation, very low capacity and injection volumes) relegated OTLC to a more restricted use than packed columns. The latter consists of metal or fused silica capillaries filled with chromatographic supports with particle sizes ranging from 1.0 to 50 µm. CapLC could be divided into micro (0.8–0.15 mm id) LC and nano (20–100 µm id) LC, depending on the inner diameter of the column, although other classifications can be found in **refs.** ***2,3***, and ***10***.

Capillary columns usually show overall performance and efficiency comparable or better than conventional columns. Jorgenson et al. reported 40,000 theoretical plates for resorcinol ($k' = 2,7$) on a 25-µm id column 33 cm of length filled with 5-µm particles. These authors observed that column efficiency improved when the column diameters diminished from 50 to 12 µm ***(11,12)***.

From: *Methods in Molecular Biology, vol. 251, HPLC of Peptides and Proteins: Methods and Protocols*
Edited by: M.-I. Aguilar © Humana Press Inc., Totowa, NJ

This improvement was explained as flow dispersion and resistance to mass transfer decreased with smaller column diameters. Wall effects make a higher contribution in capLC, than in conventional LC columns. This leads to relatively higher band broadening at high flow rates. Band broadening can be diminished using electrodriven flows, as performed in capillary electrochromatography (CEC), instead of pressure forces. CEC columns can reach efficiencies over 300,000 theoretical plates/m *(13)*, greatly exceeding the efficiency of the equivalent pressure-driven capLC. Because electrodriven flows are less influenced by particle size than pressure-driven flows, smaller particles and longer columns can be used thus increasing the efficiency that can be obtained. The separation and detection of peptide mixtures at the attomole level by CEC were described recently *(14)*. Lubman reported the analysis of protein digests by the so-called pseudoelectrochromatography, a pressurized CEC mode *(15)*. This technique is an intermediate between capLC and CEC. The preparation of CEC columns and the applications of this technique were reviewed *(4,16-18)*.

In this chapter we will focus on the preparation and handling of packed capLC columns and the corresponding instrumentation for pressure-driven separations. Nowadays, capLC is a widespread technique in laboratories involved in the analysis of peptides and proteins, especially in the proteomics area. The low flow rates employed in capLC allow both the use of expensive solvents (i.e., deuterated) and coupling with detectors that show their best performance at the nanoliter-per-minute range, such as mass spectrometers (MS) with electrospray (ESI) interfaces (ESI-MS). Several companies distribute miniaturized chromatographic systems, including pumps, automatic microinjectors, and microcolumns, as well as suitable cells or interfaces for photometric or MS detection (*see* **Note 1**). These systems can also be built using conventional laboratory equipment with minimal effort at low cost *(19,20)*.

Before Takeuchi and Ishii demonstrated the utility of fused silica capillaries *(21)*, first attempts to prepare packed capLC columns used metal or PTFE tubing *(5,8)*. Although Teflon *(22)* or PEEK lines *(23)* are still used to prepare capillary columns, fused silica is the material most authors choose. Fused silica capillary columns with internal diameters from 500 to about 50 μm can be easily prepared by slurry packing methods *(1,3,17,19,24,25)* (*see* **Subheading 3.1.**). Colon reviewed several other methods for column packing *(16)*.

The first step to column packing is to build a frit at the exit end of the fused silica capillary column. The frit can be prepared sinterizing a plug of silica particles with a flame *(1)* or it can be made pushing the capillary column through a porous membrane sheet two or three times, resulting in a stable plug at the outlet of the column. Membrane frits can also be prepared using Teflon *(24)*, PVDF *(19)*, and glass fiber sheets *(3)*. Alternatively, Lubman used a Valco microbore column fitting with a very small amount of glass wool as a frit. The thickness was 0.1 mm and it proved to be more stable than glass frits ***(15)***. A zero dead volume Valco union can also be used to hold a stainless mesh frit ***(17)***.

The sintered glass method is probably the most commonly employed. In our experience, sintered glass frits are stable when using capillaries with outer diameters bigger than 300 µm. Columns with smaller outer diameters (180 µm) are more vulnerable to the heating process that makes the capillary too fragile to handle.

After the frit is ready, the capillary column is coupled to a reservoir filled with the slurry of the packing material. The slurry is pushed under pressure into the column inlet by a flowing stream. The frit allows the solvent through the column and the solid phase fills it in a similar process to filtration. Liquid solvents ***(1,12,17,26,27)*** and supercritical fluids ***(25,28,29)*** are used for slurry-packing. Likewise, gas streams are used for dry-packing procedures ***(30,31)***. The relative advantages of these methods are the difficulty of packing and the stability and efficiency of the columns obtained.

In common practice, 20 µm is the minimum column diameter for a packed column, but most authors use a minimum id of 50–75 µm. Kennedy and Jorgensson reported the preparation by slurry-packing methods of 12 µm id capLC columns filled with 5 µm particles. Filtering the particles through an 8 µm filter was necessary to prevent the column from plugging with the larger particles in the commercial material ***(12)***.

Five micron particles are the most commonly used for column filling. Recently, MacNair et al. introduced ultrahigh pressure capLC (UHPLC) where nonporous particles of 1.0–1.5 µm are used to fill 29–100 µm id capLC columns ***(32)***. More than 200,000 theoretical plates (k′=1) can be generated using 25–50-cm-long columns. UHPLC can also produce high-speed separations (less than 100 s) with column efficiencies over 20,000 plates ***(33)***. Mac Nair demonstrated the application of this technique to analyse peptides and proteins with an ion trap MS detector ***(34)***. To take full advantage of the high speed available, Lee et al. coupled UHPLC to a fast scanning time-of-flight (TOF) MS detector ***(33)***.

Other special classes of packed columns with immobilized stationary phases ***(35)*** or a continuous porous bed (monoliths) ***(36,37)***, have been developed. The major advantage of these columns is that they do not need the end frit necessary to retain the stationary phase in conventional packed capLC columns.

The optimal flow rate for a 150-µm column is around 1 µL/min. This demands special requirements of the pumping system especially when a gradient elution is needed. Syringe pumps were used to obtain reproducible flow rates in the microlitre-per-minute range. In this sense, Lee et al. described a brilliant procedure to prepare and store the solvent gradient before the analytical step ***(38)***. They also apply a "peak parking" procedure by controlling the pressure of the syringe pump. This enables an efficient characterization of peptide mixtures by tandem mass spectrometry *(39,40)*. More simple setups are based on conventional HPLC gradient pumps with flow splitting. Commercial sources supply calibrated flow splitters that give very reproducible flows (*see* **Note 1**). Even so, most authors prepare their own splitters using a T-piece and some kind of restrictor (*see* **Subheading 3.2.1.**).

CapLC can be coupled to a wide range of detectors including UV, fluorescence, and electrochemical detectors. The analysis of peptides from biological sources is often hindered by its very low concentration in the sample. Attomole/femtomole detection limits are usually required and thus high recovery, high sensitivity separation, and detection techniques are needed. The need for an effective coupling of LC with some emerging MS techniques such as continuous flow fast atom bombardment *(41)* and ESI *(42)* has probably been the driving force towards capLC developmnent in the last decade. These MS techniques were specially suited for biopolymer analyses and offered their optimal performance at the microlitre or submicrolitre per minute flow rate. Coupling MS with LC implied either flow splitting procedures with the resulting sample loss or the development of capLC methods. In this respect, capLC/ESI-MS has become a very powerful analytical technique and, currently, most capLC/MS work is performed using ESI interfaces (*see* **Subheading 3.2.4.**). In the early 1990s, Hunt et al. sequenced subpicomole amounts of antigenic peptides bound to major histocompatibility complex (MHC) molecules using capLC columns and commercially available ESI interfaces *(43,44)*. Further downscaling of the LCMS method by coupling nanoLC to dedicated microESI interfaces resulted in detection limits down to the attomole range *(26,45–47)*. Enzymatic digest from proteomic studies can be on-line separated and sequenced using capLC/ESI-MS with automatic scanning procedures in triple quadrupole, ion trap, or hybrid quadrupole TOF mass spectrometers *(48,49)*. Protein sequence coverage higher than 50% can be routinely obtained from well defined Commassie stained spots (about 1–5 pmol total protein). Yates et al. demonstrated 99% coverage on the characterization of mutant globins (15 KDa) analyzing peptide mixtures from several enzymatic digestions of the same protein *(48)*. CapLC/ESI-MS quantitative determinations of nanomolar amounts of peptides can be performed using internal standards *(26)*. The use of gradient elution or on-line preconcentration microcolumns allows the injection of sample volumes up to 100–200 μL to be injected without deleterious effects on the system. These techniques solve the problem of small injection volumes required by capLC, improve the concentration detection limits and simplify overall sample handling. The use of selective on-line preconcentration procedures can also minimize interferences due to the complexity of the biological sample *(26,50)*. Ion exchange precolumns can be used in combination with reversed-phase capLC columns for this purpose (*see* **Subheading 3.1.**). A similar, more sophisticated approach is used in multidimensional chromatography. Jorgerson et al. demonstrated zeptomol detection limits for the analysis of peptide digest from porcine thyroglobulin using a two-dimensional (2D) anion exchange-reversed phase capLC system and fluorescence detection. The peak capacity of this 2D method was estimated to be over 2000 components per analysis ***(51)***.

In this chapter, we will describe straightforward protocols for beginners to produce capLC separations with minimal effort and economical cost. The use of packed capillary columns with membrane frits and fused silica based interfaces for UV and ESI-MS will be described. The reader is referred to the literature and the notes along the chapter for improved techniques when needed.

2. Materials

2.1. Column Preparation

2.1.1. Analytical Column

1. Fused silica capillaries of 250 × 350, 50 × 200, 75 × 200, and 30 × 150 (id × od µm) were obtained from Polymicro Technologies (Phoenix, AZ) (*see* **Note 2**).
2. Whatman GF/A glass fibre filters (Whatman International Ltd., Maidstone, UK).
3. Araldit® glue (Ceys, Ciba-Geigy).
4. Empty 100 × 1 mm HPLC column holder with the metal filters removed.
5. Bulk stationary phases 5 µm Kromasil C18 and C8 (Teknokroma, Barcelona, Spain), and Poros 20 R1 (Perseptive Biosystems, Cambridge, MA).
6. Fused silica cutter (Alltech Associates, Inc., Deerfield, IL).
7. HPLC pump (any spare pump is valid).
8. Teflon tube (300 and 100 µm id, 1/16 iod, Supelco, Bellefonte, PA).
9. Valco Union and connectors (Valco Instruments Co. Inc., Houston, TX).
10. Acetonitrile (ACN) (LiChrosolv, Merck, Darmstadt, Germany).

2.1.2. Preconcentration Columns

1. Strong-anion exchange µ-Precolumns (5 × 0.8 mm id; LC Packings, The Netherlands).

2.2. Chromatographic Equipment

2.2.1. General

1. HP-1100 system including high pressure binary HPLC pump and a solvent degasifier (Agilent Technologies, Barcelona, Spain).
2. Rheodyne injector mod 7725i with 5, 10, and 200 µL loops (Supelco).

2.2.2. Chromatographic Flow Splitting

1. Zero dead volume (ZDV) Valco Tee (Supelco).
2. HPLC columns 5 mm 150 × 2.1 mm (250 mm column), 100 × 1 (75 mm column), 35 × 3 mm (For Poros).

2.2.3. Preconcentration Switching Systems

1. Two Position Electric Actuator and VALCO valve (Valco Instruments Co. Inc., Houston, TX).
2. Auxiliary M-6000 HPLC pump (Waters, Milford, MA).

2.2.4. UV Cell Preparation (see Note 3)

1. UV detector ABI 769 A (Applied Biosystems, Foster City, CA).
2. Fused silica capillaries 250 × 350, 50 × 200, 150 × 375, and 30 × 150 (id × od).

2.2.5. ESI Interface Preparation

1. Fused silica capillaries 50 × 200, 30 × 150, 15 × 150 (id × od).
2. Hypodermic stainless steel capillary (A-M System, Inc., Carlsborg, WA).
3. Microflame torch (Microflame, Minneapolis, MN).

2.3. HPLC Gradient Systems

1. Acetonitrile 190 for UV (Romil, Teknokroma, Barcelona) (for UV detection).
2. Acetonitrile ultragradient (Romil) (for MS detection).
3. Trifluoroacetic acid (TFA) (Uvasol, Merck).
4. Acetic acid (AcOH) (Suprapur, Merck).
5. Ammonium acetate (Merck).

3. Methods

3.1. Column Preparation

3.1.1. Analytical Columns

1. The 250 or 180 μm fused silica capillaries used as columns (*see* **FSC**, **Fig. 1**) are cut to the desired length (11 cm) using a porcelain cutter.
2. The outlet line (*see* **EXC**, **Fig. 1**) is also cut from a 50 × 200 μm (250 μm id column) or a 50 × 150 (180 μm id column) capillary long enough to reach the detector (5–20 cm). The shorter the better.
3. Capillaries are flushed with ACN/water 3/1 using an HPLC pump (*see* **Note 4**).
4. A frit is produced at the end of the tube twisting the capillary end against a Whatman GF/A glass fiber filter (*see* **Note 5**). A piece of filter corresponding to the diameter of the capillary is cut out in this way, and remains inside the bore of the tube. The EXC capillary is introduced into the FSC capillary to serve as the outlet line. The frit is pushed 5 mm into the FSC by pressing with the EXC and both tubes are held together with a drop of glue (*see* **Note 6**)

   The same protocol is used for the 75 μm id × 200 od column. In this case, the od of the FSC and the EXC lines are the same and the connection is done using a 2-cm fused silica sleeve (250 μm id × 350 μm id) (*see* **FSS**, **Fig. 1**). The frit is prepared into the FSS and sandwiched between the FSC and the EXC (*see* **Note 7**).
5. The column is slurry packed with the silica particles at 1500 psi. For this purpose, the column is connected to a metallic reservoir (a 1 mm id column holder) using

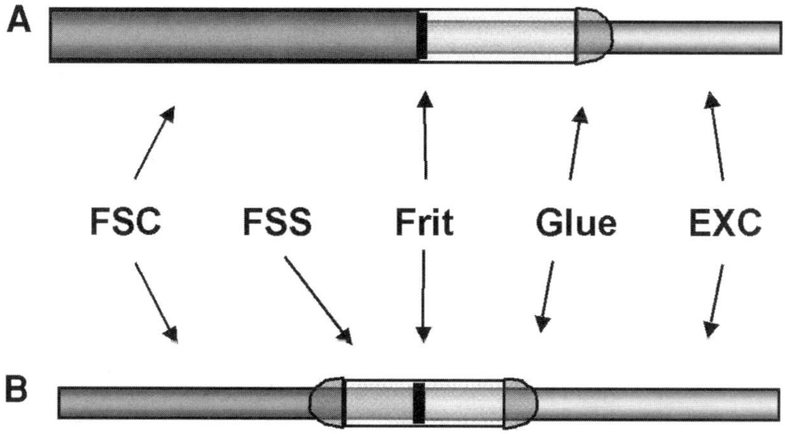

Fig. 1. General scheme for the preparation of fused silica capLC columns with internal diameter **(A)** above and **(B)** below 180 µm. *FSC*, fused silica capillary column. *FSS*, fused silica sleeve. *EXC*, fused silica exit capillary.

1/16 HPLC connectors. A Teflon sleeve is used to adapt the capillary to the ferrule diameter. One mg of stationary phase is dissolved in 4 mL of ACN/water (3/1) and 100 µL of the slurry are introduced into the reservoir with a syringe. The slurry is then pushed into the FSC connecting the reservoir to an HPLC pump (*see* **Note 8**). Flow rate is set at 50–100 µL min for a few minutes to fill the column, and then it is lowered down to approx 20 µL/min in order to compact the stationary phase at a constant pressure. After 20 min, the column can be disconnected ready to be used (*see* **Note 9**).

3.1.2. Preconcentration Columns

Preconcentration columns can be prepared in the same way as analytical columns from 250–530 id FSC tubes (*see* **Note 2** for adequate sizes). Higher column diameters can bc used when high volumes have to be preconcentrated in a reasonable time or when complex samples could produce column overloading. Capiello et al. *(23)* described how to produce convenient columns with diameters up to 0.76 mm from PEEK tubing. Conventional ZDV peak or stainless steel connectors are used as column ends with 1/16 nuts and ferrules. The membrane frit at the outlet end is introduced into the ZDV union and sealed in place by pressing the column end against it and then tightening the nut. In the examples given in this chapter, we used a commercial kit consisting of a stainless steel holder and 800 µm id cartridges. The same holder could load cartridges with id from 200 to 800 µm. Cartridges available include C18 and ion exchange materials.

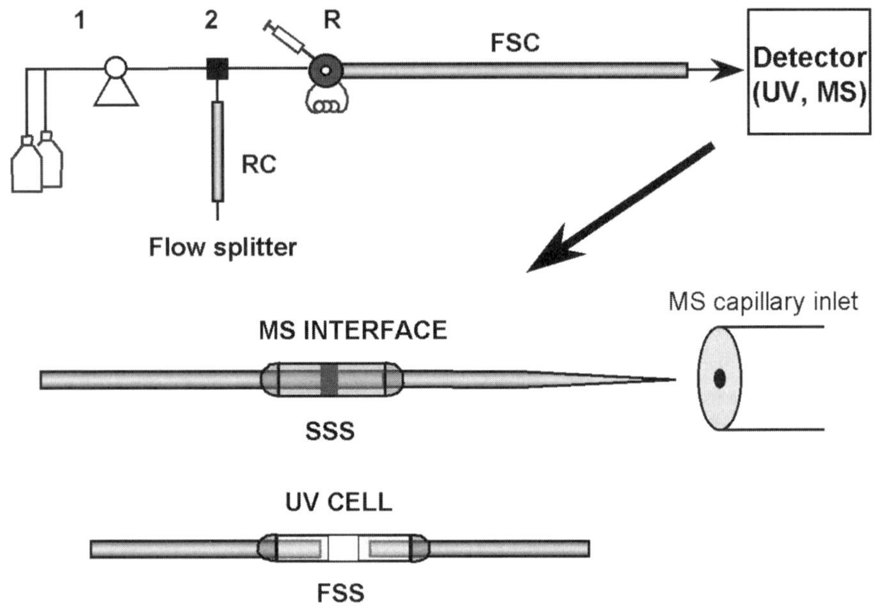

Fig. 2. Scheme of a simple capLC system and the interfaces for MS and UV detection. *1*, Gradient pump and solvents. *2*, ZDV splitting Tee. *R*, injector. *RC*, restriction column. *FSC*, fused silica capillary column. *SSS*, stainless steel sleeve. *FSS*, Fused silica sleeve.

3.2. Chromatographic Equipment

3.2.1. Column Connection

The basic scheme of a capLC system is depicted in **Fig. 2**. The main issue in the setup of capLC is to decrease the dead volumes as much as possible. Some changes can be done in a conventional HPLC pump to optimize its performance at low flow rates. We substituted a 2 × 0.1 cm precolumn for the mixer of our HPLC-1100 gradient pump and all the connections after the pump were remade with 120 μm id PEEK tube. The tubing connecting the split tee and the injector should be made of 120 μm id tube and as short as possible. The column is connected directly to the injector using a teflon sleeve with an adequate inner diameter (*see* **Fig. 3**). Switching the injector to the "load position" after the sample has been transferred to the analytical column eliminates the delay produced by the injector loop on the solvent gradient.

3.2.2. Chromatographic Flow Splitting

1. The flow splitter is built from a ZDV Valco Tee. The inlet flow from the HPLC is split in a ratio that depends on the relative flow restriction produced by the capillary column and the flow restrictor.

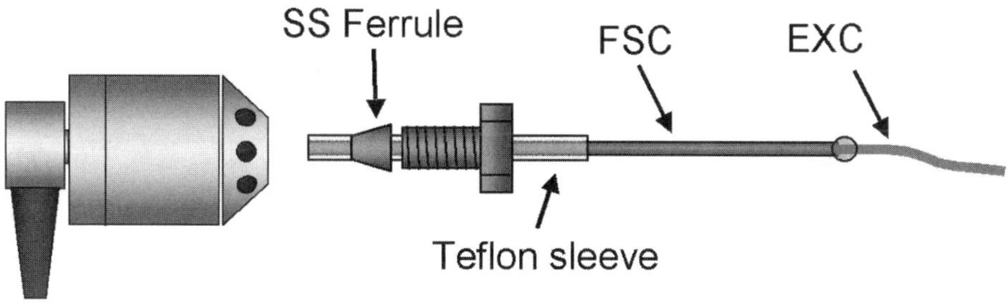

Fig. 3. Connection of the capillary column showing the use of Teflon sleeves for capillary connections. *EXC*, fused silica exit capillary. *FSC*, fused silica capillary column, *SS*, stainless steel.

2. The flow restrictor for a 250 µm id column consists of a conventional 2.1 mm id HPLC column (5 µm Nucleosil, 15 cm long). A 1 mm id column (5 µm Partisil, 10 cm long) is used in the case of 75 mm id columns (*see* **Note 10**).
3. A HPLC flow rate of 100 mL/min is typically used. After flow splitting, about 3–4 µL/min in the 250 µm column or 0.5-1 µL/min in the 75 µm column are obtained.
4. The flow at the exit of the column can be calculated collecting the eluent in a conical vial (500 µL Eppendorf tube) for 10 min and measuring its volume with a 5 or 10 µL Hamilton syringe. Alternatively, the column outlet can be introduced into the bore of the syringe to measure the volume directly from the syringe marks. Final adjustment of the flow rate is done by modifying the flow from the HPLC pump.

3.2.3. Preconcentration Switching Systems

The basic scheme of a preconcentration system is depicted in **Fig. 4**. An auxiliary HPLC pump is used to load the sample into the precolumn and to wash out not retained compounds. The auxiliary pump is permanently working at 50–200 µL min, depending on the precolumn, so that sample loading is typically made in a very fast way even with high volume samples. Sample loops from 20–200 µL are usually employed.

The precolumn is connected to a motor driven Valco switching valve. When the valve is in the "analysis" position, the precolumn is connected in series with the gradient HPLC system and the analytical column. Valve switching can be done manually or automatically using the relay contacts on the HPLC system.

3.2.4. UV Cell Preparation

1. The polyimide coating of a 250 × 350 mm (id × od) fused silica capillary is removed by carefully burning it with a flame. The capillary should be approx 2 cm long and the open window 0.5 cm (*see* **Fig. 2**).

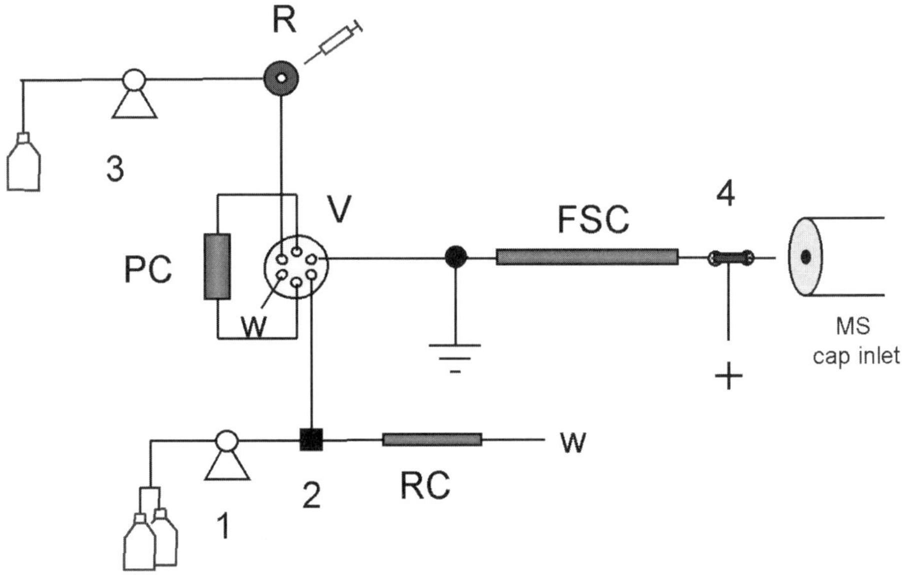

Fig. 4. Scheme of a preconcentration capLC system. *1*, Elution gradient pump and mobile phases. *2*, ZDV splitting Tee. *3*, Preconcentration pump and mobile phase. *4*, microelectrospray interface. *R*, injector. *V*, Valco electrovalve. *W*, waste. *PC*, preconcentration column. *RC*, restriction column. *FSC* fused silica capillary column.

2. Inlet and outlet capillaries are prepared from 50 × 200 µm (id × od). These capillaries are glued into the cell capillary, leaving a 5-mm space (the window area). The total volume of the flow cell is approximately 250 nL (*see* **Note 11**).
3. The flow cell is placed into the ABI CZE cell holder. The cell is used as a conventional CZE UV cell.

*3.2.5. ESI Interface Preparation (*see **Note 12***)*

1. About 10 cm of a 200 × 50 µm FS capillary is prepared. A lightweight (10 gr) is fixed to one end of the capillary. The other end is fixed to a support leaving the capillary in a vertical position with the weight hanging free. The flame of a torch is applied to the center of the line allowing the capillary to stretch under gravity. Two capillaries with sharp tips are obtained in this way (*see* **Note 13**). The needle tips are carefully cut to obtain an inner bore of about 5 µm. The bore diameter can be monitored with a stage micrometer (Graticules LTD, Tonbridge, Kent, UK) under a microscope (*see* **Note 14**).
2. The end of a 250-µm ID stainless steel capillary is twisted against a Whatman GF/A glass fiber filter to cut out a piece of filter.
3. The filter is sandwiched between two 200 × 50 µm capillaries, the electrospray needle prepared as indicated earlier, and the inlet line. These capillaries are glued into the steel capillary that serves as the electrical contact for electrospray (*see* **Fig. 2**).

4. The ESI interface was mounted into an x-y-z-manipulator from a Protana Nanoelectrospray Source Kit.
5. The ESI needle tip was carefully positioned at 3–5 mm from the inlet of a Finnigan LCQ mass spectrometer (ThermoQuest, Barcelona, Spain). For standard analyses, an electrospray voltage between 1.6 and 2.0 kV was applied. For continuous infusion experiments, the inlet capillary of the μESI was connected via an injector to a syringe pump.

3.3. HPLC Gradient Systems

The basic HPLC configurations for the examples listed below are depicted in **Figs. 2** and **4**.

3.3.1. Separation of Peptide Mixtures from Enzymatic Digestions of Proteins

A key point in the analysis of on-gel tryptic digests is the removal of salts and acrylamide contaminants from the sample before peptide separation. In general, desalting can be performed directly on the analytical column provided the peptide concentration is relatively high (clear Commassie spot). Otherwise a column switching system using a preconcentration column would be necessary (*see* **Subheading 3.3.2.**). Proper desalting involves a long washing step with aqueous solvent prior to gradient elution. Special care has to be taken to avoid the loss of very polar peptides during this step. The system described below uses a high volume loop that produces an additional delay on the gradient so that no initial 0% step is required. Examples of capLC separations are shown in **Figs. 5** and **6**.

1. Analytical column: 5 μm Kromasil C8 (250 μm × 10 cm); Injection loop, 10 μL
2. Solvent A : water (1% AcOH for MS detection or 0.05% TFA for UV/MS detection) Solvent B: ACN/water 8/2 v/v (1% AcOH for MS detection or 0.05% TFA for UV/MS detection) (*see* **Note 15**).
3. Gradient: From 0 to 55% solvent B in 60 min; from 55 to 80% B in 5 min; hold 10 min at 80% B. Back to initial conditions in 5 min. Flow rate 3 μL/min.
4. Protein digests are prepared following standard procedures ***(52)***. The final extract (about 50–150 μL) is concentrated to 20 mL in order to eliminate the extraction solvent (ACN). A 10-μL aliquot is directly injected into the system.

3.3.2. Separation of Peptides Using Preconcentration Procedures

The quantitative analysis of endothelins in human umbilical vein endothelial cell (HUVEC) supernatants is shown in **Fig. 7**. The use of ion exchange preconcentration prior to on-line reversed phase chromatography provides a very selective method for endothelin analysis. The synthetic endothelin ^{3,11}Ala-ET-1 (alaET) is used as internal standar for quantitation. Detection limits were found

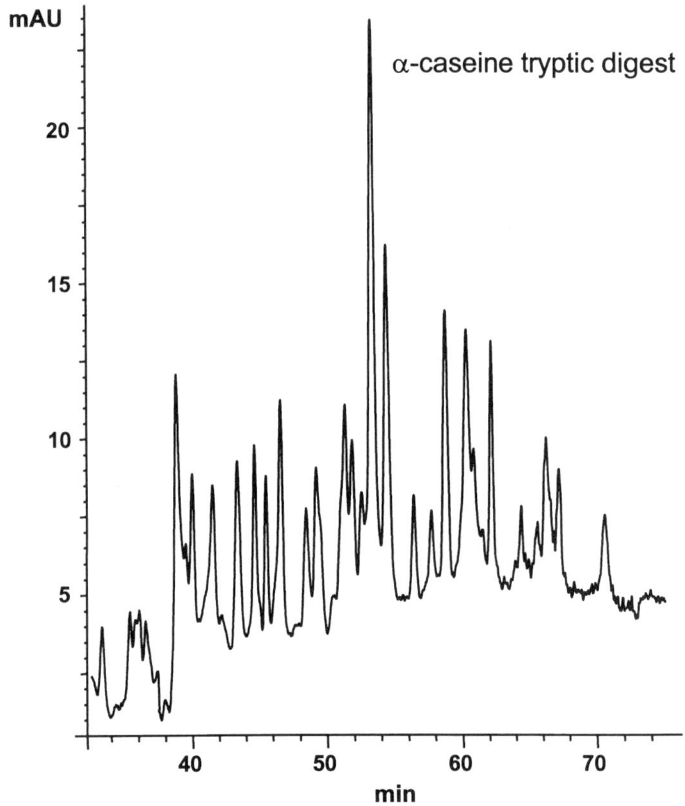

Fig. 5. CapLC-UV analysis of an in-gel tryptic digest of alpha-casein. A blue Commassie spot (1 μg protein) was excised, in-gel reduced, alkylated, and digested following conventional procedures. The peptide mixture was analyzed as indicated in **Subheading 3.3.1.** Peptides were monitored at 214 nm.

Fig. 6. *(see facing page)* Analysis of an in-gel tryptic digest of 90 KDa heat shock protein (HSP90). A protein extract from a lymphoblastoid cell line culture was processed by immunoaffinity chromatography using GST-p21Cip1-Sepharose. The p21Cip1-binding protein fraction was separated by SDS acrilamide gel electrophoresis and stained with Coomassie blue. The selected spots were excised and in-gel reduced, alkylated and digested following conventional procedures. The ThermoQuest LCQ ion trap instrument was programmed for automatic data dependent acquisition. Full MS, Zoom Scan, and Product Ion scans were made consecutively to obtain sequence information over the entire chromatographic elution. As shown in this figure, where spectral data **(right)** was acquired in the elution zone indicated by the asterisk in the chromatogram **(left)** (Retention time 33.03 to 33.14), the sequencing of even very low intensity signals was achieved.

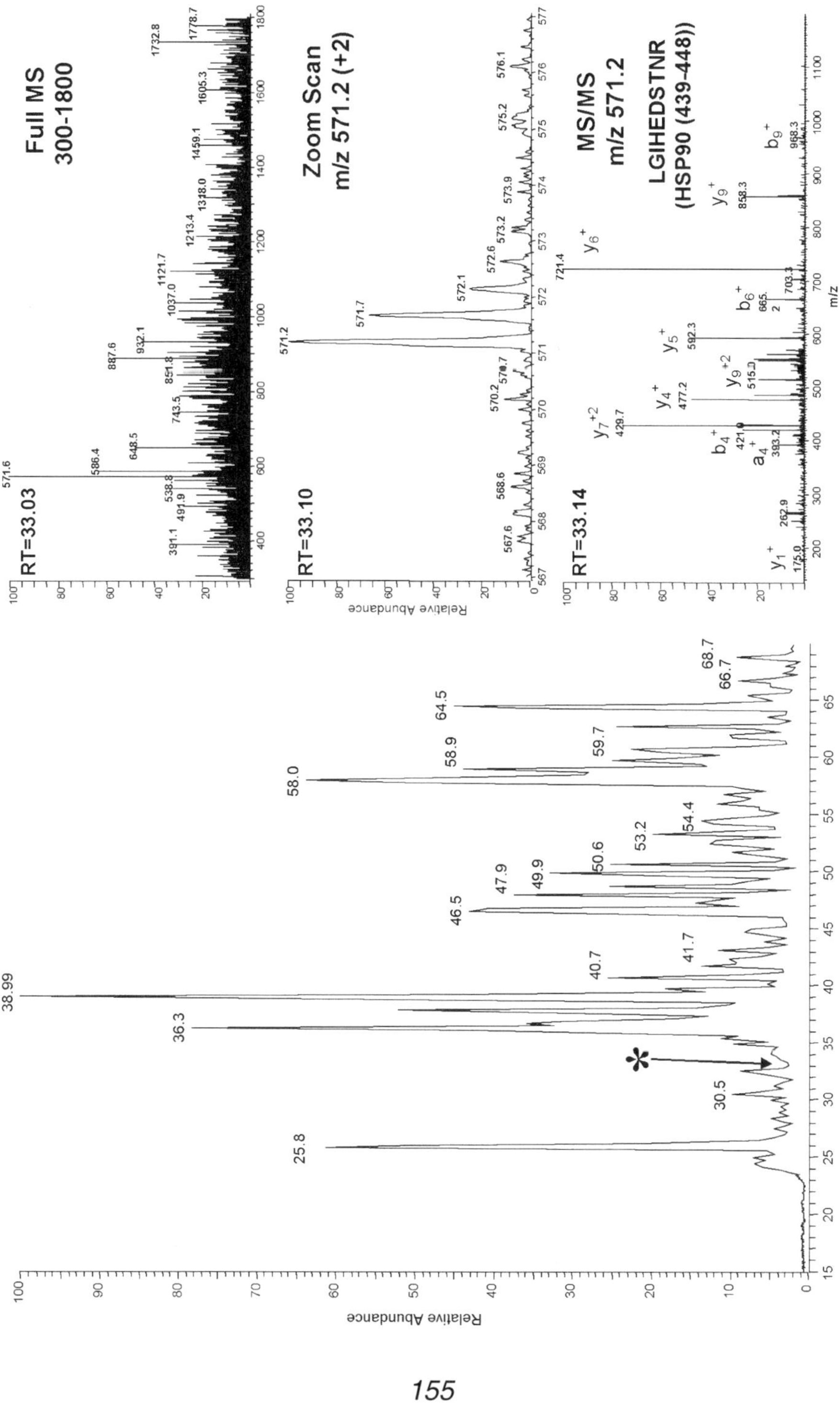
Full MS
300-1800
RT=33.03
Zoom Scan
m/z 571.2 (+2)
RT=33.10
MS/MS
m/z 571.2
LGIHEDSTNR
(HSP90 (439-448))
RT=33.14
Relative Abundance
m/z
Time (min)

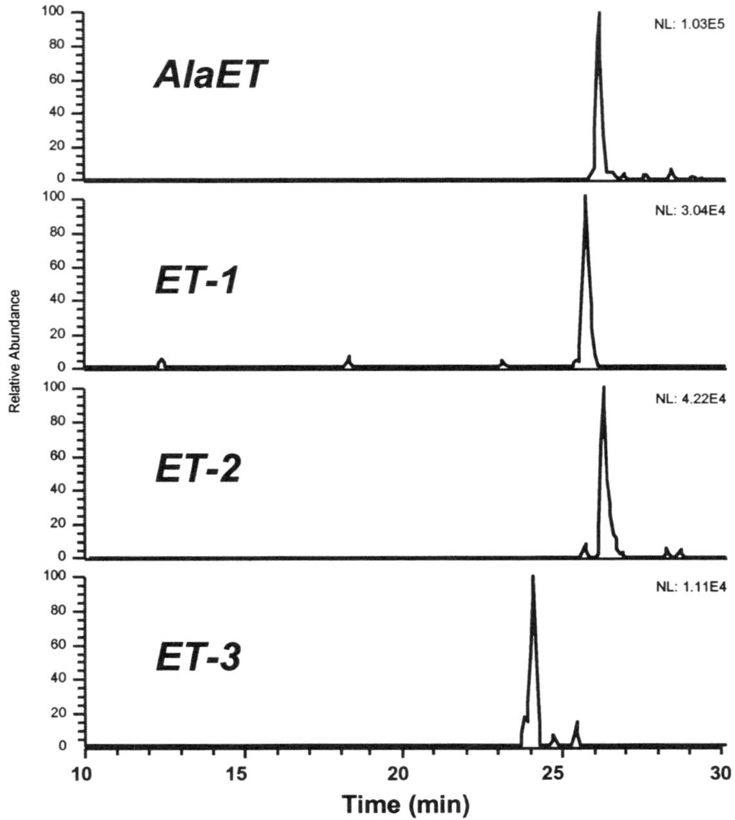

AlaET
ET-1
ET-2
ET-3
NL: 1.03E5
NL: 3.04E4
NL: 4.22E4
NL: 1.11E4
Relative Abundance
Time (min)

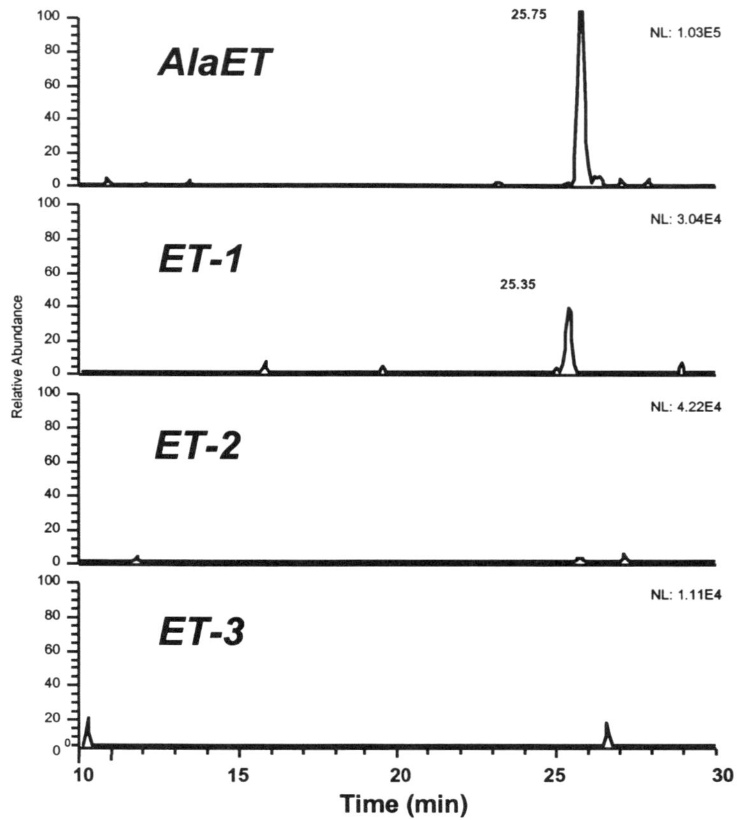

AlaET
ET-1
ET-2
ET-3
25.75
25.35
NL: 1.03E5
NL: 3.04E4
NL: 4.22E4
NL: 1.11E4
Relative Abundance
Time (min)

down to 30 pmol/L (1.5 fmol on column) and the method was linear from 50 to 2000 pmol/L ***(26)***.

1. Chromatographic Columns:
   Preconcentration precolumn: SAX μ-Precolumn (800 μm × 0.5 cm).
   Analytical column: 5 μm Kromasil C18 (75 μm × 10 cm).
   Injection loop 200 μL.
2. Preconcentration solvent: 10 mmol/L ammonium acetate, pH 7.0 (10:90, v/v). Flow rate 200 μL/min.
3. Gradient solvents:
   Solvent A: ACN:water:AcOH:TFA (10:90:1:0.05, v/v).
   Solvent B: ACN:MeOH:water:AcOH:TFA (45:45:10:1:0.05, v/v).
   See **Table 1** for gradient program. Flow rate 550 nL/min.
4. HUVEC supernatants (500 μL) are precipitated with ACN (1 mL). The material is centrifuged and the supernatant evaporated close to dryness. Just before injection, 100 μL of MeOH:water (50:50 v/v) and 1.9 ml of the preconcentration mobile phase are added. A 200-μL aliquot is injected into the analytical system.

4. Notes

1. LC Packings (LC Packings (The Netherlands) BV) were probably the first company to distribute capillary columns and other instrumentation for capLC since the early 1990s. A capillary-perfusion Tool Kit is available from Perseptive Biosystems for the preparation of PEEK capLC columns (ref. 5-2218-00-0000). Low flow pumps and microinjectors for capLC are currently available from major HPLC companies.
2. Other diameters can be selected depending on the application. For example: 530 × 700/150 × 375; 180 × 350/30 × 150 (id × od column / id × od exit line).
3. Any other UV detector provided with cell holders for CZE could be used with homemade fused silica cells. Otherwise, commercial cells can be purchased from LC Packings (*see* **Note 11**). We routinely use a HP1100 UV detector with a LC-Packings U-Z View (CAP) for high sensitivity detection.
4. Capillaries can be deactivated by treatment with *bis*-trimethylsilyl trifluoroacetamide (BSTFA) (Pierce, Rockford, IL). This treatment eliminates the free silanol groups in the silica surface and minimizes peptide adsorption. The capillary is filled with the reagent using a 50-μL glass syringe. After 10–15 min at room temperature, acetonitrile is flushed to remove excess reagent and to clean the capillary. BSTFA is a volatile, strong derivatizing agent that rapidly reacts with moisture. Its handling has to be done in a fume hood. BSTFA is sold in ampoules

---

Fig. 7. *(see opposite page)* Analysis of HUVEC supernatants by PC-capLC-ESIMS/MS. **(Left)** HUVEC supernatant spiked with 200 p*M* endothelins and **(right)** HUVEC supernatant. (200 μL injection; internal standard 2 n*M* AlaET). ET, endothelin; AlaET, [3,11 Ala]-ET-1 ***(26)***.

Table 1
Time Event of the SAX PC capLC/ESI-MS Method

Time (min)	Concentration of solvent B	Flow rate (mL/min) of LC pump	Valve V position
0	10%	0.4	Preconcentration
10	10%	0.4	
10.1	20%	0.3	
15			Analysis
20	60%	0.3	
20.1	80%	0.3	
25	80%	0.3	
25.1	10%	0.3	

that, after opening, can be stored inside a small flask in the refrigerator for several weeks.

5. Alternatively, frits made of sintered glass can be built as follows: The end of the fused capillary is burned with a flame to remove a section of a few mm of the polyimide coating. Any black residue is carefully wiped out. The capillary is connected to a 500-µL syringe provided with a Teflon connector and flushed with ACN. After ACN removal, the frit is prepared by carefully tapping the tip of the column into a pile of 5 µm silica particles, filling approx 1 mm of the capillary section. This end is then heated with the flame of a microtorch to sinter the silica particles. The resulting frit is usually able to withstand high pressures. However, this procedure requires some practice. The quality of the frit should be tested by flushing some solvent before packing the column. We made stable frits when using capillaries with outer diameters greater than 300 µm. In our experience, columns with smaller outer diameters (180 µm) are more vulnerable to the heating process making the capillary too fragile to handle. Special care needs to be taken to connect the column outlet to the transfer line with the teflon connector (use a preformed connector).
6. Cyanoacrylate glues are too liquid and tended to spread out contaminating the system. Epoxy resins, such as Araldit, are more suitable. The glue must fill the area between the coaxial capillaries, otherwise the connection tends to fail when subjected to pressure. Owing to glue viscosity, this process requires the inner capillary to be pulled 1 mm in and out just after the glue is deposited on the external part of the connection. A gentle warming of the capillary connection is also helpful.
7. This is the general procedure for preparing columns with inner diameters below 150 µm.
8. We commonly use a Vibro-Graver (Burguess Vibro-Crafters, Inc, Graystake, ILL) against the reservoir to prevent the aggregation of particles (some authors immerse the whole system in an ultrasonic bath for this purpose). We do not observe differences in the column quality whether the slurry was previously sonicated or not.

9. This is a very simple procedure that yields reasonable efficiency. Several other solvent mixtures were tested for slurry packing including acetone, methanol, acetonitrile, carbon tetrachloride/paraffin, and n-heptane. In most cases the solvent used for particle suspension was different from the packing (flushing) solvent. Surfactant additives such as sodium lauryl sulphate or Tween were employed to improve particle dispersion. Other packing methods use gases (dry method) or supercritical fluids to pack the particles. A description of available methods can be obtained from **refs. 1, 12, 16, 17, 19, 24–31**.
10. When an old LC column is not available, a 50–100 cm section of fused silica capillary can also be used as a restrictor. The split ratio is controlled selecting the appropriate length and internal diameter of the capillary. CapLC columns packed with perfusive chromatography material such as POROS produce a much lower restriction. Thus, the split restrictor must be prepared accordingly. We use a Hypersil C4 0.3 × 3.5 cm column (5 µm, 300 Å particle) as a restrictor for a 250-µm id column (10-cm long) filled with POROS R1.
11. LC Packing offers optimized UV cells for capLC that can be adapted to most commercial detectors. Available cells are prepared to work at three different flow rate ranges: 10–100, 1–10, and 0.1–1 µL/min (Micro, Capillary and NanoLC cells, respectively). Cell volumes are 140 nL (micro), 35 nl (capillary), and 3 nL (nanoLC).
12. The conventional electrospray interface of the Finnigan instruments can be used instead of this dedicated interface when using capLC columns with inner diameter larger than 180 µm at flow rates higher than 3–5 µL/min. When working at flow rates close to or lower than 5 µL/min, sensitivity can be improved by reducing the default distance between the needle tip and the MS entrance (typically 3 cm) to 5 mm.
13. Pulled ESI needles can be prepared with a micropipette puller device. However, some glass pullers do not produce enough energy to fuse the silica. Most authors use laser pullers for this purpose (Sutter Instruments [Phoenix, CA]).
14. Stainless steel, fused silica and glass capillaries have been used as ESI needles. Some changes are usually made to the capillary tip in order to optimize spray formation. Fused silica capillaries with small id are commonly etched with HF ***(45)*** or the tip is polished with sand paper, resulting in very sharp ESI tips. Electropolishing has also been used with metal capillaries to produce similarly sharp tips ***(53)***. ESI tips have been constructed with id values down to 10 µm. Spray tips as small as a few µm can be prepared from glass or fused silica capillaries using a micropipette puller ***(54,55)***. Valaskovic et al. ***(55)*** obtain reproducible tip diameters down to 2 µm by first pulling capillaries and later etching them with HF. These tips can be operated at flow rates from 0.1–20 nL/min, depending on the diameter. New Objetive (Cambridge, MA) distribute glass or fused silica tapered ESI tips with different internal diameters.
15. Peptide separations are greatly enhanced by the ion pairing effect of TFA. However, TFA is incompatible with ESI ionization and a higher concentration than 0.2% approx can eliminate the ESI signal completely. For this reason, TFA should

be avoided from capLC/ESI-MS eluents and replaced by formic or acetic acids to maximize sensitivity. In some cases, the use of mixtures containing 1% AcOH and a small amount of TFA (0.02–0.05%) may provide a good balance between chromatographic separation and signal intensity. When on-line UV detection is required, AcOH cannot be used at the concentration indicated earlier because of its high absorbance at 214 nm. In this case, it is advisable to use the minimal amount of TFA (0.05%) to allow both separation and compatibility with MS.

Acknowledgments

The sample analyzed in **Fig. 6** was provided by Dr. Oriol Bachs (Biologia Cellular i Anatomia Patologica Department, School of Medicine, IDIBAPS, Barcelona, Spain). HUVEC culture supernatants were provided by Dr. Ginés Escolar (Hospital Clinic, Barcelona, Spain).

References

1. Tomer, K. B., Moseley, M. A., Deterding, L. J., and Parker, C. E. (1994) Capillary liquid chromatography/mass spectrometry. *Mass Spectrom. Rev.* **13,** 431–457.
2. Visser, J. P. C., Claessens, H. A., and Cramers, C. A. (1997) Microcolumn liquid chromatography: instrumentation, detection and applications. *J. Chromatogr. A.* **779,** 1–28.
3. Abian, J., Oosterkamp, A. J., and Gelpi, E. (1999) Comparison of conventional, narrow bore and capillary liquid chromatography/mass spectrometry for electrospray ionization mass spectrometry: Practical considerations. *J. Mass Spectrom.* **34,** 244–254.
4. Krull, I. S., Sebag, A., and Stevenson, R. (2000) Specific applications of capillary electrochromatography to biopolymers, including proteins, nucleic acids, peptide mapping, and so forth. *J. Chromatogr. A.* **887,** 137–163.
5. Horvarth, C. G., Preiss, B. A., and Lipski, S. R. (1967) Fast liquid chromatography: An investigation of operating parameters and the separation of nucleotides on pellicular ion exchangers. *Anal. Chem.* **39,** 1422–1428.
6. Scott, R. P. W. and Kucera, P. (1979) Mode of operation and performance characteristics of microbore columns for use in liquid chromaatography. *J. Chromatogr.* **169,** 51–72.
7. Hirata, Y. and Novotny, M. (1979) Techniques of capillary liquid chromatography. *J. Chromatogr.* **186,** 521–528.
8. Ishii, D., Asai, K., Hibi, K., Jonokuchi, T., and Nagaya, M. (1977) A study of micro-high performance liquid chromatography. I.Development of technique for miniaturization of high performance liquid chromatography. *J. Liqu. Chromatogr.* **144,** 157–168
9. Ishii, D., Tsuda, T. and Takeuchi, T. (1979) Studies of open-tubular micro-capillary liquid chromatography. IV. Soda lime glass columns treated with alkaline solution. *J. Chromatogr.* **185,** 73–78

10. Basey, A. and Oliver, R.W.A. (1982) Nomenclature of liquid chromatography. On the description of small diameter columns used in liquid chromatography. *J. Chromatogr.* **251,** 265–268
11. Kennedy, R.T. and Jorgenson, J.W. (1989) Preparation and Evaluation of Packed Capillary Liquid Chromatography Columns with inner diamenters from 20 to 50 μm. *Anal. Chem.* **61,** 1128–1135.
12. Hsieh, S. and Jorgenson, J.W. (1996) Preparation and evaluation of slurry-packed liquid chromatography microcolumns with inner diameters from 12 to 33 μm. *Anal. Chem.* **68,** 1212–1217.
13. Spikmans, V., Lane, S. J., Tjaden, U. R., and van der Greef, J. (1999) Automated capillary electrochromatography tandem mass spectrometry using mixed mode reversed-phase ion exchange chromatography columns. *Rapid Commun Mass Spectrom.* **13,** 141–149.
14. Gucek, M., Gaspari, M., Walhagen, K., Vreeken, R. J., Verheij, E. R., and van der Greef, J. (2000) Capillary electrochromatography/nanoelectrospray mass spectrometry for attomole characterization of peptides. *Rapid Comm. Mass Spectrom.* **14,** 1448–1454
15. Wu, J.-T., Huang, P., Li, M. X., and Lubman, D. M. (1997) Protein digest analysis by pressurized capillary electrochromatography using an ion trap storage/reflectron time-of-flight mass detector. *Anal. Chem.* **69,** 2908–2913.
16. Colon, L. A., Maloney, T. D., and Fermier, A. M (2000) Packing columns for capillary electrochromatography. *J. Chromatogr. A.* **887,** 43–53.
17. Angus, P. D. A., Demarest, C. W., Catalano, T., and Stobaugh, J. F. (2000) Aspects of column fabrication for packed capillary electrochromatography. *J. Chromatogr. A.* **887,** 347–365.
18. Pyell, U. (2000) Advances in column technology and instrumentation in capillary electro-chromatography. *J. Chromatogr. A.* **892,** 257–278.
19. Davis, M. T. and Lee, T. D. (1992) Analysis of peptide mixtures by capillary high performance liquid chromatography: A practical guide to small-scale separations. *Protein Sci.* **1,** 935–944.
20. Grimm, R., Serwe, M., and Chervet, J.-P. (1997) Capillary LC using conventional HPLC Instrumentation. *LC/GC* **15,** 960–965.
21. Takeuchi, T. and Ishii, D. (1982) Micro-HPLC with long micro-packed flexible fused silica columns. *J. Chromatogr.* **238,** 409–418
22. Bateman, K. P., White, R. L., and Thibault, P. (1998) Evaluation of absoprtion preconcentration/CZE/nanoelectrospray mass spectrometry for peptide and glycoprotein analyses. *J. Mass Spectrom.* **33,** 1109–1123.
23. Cappiello, A., Palma, P., and Mangani, F. (1991) New materials and packing techniques for micro-HPLC packed capillary columns. *Chromatographia* **32,** 389–391.
24. Robins, R. H. and Guido, J. E. (1997) Design, construction and application of a simple packed capillary high performance liquid chromatography/electrospray mass spectrometry system. *Rapid Commun. Mass Spectrom.* **11,** 1661.

25. Roulin, S., Dmoch, R., Carney, R., et al. (2000) Comparison of different packing methods for capillary electrochromatography columns. *J. Chromatogr. A.* **887,** 307–312.
26. Oosterkamp, A. J., Gelpi, E., and Abian, J. (1998) Quantitative peptide bioanalysis using column-switching nano liquid chromatography/mass spectrometry. *J. Mass Spectrom.* **33,** 976–983.
27. Zimina, T. M., Smith, R. M., Myers, P., and King, B. W. (1995) Effects of kinematic viscosity of the slurry on the packing efficiency of peek microbore columns for liquid chromatography. *Chromatographia* **40,** 662–668.
28. Malik, A., Li, W., and Lee, M.L. (1993) Preparation of long packed capillaaary columns using carbon dioxide slurries. *J. Microcol. Sep.* **5,** 361–369.
29. Tong, D., Bartle, K. D., Clifford, A. A., and Edge, A. M. (1995) Theoretical studies of the preparation of packed capillary columns for chromatography. *J. Microcol. Sep.* **7(3),** 265–278.
30. Crescentini, G., Bruner, F., Mangani, F., and Yafeng, G. (1988) Preparation andd Evaluation of dry packed capillary columns for high performance liquid chromatography. *Anal. Chem.* **60,** 1659–1662.
31. Guan, Y., Zhou, L., and Shang, Z. (1992) Dry packed capillary columns for micro HPLC. *J. High Resolution Chromatogr.* **15,** 434–436.
32. MacNair, J. E., Patel, K. D., and Jorgenson, J. W. (1999) Ultrahigh-Pressure reversed phase capillary liquid chromatography: isocratic and gradient elution using columns packed with 1.0 μm particles. *Anal. Chem.* **71,** 700–708.
33. Wu, N., Collins, D. C., Lippert, J. A., Xiang, Y., and Lee, M.L. (2000) Ultra high pressure liquid chromatography/time of flight mass spectrometry for fast separations. *J. Microcol. Sep.* **12,** 462–469.
34. MacNair, J. E., Opitek, G. J., Jorgenson, J. W., and Moseley III ,M.A. (1997) Rapid Separation and characterization of protein and peptide mixtures using 1.5 μm diameter non-porous silica in packed capillary liquid chromatography/mass spectrometry. *Rapid Comm. Mass Spectrom.* **11,** 1279–1285.
35. Chirica, G. . and Remcho, V. (2000) Fritless capillary columns for HPLC and CEC prepared by immobilizing the statuionary phase in organic polymer matrix. *Anal. Chem.* **72,** 3605–3610.
36. Gusev, I., Huang, X., and Horváth, C. (1999) Capillary columns with in situ formed porous monolithic packing for micro high-performance liquid chromatography and capillary electrochromatography. *J. Chromatogr. A.* **855,** 273–290.
37. Hayes, J. D. and Malik, A., (2000) Sol-gel Monolithic columns with reversed electroosmotic flow for capillary electrochromatography. *Anal. Chem.* **72,** 4090–4099.
38. Davis, M. T., Stahl, D. C., and Lee, T. D. (1995) Low flow high performance liquid chromatgography solvent delivery system designed for tandem capillary liquid chromatography-mass spectrometry. *J. Amer. Soc. Mass Spectrom.* **6,** 571–577.
39. Davis, M. T. and Lee, T. D. (1997) Variable flow liquid chromatography tandem mass spectrometry and the comprehensive analysis of complex protein digest mixtures. *J. Amer. Soc. Mass Spectrom.* **8,** 1059–1069.

40. Davis, M. T. and Lee, T. D. (1998) Rapid protein identification using a microscale electrospray LC/MS system on an ionmm trap mass spectrometer. *J. Amer. Soc. Mass Spectrom.* **9,** 194–201.
41. Caprioli, R. M., Fan, T., and Cotrell, J.S . (1986) Continuous-Flow sample probe for fast atom bombardment mass spectrometry. *Anal. Chem.* **58,** 2949.
42. Yamashita, M. and Fenn, J. B. (1984) *J. Phys. Chem.* **88,** 4451.
43. Hunt, D. F., Henderson, R. A., Shabanowitz, J., et al. (1992) Characterization of peptides bound to Class I MHC molecule HLA-A2.1 by mass spectrometry. *Science* **255,** 1261.
44. Henderson, R.A., Michel, H., Sakaguchi, K., et al. (1992) HLA-A2.1_Associated peptides from a mutant cell line: a second PAthway of antigen presentation. *Science* **255,** 1264.
45. Emmett, M. R. and Caprioli, R.M . (1994) Micro-electrospray: mass spectrometry: ultra-high-sensitivity analysis of peptides and Proteins. *J. Amer. Soc. Mass Spectrom.* **5,** 605–613.
46. Chervet, J. P., Ursem. M., and Salzmann, J. P. (1996) Instrumental requirements for nanoscale liquid chromatography. *Anal. Chem.* **68,** 1507–1512.
47. Vanhoutte, K., Hoes, I., Lemiere, F., et al. (1997) Development of a nanoscale liquid chromatography/electrospray mass spectrometry methodology for the detection and identification of DNA adducts. *Anal. Chem.* **69,** 3161–3168.
48. Gatlin, C. L., Eng, J. K., Cross, S. T., Detter, J. C., and Yates III, J.R. (2000) Automated identification of aminoacid sequence variations in proteins by HPLC/microspray tandem mass spectrometry. *Anal. Chem.* **72,** 757–763.
49. Devreese, B., Vanrobaeys, F., and Beeumen, J. V. (2001) Automated nanoflow liquid chromatography/tandem mass spectrometric identification of proteins from Shewanella putrefaciens separated by two-dimensional polyacrylamide gel electrophoresis. *Rapid Commun. Mass Spectrom.* **15,** 50–56.
50. Van der Heeft, E., Jan ten Hove, G., Herberts, C. A., Meiringg, H. D., van Els, C. A. C. M. and de Jong, A. P. J. M. (1998) A microcapillary column switching HPLC-Electrospray ionization MS system for the direct identification of peptides presented by major histocompatibility complex class I molecules. *Anal Chem.* **70,** 3742–3751.
51. Holland, L.A. and Jorgenson, J.W. (2000) Characterization of a comprehensive two-dimensional anion exchange-perfusive phase liquid chromatography system for improved separation of peptides. *J. Microcolumn Separ.* **12,** 371–377.
52. Shevchenko, A., Wilm, M., Vorm, O., and Mann, M. (1996) Mass Spectrometric sequencing of proteins from silver-stained polyacrilamide gels. *Anal. Chem.* **68,** 850–858.
53. Chowdhury, S. K. and Chait, B. T. (1991) Method for the electrospray Ionization of highly conductive Aqueous solutions. *Anal. Chem.* **63,** 1660–1664.
54. Wilm, M.S. and Mann, M. (1994) Electrospray and Taylor-Cone theory. Dole's beam of macromolecules at last?. *Int. J. Mass Spectrom. Ion Processes* **136,** 167–180.

55. Valaskovic, G. A., Keheller, H. D., Little, P., Aaserud, D. J., and McLafferty, F.W. (1995) Attomole-sensitivity electrospray source for large-molecule mass spectrometry. *Anal. Chem.* **67,** 3802–3805.

11

Micropreparative HPLC of Peptides and Proteins

Edouard C. Nice and Marie-Isabel Aguilar

1. Introduction

The introduction of high-performance liquid chromatography (HPLC) for the analysis and separation of peptides and proteins in the late 1970s offered unrivalled advantages in terms of speed, resolution, sensitivity, and recovery when used for the purification of low microgram levels of complex mixtures of peptides and proteins *(1,2)*. However, the conventional HPLC columns of typically 4.6 mm id and operated at flow rates of 1 mL/min resulted in peak volumes of approx 1 mL or even larger. The resultant sample concentrations (μg/mL) were not ideally suited to subsequent manipulations, and at such low concentrations, losses caused by nonspecific adsorption on either the chromatographic support or associated equipment (e.g., syringes, sample vials, recovery vials) were commonplace *(3,4)*. Attempts to reduce peak volumes by operating at lower flow rates were shown to be associated with poor recoveries of, in particular, hydrophobic proteins *(5,6)*.

It can be shown that eluant peak volumes from chromatographic columns are directly proportional to the column length and inversely proportional to the square of the column diameter *(7,8)*. Thus, if a 4.6 mm id column operated at a flow rate of 1 mL/min is compared with a 1 mm id microbore column, the flow rate for the 1 mm id column would need to be reduced by $(4.6/1)^2$, i.e., 21-fold, to approx 50 μL/min to operate at the same linear velocity as shown schematically in **Fig. 1**. Under these conditions, if the same mass of protein is loaded on to each column, and the chromatographic efficiency of each column is identical, then samples will be recovered in concomitantly reduced peak volumes (50 μL compared to 1 mL for a peak 1 min wide). For concentration-

From: *Methods in Molecular Biology, vol. 251, HPLC of Peptides and Proteins: Methods and Protocols*
Edited by: M.-I. Aguilar © Humana Press Inc., Totowa, NJ

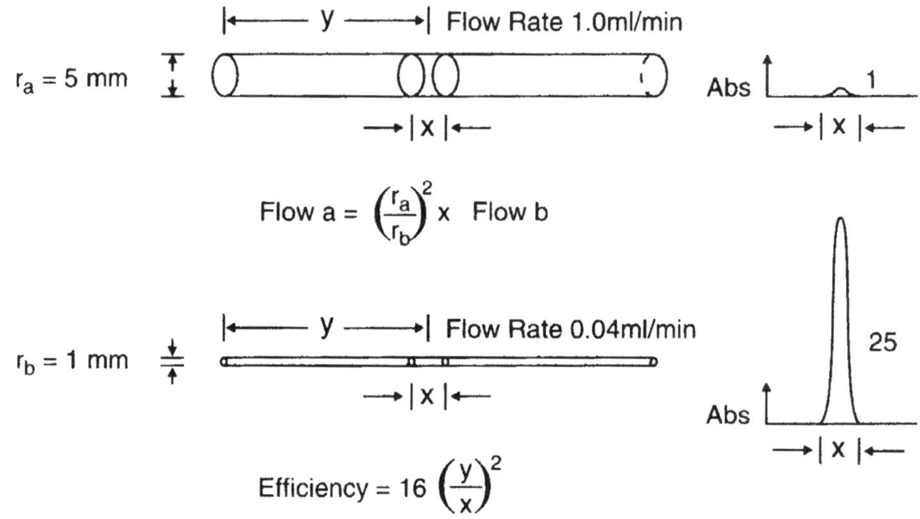

Fig. 1. Theory of micropreparative HPLC.

dependent detectors, this increase in sample concentration will give a directly proportional increase in the detection signal if the flow cell geometry is conserved. Nanogram quantities of peptides and proteins can thus be readily detected using small-bore columns (*see* **Fig. 2**) with peak elution volumes of 100 μL or less.

Sample volumes are not limiting: large sample volumes can be loaded onto interactive supports (e.g., reversed phase, ion-exchange, hydrophobic interaction, or affinity) under appropriate mobile phase conditions (trace enrichment) ***(9–14)***. Suitable instrumentation requires careful consideration with regard to minimised pre- and postcolumn dead volumes, accurate gradient formation at low flow rates, detector flow cell geometry, and sample recovery.

These techniques are ideally suited for the purification and concentration of trace protein or peptide components in bulk biological samples prior to application to highly specific and sensitive downstream analytical techniques such as microsequence analysis, mass spectrometry, biosensor analysis, or biological assay ***(3,4,14–17)***. They also facilitate the recovery of low levels of material following chemical or enzymatic modification. The techniques can be readily applied to the development of large-scale processes, whereby multidimensional purification protocols can be rapidly developed using minimum amounts of sample, solvent, and column packing materials and then scaled up appropriately.

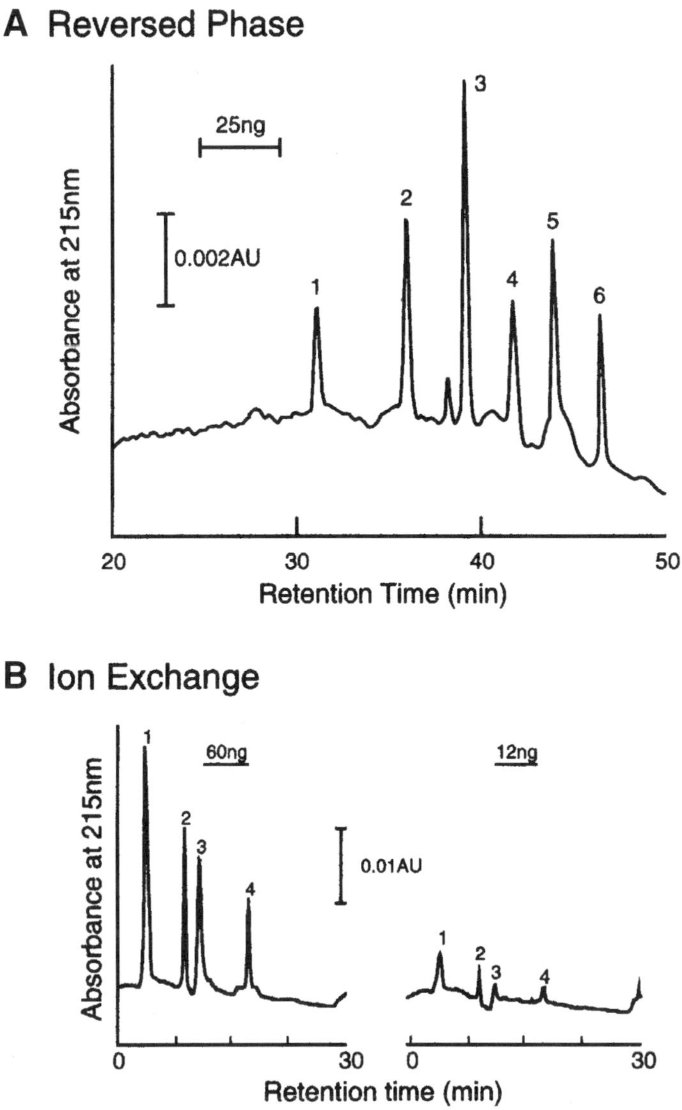

Fig. 2. Separation of low level protein standards by micropreparative RP- and anion-exchange HPLC. **(A)** Micropreparative RP-HPLC. Column: Brownlee RP-300 column, 30 cm × 2.1 mm id, and a linear gradient from 0–60% acetonitrile in 0.15% TFA over 60 min at a flow rate of 0.1 mL/min. **(B)** Micropreparative anion-exchange HPLC using a Mono-Q PC 1.6/5 anion-exchange column (Pharmacia), 50 cm × 1.6 mm id, using a linear gradient from 0–1 *M* sodium chloride in 20 m*M* Tris-HCl, pH 7.5 over 50 min at a flow rate of 0.1 mL/min. Reproduced with permission from **ref.** ***12*** © John Wiley & Sons, Ltd. Reprinted with permission.

2. Materials

2.1. Equipment

1. Suitable instrumentation for micropreparative HPLC requires (a) precise, pulseless pumping at low flow rates (*see* **Note 1**); (b) the ability to generate accurate and reproducible gradients accurately at low flow rates (*see* **Note 2**); (c) the minimization of dead volumes throughout the instrument (*see* **Note 3**); and (d) the reliable recovery of small volume fractions (*see* **Notes 4** and **5**).
2. Suitable instrument: *see* **Note 1**.
3. Solvent filtration apparatus equipped with a 0.22-μm Teflon filter.
4. Sample filters, 0.22 μm porosity.

2.2. Buffers

2.2.1. Anion-Exchange Chromatography

1. Buffer A : 20 m*M* Tris-HCl, pH 7.5.
2. Buffer B: 1 *M* NaCl in 20 m*M* Tris-HCl, pH 7.5.

2.2.2. Reversed-Phase Chromatography

1. Buffer A: 0.15% (v/v) aqueous trifluoroacetic acid (TFA).
2. Buffer B: 60% acetonitrile (HPLC grade) / 40% water (Milli-Q) containing 0.125% (v/v) aqueous TFA.

2.3. Columns

1. Mono-Q PC 1.6/5 anion-exchange column (50 × 1.6 mm id) (Amersham Biosciences).
2. Brownlee RP-300 (30 × 2.1 mm id) reversed-phase column (Applied Biosystems).

3. Methods

The following subheadings describe a method in which an initial screen of a range of chromatographic conditions is performed, followed by a detailed description of the loading and elution of a sample protein from a micropreparative column.

3.1. Solvent Preparation

Filter all solvents through a 0.22-μm filter before use. This removes particulates that could block solvent lines or the column and also serves to degass the solvent. If the HPLC instrument is not installed with on-line degassing capability, check your instrument requirements to assess whether further degassing is required. Failure to do this can result in spikes on the chromatogram due to bubble formation in the detector.

Table 1
Standard Loading and Eluting Buffers for Minicolumns Used in Column Screening

Column	Loading buffer	Eluting buffer
Anion-exchange	20 m*M* Tris-HCl, pH 7.5	20 m*M* Tris-HCl, pH7.5; 1 *M* NaCl
Cation exchange	20 m*M* sodium acetate, pH 5.0	20 m*M* sodium acetate, pH 5.0, 1 *M* NaCl
Ligand dye	20 m*M* sodium phosphate, pH 7.0	20 m*M* sodium phosphate, pH 7.0, 2 *M* NaCl
Reversed phase	0.1% TFA in water	0.1% TFA in water/ 50% acetonitrile.
Hydrophobic interaction	10 m*M* phosphate buffer, 2 *M* ammonium sulphate, pH 7.0	10 m*M* phosphate buffer, pH 7.0
Hydroxyapatite	10 m*M* sodium phosphate, pH 6.8	400 m*M* sodium phosphate, pH 6.5

3.2. Initial Column Screen

To assist in the design and optimization of a multidimensional micropurification protocol for a new protein sample, an initial screen of a series of column packings with alternative selectivities is advantageous, especially if a sensitive and specific bioassay is available to monitor the eluant fractions (e.g., mitogenic assays for many growth factors, dot blot assays). The method uses either small disposable columns (e.g., Biorad Econocolumn) or small tap-packed cartridges made using HPLC tubing and fittings, which can be attached to an HPLC pump for batchwise elution. Typically the column is packed with 0.5 mL gel volume. The experimental conditions are defined to determine whether a specific packing selectivity can retain the protein of interest under the loading conditions used, and whether the activity can be recovered batchwise in good yield using standard chromatographic buffers.

1. Dilute the sample 1:5 with the loading buffer (typically 100–200 µL sample providing the downstream assay is suitably sensitive). Apply the sample to each of the following column types using the standard loading conditions for each column as listed in **Table 1**.
2. Collect the load fraction.

3. Wash with loading buffer and collect 5 × 0.5 mL fractions.
4. Elute with eluting buffer and collect 5 × 0.5 mL fractions.
5. Assay fractions according to protocols defined for target protein to identify the packing and chromatographic conditions, which resulted in adsorption and recovery of the protein.
6. Analyze active fractions by gel electrophoresis to determine protein heterogeneity of recovered fractions.
7. Determine which column resulted in the best purification and integrate these columns into an optimized sequence of purification steps.

3.3. Sample Preparation

It may be necessary to adjust certain sample parameters (e.g., pH, solvent conditions, salt or organic solvent concentration to ensure that the protein of interest can be trace-enriched onto the column. This is especially true when going from one chromatographic system to another, i.e., multidimensional HPLC. In case of precipitation, centrifuge to remove sediment before taking the clarified sample up carefully into the loading syringe. Sample dilution is best done directly in the syringe to minimize losses.

3.4. Column Equilibration and Blank Run

1. Connect the guard and the column to the solvent delivery system according to the HPLC system requirements and equilibrate with Buffer A.
2. Once a stable baseline is obtained, perform blank run, using same injection conditions/volumes as for final sample.

3.5. Sample Injection and Analysis

This section describes large volume trace enrichment, either via pump, or large volume (e.g., 2 mL) injection loop, typically at higher flow rates to minimize loading time (*see* **Note 6**).

3.5.1. Sample Loading Directly Through Pumps

1. Reduce dead volumes by removing the pulse dampener if present.
2. Remove frits from the check valve to avoid blocking of the frits.
3. Place sample into a 50 mL Falcon tube. The conical shape of these tubes facilitates loading of the last few drops of sample.
4. Insert the solvent inlet tube into the sample (remove any filters that are in line), set flow to 1–2 mL/min, depending on back pressure and load sample. It is worth watching the back pressure during this operation since the back pressure can increase as the sample loads.
5. Add 20 mL of loading buffer into the Falcon tube and pump buffer at 1–2 mL/min (depending on the back pressure limits of the column) until sample has loaded completely onto the column, monitoring the UV output to check the eluant (*see* **Note 5**).

6. When the baseline has returned to zero (i.e., washing has completed), immediately shut down the flow rate to 0 mL/min.
7. Ensure that the loading loop is out of line.
8. Put the primary solvent line back into the solvent bottle
9. Set to flow rate compatible with the microcolumn that is being used (typically 50–100 µL/min) and wait for baseline to stabilize.
10. Start the gradient to elute the sample.

3.5.2. Sample Loading Via Syringe and Injection Loop

1. Install a 2-mL loop into the injection valve and switch loop out of line (into the load position). Set flow rate at 1–2 mL/min depending on back pressure.
2. Fill a 1-mL syringe with sample and load into the loop.
3. Switch loop in line (inject position) to load sample onto the column and monitor the baseline.
4. For multiple injections, repeat above steps, allowing sufficient time between injections for the loop to empty before switching back out of line.
5. When the total sample has been loaded, reduce the flow rate to the operating flow rate.
6. Start gradient. Linear 50-min gradient from 0–50% acetonitrile (reversed phase) or 0–1 *M* NaCl (ion-exchange). Hold for 10 min at the top of the gradient to account for any dead volume between where the gradient is generated and the column.

If using a large volume sample loop for injection, remember to switch it out of line after loading, otherwise a large dead volume will be introduced into system (e.g., a 2 mL loop takes 20 min to clear at 100 µL/min). **Figure 3** shows the chromatograms obtained following micropreparative ion-exchange and reversed phase isolation of the human EPH-like receptor tyrosine kinase ligand (HEK) ***(14)***.

4. Notes

1. Most conventional HPLC systems using high-pressure mixing can be readily modified for use with narrow-bore or microbore columns and, using the simple modifications described in Notes 2 and 3 will operate smoothly at low flow rates. However, there are a number of commercially available microLC instruments including the Amersham Biosciences SMART Micropreparative HPLC and Ettan MicroLC system, the Agilent 1100 Capillary LC system, the Waters Cap LC system and LC Packings UltiMate.
2. The chief concern with some conventional instruments is the dead volume of the dynamic mixer. A number of static mixer designs have been investigated but none were capable of adequately mixing the gradients required for protein elution (particularly *N*-propanol gradients, which had been found ideal for the recovery of more hydrophobic proteins from reversed phase supports). One approach to using large volume mixers was to form the gradient within the mixing chamber by direct displacement with 100% Buffer B at low flow rate ***(18)***. In this way, an exponential gradient of approx 3.6 mL was generated that gave excellent separation of

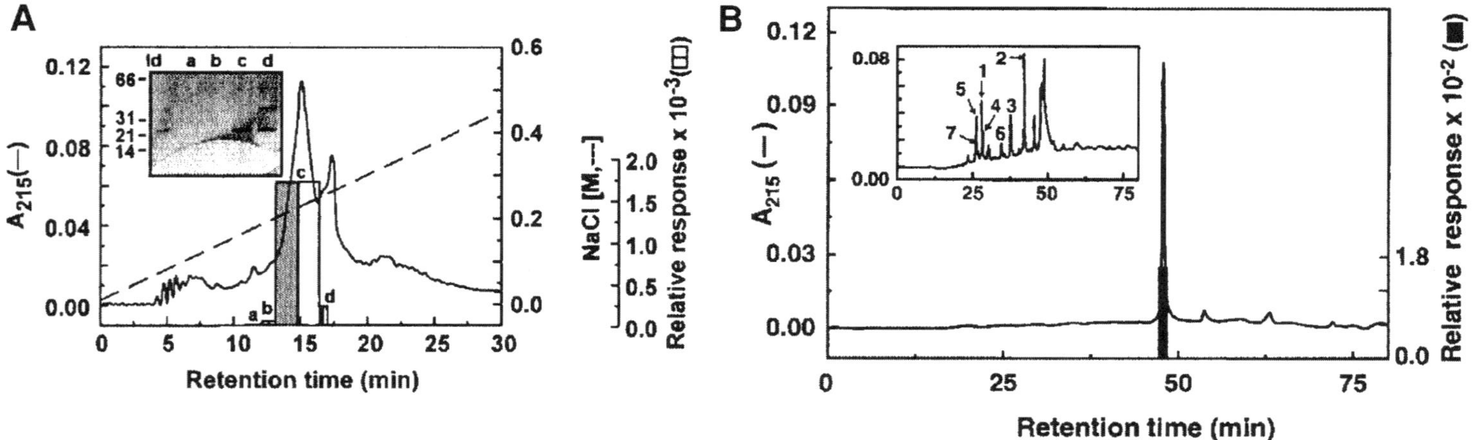

Fig. 3. Micropreparative ion-exchange chromatography of the HEK ligand. **(A)** Combined active fractions from a size exclusion column were adjusted to 50 m*M* NaCl and fractionated on a µ-Mono-Q column with a 40-min gradient (dashed line) of 0–600 m*M* NaCl in 20 m*M* Tris-HCl (pH 8.5) at a flow rate of 100 µL/min. Elution was monitored at 215 nm (solid line). **(B)** RP-HPLC of the HEK ligand and derived tryptic fragments. Approx 0.6 µg of µ-Mono-Q-purified HEK ligand was applied to a 1-mm idD RP300 HPLC column and eluted with a 0–60% CH3CN gradient in 0.09% TFA at a flow rate of 100 µL/min and elution monitored at 215 nm. Inset: Tryptic map of the reduced and alkylated HEK ligand was produced on the same narrow bore RP-HPLC column. Reproduced from **ref. *14***.

a number of protein standards. An alternative approach to enable gradients of any shape to be generated by mixing of metered volumes of Buffers A and B was to reduce the volume of the mixer by the use of a Teflon insert. Another alternative is to replace the standard mixer with a low volume alternative (e.g., that available from Gilson which has a volume of 60 µL).

In contrast, low-pressure mixing systems, which employ valve mixing blocks to generate the gradient prior to a single high pressure pump, are not easy to modify, primarily because of excessive instrumental dead volumes inherent to the system, which could not be easily reduced.

3. By removing the pulse dampers from the instrument and some simple replacement and rerouting of the stainless steel solvent lines to minimize the volumes between the pumps, the mixer, injector, column, and detector, the instrumental dead volume can be significantly reduced.
4. Recovery can be easily performed by manual fraction collection, allowance being made for the dead volume between the detector flow cell and the collection port, which is significant when operating at low flow rates (<200 µL/min). Manual collection involves holding the tip of the collection device either against the side of the collection tube (we usually use 0.5- or 1.5-mL Eppendorf tubes which have particularly good recovery and low level background contamination properties) or under the liquid meniscus to prevent droplet formation. The drop size (typically around 10 µL) becomes extremely significant in terms of both resolution and overall recovery when dealing with peak volumes of 20–100 µL.
5. The requirement for special low volume flow cells in the detector is often overstated. In practice, we found that detector flow cell volumes of 4.6 µL were suitable for peak volumes >50 mL. This is in agreement with published data that demonstrated that flow cell volumes of up to 10% of the peak volume are acceptable ***(19)***. It should be borne in mind that small volume flow cells often have reduced optical path lengths that proportionately reduce the overall sensitivity of detection. Flow cell design however is important since poorly designed flow cells can cause turbulence and hence poor sample clearance and peak mixing.
6. When purifying trace components from complex mixtures, it is imperative that the total sample be loaded if optimum benefits of microcolumns are to be achieved. Proteins and peptides bind very strongly to interactive supports below critical secondary solvent concentrations (*see* **Fig. 4**). Under these conditions, trace enrichments of large sample volumes directly onto micropreparative columns is possible before recovering retained proteins or peptides in small volumes at high concentration by gradient or step elution. An important practical point is to ensure that the sample is in a buffer that will allow trace enrichment. Samples may require dilution to reduce the salt or organic solvent concentration used for elution at the previous chromatographic step. Samples are typically loaded at high flow rate by means of a large injection loop (up to 2 mL), often using multiple injections, or by direct loading via the primary or auxillary pump. Large volume injection loops should be switched out of the main flow path following sample loading to minimize the system volume between the pumps and the column, which would other-

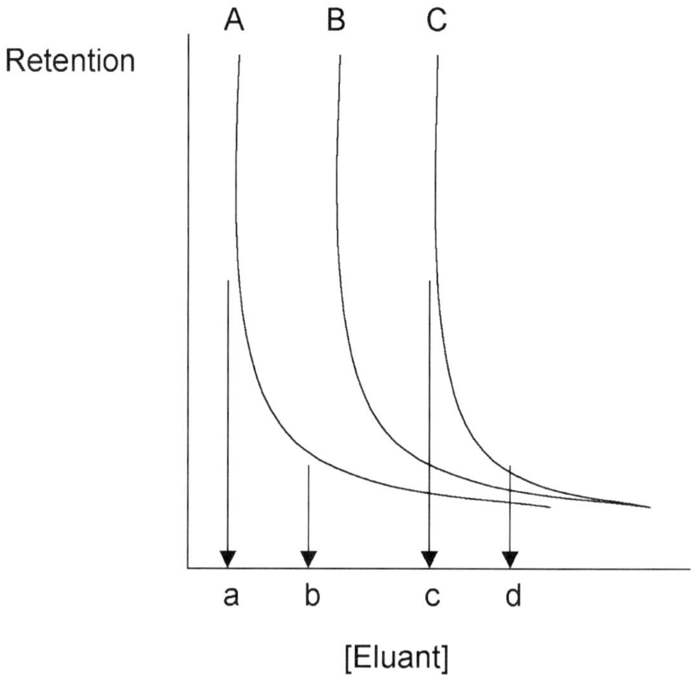

Fig. 4. Relationship between retention and secondary solvent concentration for the separation of peptides and proteins in interactive modes of chromatography. All three proteins *(A–C)* are retained at concentration of eluant *a*, while protein A can be selectively eluted by increasing the eluant concentration to *b*. Similarly, after elution of *A* and *B* using eluant concentration level *c*, protein C can be eluted at eluant concentration *d*.

wise result in unacceptable long delays in the gradient delivery. We have frequently loaded samples in volumes of up to 100 mL onto 1 mm id reversed-phase columns and recovered the retained proteins in peak volumes of <100 μL (>1000-fold concentration).

References

1. High resolution separation and analysis of biological macromolecules: Part A Fundamentals. (1996) *Methods in Enzymol.* **270,** entire volume.
2. High resolution separation and analysis of biological macromolecules: Part B Applications. (1996) *Methods in Enzymol.* **271,** entire volume.
3. Nice, E. C., Simpson, R. J., and Nicola, N. A. (1987) A micropreparative chromatographic strategy for the purification and sequence analysis of murine granulocyte-colony stimulating factor (mG-CSF). *Chromatographia* **24,** 449–468.
4. Grego, B., Van Driel, I. R., Goding, J. W., Nice, E. C., and Simpson, R. J. (1986) Use of microbore high-performance liquid chromatography for purifying sub-

nanomole levels of polypeptides for microsequencing. Structural studies on the murine plasma cell antigen PC-1. *Int. J. Pept. Protein Res.* **27,** 201–207.

5. Cooke, N. H., Archer, B. G., O'Hare, M. J., Nice, E. C., and Capp, M. W. (1982) Effect of chain length and carbon load on the performance of alkyl-bonded silicas for protein separations. *J. Chromatogr.* **255,** 115–123.
6. O'Hare, M. J., Capp, M. W., Nice, E. C., Cooke, N. H., and Archer, B. G. (1982) Factors influencing chromatography of proteins on short alkylsilane-bonded large pore-size silicas. *Anal. Biochem.* **126,** 17–28.
7. Kucera, P. (1980) Design and use of short microbore columns in liquid chromatography. *J. Chromatogr.* **198,** 93–109.
8. Scott, R. P. and Kucera, P. (1979) Use of microbore columns for the separation of substances of biological origin. *J. Chromatogr.* **185,** 27–41.
9. Alexson, S. E., Finlay, T. H., Hellman, U., Svensson, L. T., Diczfalusy, U., and Eggertsen, G. (1994) Molecular cloning and identification of a rat serum carboxylesterase expressed in the liver. *J. Biol. Chem.* **269,** 17,118–17,124.
10. Poulter, L., Barratt, D., Scott, C. W., and Caputo, C. B. (1993) Locations and immunoreactivities of phosphorylation sites on bovine and porcine tau proteins and a PHF-tau fragment. *J. Biol. Chem.* **268,** 9636–9644.
11. Nice, E. C., Catimel, B., Lackmann, M., et al. (1997) Strategies for the identification and purification of ligands for orphan biomolecules. *Lett. Peptide Sci.* **8,** 107–120.
12. Nice, E. C. (1996) Micropreparative HPLC of proteins and peptides: principles and applications. *Biopolymers (Peptide Science)* **40,** 319–341.
13. Nice, E. C., Fabri, L., Hammacher, A., Holden, J., Simpson, R. J., and Burgess, A. W. (1992) The purification of a Rap1 GTPase-activating protein from bovine brain cytosol. *J. Biol. Chem.* **267,** 1546–1553.
14. Lackmann, M., Bucci, T., Mann, R. J., et al. (1996) Purification of a ligand for the EPH-like receptor HEK using a biosensor-based affinity detection approach. *Proc. Natl. Acad. Sci. USA* **93,** 2523–2527.
15. Catimel, B., Weinstock, J., Nerrie, M., Domagala, T., and Nice, E. C. (2000) Micropreparative ligand fishing with a cuvette-based optical mirror resonance biosensor. *J. Chromatogr. A* **869,** 261–273.
16. Catimel, B., Rothacker, J., and Nice, E. (2001) The use of biosensors for microaffinity purification: an integrated approach to proteomics. *J. Biochem. Biophys. Meth.* **49,** 289–312.
17. Moritz, R. L., Ritter, G., Catimel, B., et al. (1998) Micro-sequencing strategies for the human A33 antigen, a novel surface glycoprotein of human gastrointestinal epithelium. *J. Chromatogr. A.* **798,** 91–101.
18. Nice, E., Lackmann, M., Smyth, F., Fabri, L., and Burgess, A. W. (1994) Synergies between micropreparative high-performance liquid chromatography and an instrumental optical biosensor. *J. Chromatogr. A* **660,** 169–185.
19. Kirkland, J. J., Yau, W. W., Stoklosa, H. J., and Dilks, C. H., Jr. (1977) Sampling and extra-column effects in high-performance liquid chromatography; influence of peak skew on plate count calculations. *J. Chromatogr. Sci.* **15,** 303–316.

12

Multidimensional HPLC Purification of Proteins

Edouard C. Nice and Marie-Isabel Aguilar

1. Introduction

Purification of trace compounds (e.g., growth factors, receptors) from bulk biological samples typically requires high purification factors to achieve purification to homogeneity ***(1–9)***. The sequential use of microcolumns of varying selectivity, assuming that they have good recovery characteristics, allows very high purification factors to be achieved. As described in Chapter 11, the use of short narrow-bore (2.1 mm id), microbore (1 mm id), or capillary (<1 mm id) columns allows the recovery of purified proteins and peptides in reduced volume at increased concentration compared with larger bore columns. Indeed, as shown in **Table 1**, it can be calculated that purification factors far in excess of those achieved by two-dimensional polyacrylamide gel electrophoresis can be readily achieved using multidimensional purification protocols, with the added advantage that the purified sample is in a form directly compatible with downstream analysis. Successful micropreparative HPLC requires minimal losses of both mass and biological activity during the chromatographic purification and other associated nonchromatographic sample manipulation (e.g., sample dilution, pH adjustment, storage, chemical manipulation) which we refer to as micromanipulation ***(3,6,7)***. By maintaining high overall recovery throughout the procedure, it is possible to take a sample though successive chromatographic steps (i.e., multidimensional purification) and still have sufficient material for structural and/or biological analysis. This chapter illustrates the potential of short micropreparative columns with different selectivity by the sequential use of size exclusion, anion-exchange and reversed-phase (RP) columns for the final purification of a sample of murine epidermal growth factor (mEGF) isolated from mouse salivary glands ***(4)***. By using the columns in this sequence,

From: *Methods in Molecular Biology, vol. 251, HPLC of Peptides and Proteins: Methods and Protocols*
Edited by: M.-I. Aguilar © Humana Press Inc., Totowa, NJ

Table 1
Potential Peak Resolution in Multidimensional HPLC

Chromatographic mode	Peak capacity[a]	Cumulative resolution
Hydrophobic interaction	50	50
Size exclusion	10	500
Ion-exchange	50	25,000
Reversed phase	100	2,500,000

[a]Peak capacity is defined as the number of peaks that can be theoretically resolved in a typical chromatographic run.

it is possible to reinject the recovered sample without prior dilution because the buffers used for size exclusion and anion-exchange stages are compatible with subsequent trace enrichment onto the anion exchange and RP columns. Thus by careful consideration of the order in which columns are used in a multidimensional purification scheme, the need to manipulate the sample between stages can be avoided, or minimised, thereby minimising sample handling and dilution and, hence, improving overall recoveries. The differing selectivities obtainable with each column used in the protocol described in this chapter resulted in purification of the mEGF from minor contaminants at each stage.

2. Materials

2.1. Equipment

1. Pharmacia SMART system equipped with automatic fraction collection. This method can also be performed with other instruments with appropriate features as outlined in Chapter 11.
2. Solvent filtration apparatus equipped with a 0.22-µm Teflon filter.
3. Sample filters, 0.22-µm porosity.

2.2. Buffers

1. Size exclusion chromatography.
   a. 1% w/v NH_4HCO_3 / 0.02% v/v Tween-20 (*see* **Note 1**).
2. Anion-exchange chromatography.
   a. Buffer A: 20 m*M* Tris-HCl (pH 7.5).
   b. Buffer B: 1 *M* NaCl in 20 m*M* Tris-HCl (pH 7.5).
3. RP chromatography.
   a. Buffer A: 0.15% (v/v) aqueous trifluoroacetic acid (TFA).
   b. Buffer B: 60% acetonitrile (HPLC grade) / 40% water (Milli-Q) containing 0.125% (v/v) aqueous TFA (*see* **Note 2**).

2.3. Columns

1. Superose 12PC 3.2/30 (300 × 3.2 mm id) (Amersham Biosciences).
2. Mono-Q PC 1.6/5 anion-exchange column (50 × 1.6 mm id) (Amersham Biosciences).
3. Brownlee RP-300 (30 × 2.1 mm id) RP column (Applied Biosystems).

3. Methods

3.1. Sample Preparation

Dissolve sample in 100 µL of Buffer A (*see* **Note 3**). If there is some undissolved material, centrifuge the sample in an Eppendorf tube (*see* **Note 4**).

3.2. Solvent Preparation

Filter all solvents through a 0.22-µm filter before use. This removes particulates that could block solvent lines or the column and also serves to degas the solvent. If the HPLC instrument is not installed with on-line degassing capability, check with your instrument requirements to assess whether further degassing is required.

3.3. Size-Exclusion Chromatography

1. Connect the Superose 12 column to the solvent delivery system according to the HPLC system requirements and equilibrate using 1% w/v NH_4HCO_3 / 0.02% w/v Tween-20 at a flow rate of 100 µL/min, a column temperature of 25°C and detection at 215 nm.
2. Once a stable baseline is obtained, inject 4.5 µg of mEGF (750 pmol) (either manually or via an automatic injector).
3. Collect the main eluting protein peak for injection onto the micropreparative anion-exchange column (*see* **Note 4**).

3.4. Anion-Exchange Chromatography

1. Connect the Mono-Q column to the solvent delivery system according to the HPLC system requirements and equilibrate using Buffer A at a flow rate of 100 µL/min, a column temperature of 25°C and detection at 215 nm.
2. After controlled equilibration (*see* **Note 5**), inject the fraction from the size exclusion column onto the anion-exchange column (either manually or via an automatic injector).
3. Elute the sample with a linear gradient from 0–100% Buffer B over 50 min.
4. Collect the main eluting protein peak for injection onto the micropreparative RP column.

3.5. RP Chromatography

1. Connect the RP column to the solvent delivery system according to the HPLC system requirements and equilibrate using Buffer A at a flow rate of 2 mL/min, a column temperature of 25°C and detection at 215 nm.

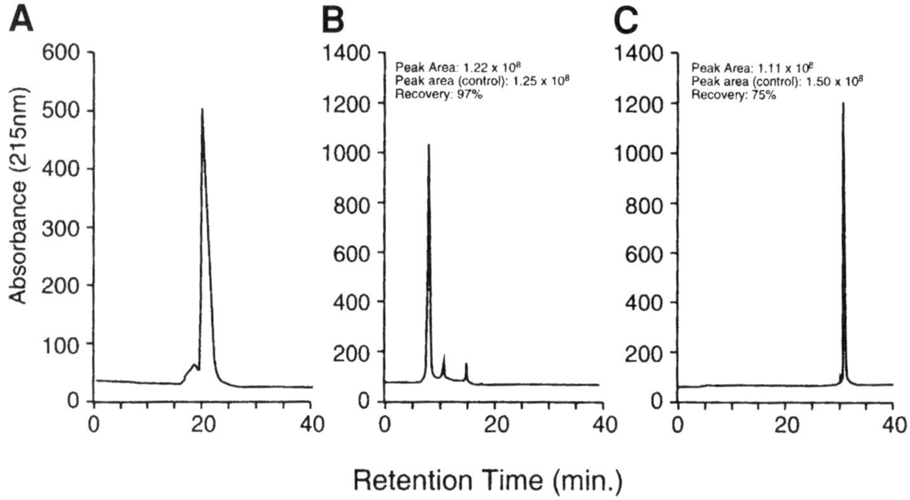

Fig. 1. Mulitdimensional micropreparative HPLC of mEGF. (**A**) Size exclusion chromatography. (**B**) Anion-exchange chromatography. (**C**) RP chromatography. The operating conditions for each system are given in the text. Reproduced with permission from **ref. *4*** © 1993 John Wiley & Sons.

2. After controlled equilibration, reduce the flow rate to 100 μL/min and inject the fraction from the anion-exchange column onto the reversed phase column (either manually or via an automatic injector).
3. Elute the sample with a linear gradient from 0–100% Buffer B over 60 min.
4. Collect the main eluting protein peak.

Following micropurification, the sample is suitable for a number of downstream manipulations such as bioassays, microsequencing, mass spectrometry or protein interaction studies using optical biosensors ***(8–15)***. **Figure 1** shows typical chromatograms obtained using the protocols outlined earlier for isolation of mEGF. The integrated peak area (Panels **A**, **B**, and **C**) and the calculated recovery between stages (Panels **B** and **C**) are indicated in the figure. Using the conditions defined, the mEGF should elute at a retention time close to those shown. However, as outlined in Chapters 2–5, there are a number of operating parameters that can be changed in order to manipulate the resolution of peptide and protein mixtures in size exclusion, ion-exchange, and RP HPLC.

4. Notes

1. Tween-20 helps prevent sample losses by nonspecific adsorption. The bicarbonate buffer is a volatile buffer system BUT is not compatible with many silica-based columns.

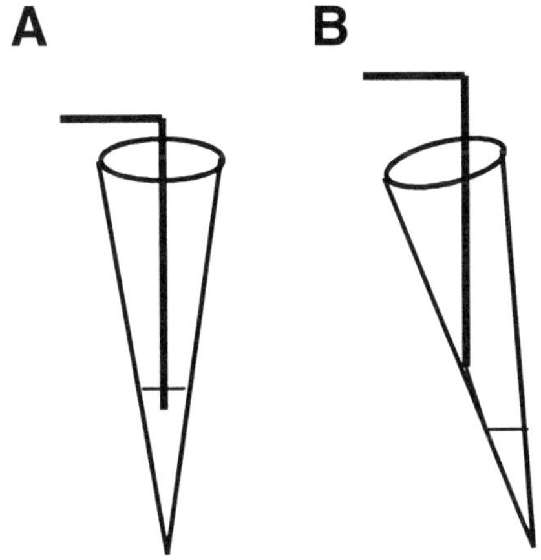

Fig. 2. Split-drop methods for manual fraction collection in micropreparative chromatography.

2. The slight differences in TFA concentration help minimize baseline drift: in fact, the TFA can be carefully titrated into the solvents, using the detector to see the baseline differences at 0 and 100% B until the solvents are balanced.

   For size exclusion chromatography, any buffer can be used. Typically, the sample is made up in 1% w/v NH_4HCO_3, but other buffers can be used since the sample will be buffer-exchanged in the column. For downstream bioassays, isotonic saline or PBS are particularly useful. If BIAcore assays are being used, then the BIAcore running buffer can be used (typicaly HEPES) to eliminate refractive index changes ***(15)***.
3. Eppendorf tubes are ideal and show low nonspecific losses compared with some other brands. Avoid filters for clarifying samples as the large surface areas can cause significant sample loss when working at low levels.
4. With the SMART system, fraction collection can be performed automatically. If fractions are collected manually on other instruments, allowance must be made for the dead volume between the detector and the outlet. When collecting fractions manually, to reduce sample crossover bewteen fractions use the split drop methods obtained holding the collection tube either just under the surface (*see* **Fig. 2**) or against the side wall of the tube. When operating at low flow rates the volume of a single drop (typically 5–10 µL) becomes significant.
5. When using gradient elution methods, it is important that the same volume of solvent is used for equilibration between each run. Also, the "first run" syndrome (where a couple of runs are often required before reproducible chromatograms are observed) can be avoided by leaving the column in 100% buffer B and re-equilibrating from there, using the defined volume of solvent.

References

1. Nice, E. C., Lloyd, C. J., and Burgess, A. C. (1984) The role of short microbore HPLC columns for protein separation and trace enrichment. *J. Chromatogr.* **296,** 153–170.
2. Nice, E. C. and Simpson, R. J. (1989) Micropreparative high-performance liquid chromatography of proteins and peptides. *J. Pharm. Biomed. Anal.* **7,** 1039–1053.
3. Nice, E. C. (1990) Micropreparative LC of proteins. *Nature* **348,** 462–463.
4. Nice, E. C., Fabri, L., Hammacher, A., Andersson, K., and Hellman, U. (1993) Micropreparative ion exchange high performance liquid chromatography: applications to microsequence analysis. *Biomed Chromatogr.* **7,** 104–111.
5. Lackmann, M., Bucci, T., Mann, R. J., et al. (1996) Purification of a ligand for the EPH-like receptor HEK using a biosensor-based affinity detection approach. *Proc. Natl. Acad. Sci. USA* **93,** 2523–2527.
6. Nice, E. C. and Catimel, B. (2000) Miniaturization of HPLC separations and equipment in protein and peptide analysis. *Encyclopedia of Analytical Chemistry* (Meyers, R. A., ed.), Wiley, Chichester, UK, **7,** 5823–5845.
7. Nice, E. C. (1996) Micropreparative HPLC of proteins and peptides: principles and applications. *Biopolymers (Peptide Science)* **40,** 319–341.
8. Alexson, S. E., Finlay, T. H., Hellman, U., Svensson, L. T., Diczfalusy, U., and Eggertsen, G. (1994) Molecular cloning and identification of a rat serum carboxylesterase expressed in the liver. *J. Biol. Chem.* **269,** 17,118–17,124.
9. Poulter, L., Barratt, D., Scott, C. W., and Caputo, C. B. (1993) Locations and immunoreactivities of phosphorylation sites on bovine and porcine tau proteins and a PHF-tau fragment. *J. Biol. Chem.* **268,** 9636–9644.
10. Simpson, R. J., Moritz, R. L., Nice, E. E., and Grego, B. (1987) A high-performance liquid chromatography procedure for recovering subnanomole amounts of protein from SDS-gel electroeluates for gas-phase sequence analysis. *Eur. J. Biochem.* **165,** 21–29.
11. Simpson, R. J., Moritz, R. L., Begg, G. S., Rubira, M. R., and Nice, E. C. (1989) Micropreparative procedures for high sensitivity sequencing of peptides and proteins. *Anal. Biochem.* **177,** 221–236.
12. Catimel, B., Weinstock, J., Nerrie, M., Domagala, T., and Nice, E. C. (2000) Micropreparative ligand fishing with a cuvette-based optical mirror resonance biosensor. *J. Chromatogr. A* **869,** 261–273.
13. Catimel, B., Rothacker, J., and Nice, E. (2001) The use of biosensors for microaffinity purification: an integrated approach to proteomics. *J. Biochem. Biophys. Methods* **49,** 289–312.
14. Moritz, R. L., Ritter, G., Catimel, B., et al. (1998) Micro-sequencing strategies for the human A33 antigen, a novel surface glycoprotein of human gastrointestinal epithelium. *J. Chromatogr. A* **798,** 91–101.
15. Nice, E., Lackmann, M., Smyth, F., Fabri, L., and Burgess, A. W. (1994) Synergies between micropreparative high-performance liquid chromatography and an instrumental optical biosensor. *J. Chromatogr. A* **660,** 169–185.

13

Analytical High-Performance Liquid Chromatography

Kálmán Benedek

1. Introduction

High-performance liquid chromatography (HPLC) has become the most significant analytical technique of the last 30 years. Whereas HPLC has revolutionized analytical chemistry in general, its significance is mostly recognized in the development of modern biochemistry/biotechnology. Indeed, the scientific accomplishments of many biological disciplines can be largely credited to the development of modern HPLC *(1)*. The popularity and success of the technique has arisen from synergistic improvements in surface chemistry, column technology, instrumentation, and software developments. HPLC is now used for biopolymer purification at the research level, large-scale purifications, and analysis at the development level. In particular, the efficiency, speed, and recovery accomplished by HPLC allowed never seen development of modern biotechnology and pharmaceutical R&D.

The first benefit gained by the application of HPLC methods is the *speed* of separation as compared to traditional low-pressure chromatography. Separations traditionally performed over a period of a few hours can be performed routinely under 60 min. Ultrafast separations can be performed under 1 min, but these techniques are for very special routine analysis and usually require the guidance of an HPLC specialist. The second major improvement modern HPLC can provide is increased *resolution* and the third important factor is the very low detection limit and increased *sensitivity*. State-of-the-art separations in the order of 60 s for samples of nanograms quantities and very complex mixtures can be performed. With the improvements in speed, resolution and detection came the discovery of new molecules and biological phenomena and the

From: *Methods in Molecular Biology, vol. 251, HPLC of Peptides and Proteins: Methods and Protocols*
Edited by: M.-I. Aguilar © Humana Press Inc., Totowa, NJ

increased speed and precise temperature control of separations also revealed new kinetic and thermodynamic regimes of studies.

One of the most widely used applications of analytical HPLC is peptide mapping, i.e., the analysis of enzymatic digests of peptides and proteins. The most important enzyme routinely used for peptide mapping is trypsin. Tryptic digestion combined with reversed-phase (RP) HPLC and mass spectrometry has become the method of choice to establish protein composition, side-chain modifications, and production lot reproducibility and of cellular changes initiated by the methods of molecular biology (*see* also Chapter 16).

Tryptic mapping is a two-step procedure involving the tryptic digestion of the protein sample followed by the analysis of the digested products by RP chromatography. The separation development and optimization of a peptide map is not a trivial task. The numbers of expected peaks are usually calculated from the sequence of the protein. A typical digest can contain 20–30 different oligopeptides and the goal is to have baseline separation of the peaks. The UV trace of the individual peak areas are integrated and a mass balance can be established. The most frequently used method applies acetonitrile gradient at low pH (0.1% trifluoroacetic acid [TFA]). Chapter 2 in this book on RP chromatography provides a comprehensive description of the use of reversed phase HPLC, whereas Chapter 16 demonstrates the use of mass spectrometry in proteolytic peptide mapping. The focus of this chapter is to outline an approach to the design of separation procedures for a complex mixture of tryptic peptides by analytical RP-HPLC through the combination of experimental scouting experiments and the application of optimization software.

2. Materials

2.1. Chemicals

1. Acetonitrile (CH_3CN), HPLC grade.
2. Milli-Q water.
3. TFA.

2.2. Equipment and Supplies

1. HPLC solvent delivery system with binary gradient capability and a UV detector (214 nm).
2. RP octadecylsilica (C18) column, 4.6 mm id × 250 mm length (*see* **Note 1**), 5 μm particle size, 300 A pore size.
3. C18 guard column.
4. Solvent filtration apparatus equipped with a 0.22-μm Teflon filter.
5. Sample filters, 0.22 μm porosity.
6. Buffer A: 0.1% (v/v) TFA in water (*see* **Note 2**).
7. Buffer B: 100% CH_3CN containing 0.1% (v/v) TFA.

3. Methods

3.1. Sample Preparation

Dissolve 1 mg of sample in 1 mL of Buffer A. If there is some undissolved material, filter the sample through a 0.22-µm filter.

3.2. Solvent Preparation

Filter all solvents through a 0.22-µm filter before use. This removes particulates that could block solvent lines or the column and serves to degas the solvent. If the HPLC instrument is not installed with on-line degassing capability, check with your instrument requirements to assess whether further degassing is required.

3.3. Column Equilibration and Blank Run

1. Connect the guard and the column to the solvent delivery system according to the HPLC system requirements and equilibrate under the following initial conditions:
   a. Solvent: 100% Buffer A
   b. Flow rate: 1 mL/min
   c. Detection wavelength: 214 nm
   d. Temperature: thermostated (*see* **Note 3**)
2. Once a stable baseline is obtained, inject 10 µL of Milli-Q water (either manually or via an automatic injector). It is generally advisable to perform two to three blank runs to ensure proper equilibration of the column.

3.4. Sample Injection and Analysis

Once a stable baseline is obtained, inject 10 µL of the sample (either manually or via an automatic injector) and use a linear gradient from 0–100% buffer B over 60 min to elute the sample.

Figure 1 displays a chromatogram of a bovine serum albumin (BSA) tryptic digest. The tryptic map is a typical busy chromatogram. The number of peaks is astonishing and to attempt better resolution via trial and error could be a very time-consuming task. Using mass spectrometry as detection can help the identification of the peaks and establish peak purity. The problem is that many research laboratories are not equipped with a mass spectrometer.

3.5. Computer-Aided Separation Optimization

Optimization of separations and sample recovery and reproducibility for 20–30 peaks is a significant task and the optimization by experimental trial and error method can be extremely time consuming. A software package called DryLab *(2–4)* developed by LC Resources is a powerful optimization software for separation optimization. The software requires the collection of four chro-

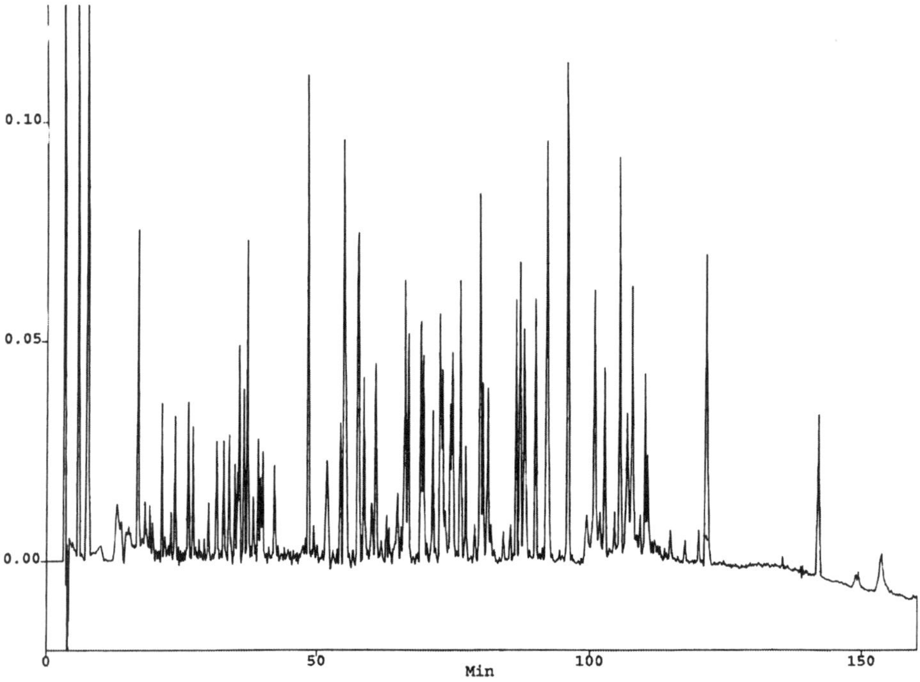

Fig. 1. Tryptic map of BSA on a RP column. Courtesy of Jerry Higgins at Higgins Analytical (Mountain View, CA).

matograms; a short (30 min) and a long (120 min) gradient at two temperatures, 20 and 60°C, respectively. After integration of the peaks, the elution times and peak areas are entered into the software. The peaks can be identified based on their integrated peak area, which remains constant during the different runs. The change in elution time as a function of gradient slope is a parameter which was first identified and studied in detail by Lloyd Snyder for separation optimization. These studies formed the foundation of DryLab, a computer-aided separation optimization software. Peak elution times can change during chromatography at different temperatures and is an important characteristic of large polypeptides. The effect of temperature on the elution time is an important variable, which can be used for resolution optimization. DryLab includes the role of temperature on the resolution of complex chromatograms as well as a series of other chromatographic parameters, such as column geometry (column length and width), particle and pore size, and flow rate. The effect of all these parameters on separation would normally be studied experimentally and in a traditional analytical laboratory a peptide map method development usually required 4–6 wk. With DryLab however, the process can be finished in 5–10 d. It is, there-

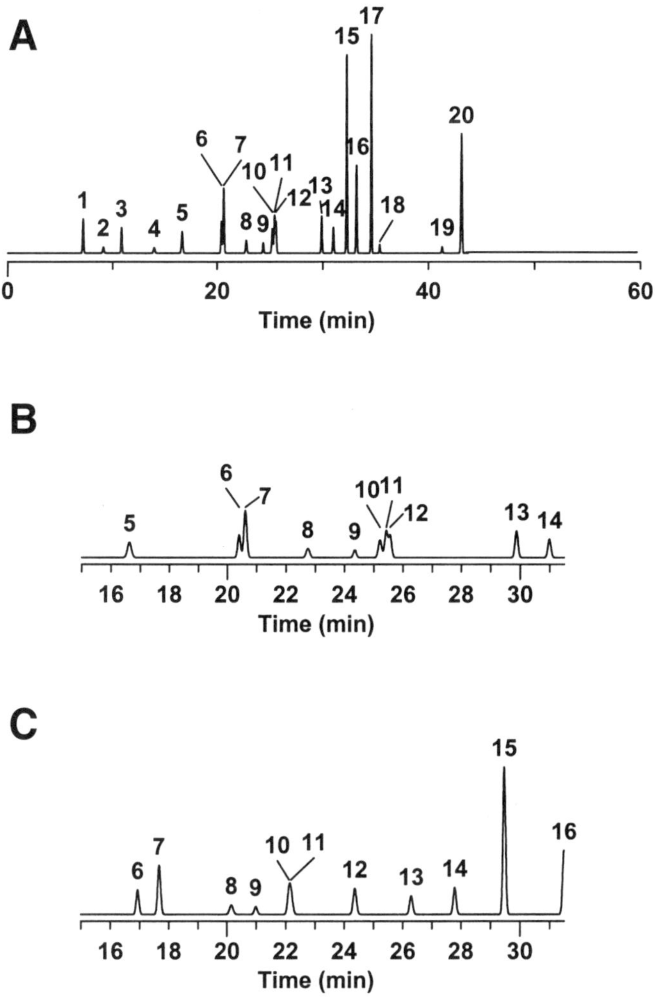

Fig. 2. Effect of temperature on the separation of tryptic peptide, a simulated chromatogram. **(A)** Conditions: column: 4.6 × 150 mm, particle size: 5 μm, flow rate: 1 mL/min, gradient time: 60 min, temperature: 20°C. **(B)** Enlargement of part of the chromatogram displayed in A. **(C)** Conditions: same as A, except gradient time: 60 min, temperature: 60°C.

fore, easy to calculate that the time and material savings accomplished by the computer-aided optimization makes the use of the software very economical.

To illustrate the use of the software, **Fig. 2A** shows a simulated chromatogram of a tryptic digest. A 60-min gradient was used and the chromatogram was calculated for 20°C. The peaks are unevenly distributed and, as the insert

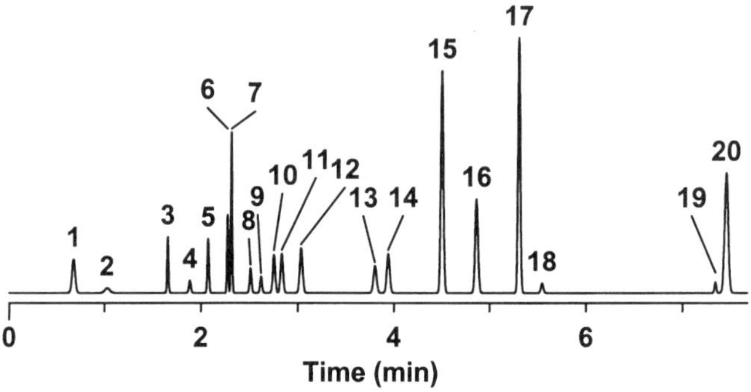

Fig. 3. Fast chromatography. Conditions: column: 2.1 × 150 mm, particle size: 3 µm, flow rate: 2 mL/min, gradient time: 12 min, temperature: 43°C.

Fig. 2B emphasizes, some of the peaks are overlapping and resolution would be considered insufficient for quantitative evaluation. The distance between some other peaks are also too large, which results in significant time wastage. **Figure 2C** displays the same sample simulation at 60°C. The resolution for most peaks is better, however peaks 10 and 11 are overlapping. **Figure 3** illustrates a model of a fast separation using higher flow rates and optimal column and temperature conditions. **Figure 4** displays the DryLab window. The upper-right side shows the resolution map, which is easily calculated by the software for all parameters. Chromatograms, which typically run between 30–60 min, can be calculated and displayed on the lower-right portion of the window in seconds. The column parameters, which can be varied, include column length and diameter, stationary phase characteristics, and so on without the need to purchase the columns and perform all the experiments. Nevertheless, although optimization software can increase productivity and efficiency, it is important to note that experimental work cannot be eliminated from method development regardless of the superiority of the computer-aided designs.

4. Notes

1. Traditionally 4.6-mm id and 15- or 25-cm columns are used. As a starting operational condition they are still recommended for use, however, in many cases a smaller bore (2.1-mm id) and shorter column length (5 cm) can provide almost identical chromatograms in a shorter period of time. (*See* Chapter 11 and Higgins Analytical Application Notes *[5]*.)

   The scope lies in the wide range of operating parameters that can be changed in order to manipulate the resolution of peptide and protein mixtures in RP-HPLC. These parameters include the immobilized ligand, the column packing geometry, the column dimensions, the ionic additive, the organic solvent, the mobile phase

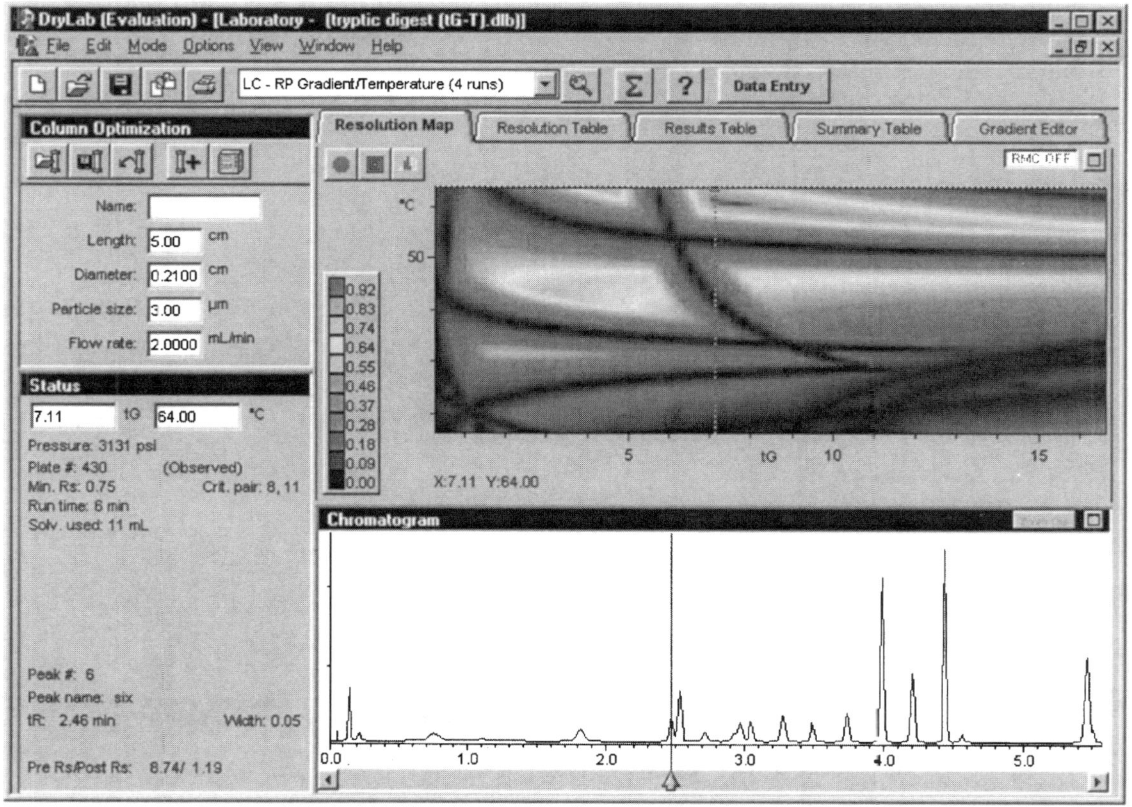

Fig. 4. DryLab Window for resolution optimization using gradient elution.

flow rate, the gradient time and gradient shape, and the operating temperature (*see* also Chapter 2).

2. Peptide mapping is usually performed in TFA containing solvents. Historically, the analyzed peptides had to be collected, evaporated, and analyzed for amino acid content. Consequently, a volatile solvent was the acid of choice. Modern-day liquid chromatography–mass spectrometry applications are also compatible with TFA, which has become a standard acid additive of mobile phases. Other acid additives which can be used and also may provide useful changes in resolution include phosphoric acid and heptafluorobutyric acid.
3. Thermostated temperature is critical in any analytical application of HPLC and in peptide mapping specifically. The retention times can be significantly altered at different temperatures ***(6)***. A systematic evaluation of temperature on analytical separations is highly recommended and can be used to alter resolution and decrease separation time ***(7,8)***.

References

1. Horváth, C. S., Preiss, B. A., and Lipsky, S. R. (1967) Fast liquid chromatography: an investigation of operating parameters and the separation of nucleotides on pellicular ion exchangers. *Anal. Chem.* **39,** 1422–1428.
2. Quarry, M. A., Grob, R. L., and Snyder, L. R. (1986) Prediction of precise isocratic retention data from two or more gradient elution runs. Analysis of some associated errors. *Anal. Chem.* **58,** 907–917.
3. DryLab bibliography, www.lcresources.com.
4. http://www.molnar-institut.com/.
5. http://www.higanalyt.com.
6. Chloupek, R. C., Hancock, W. S., and Snyder, L. R. (1992) Computer simulation as a tool for the rapid optimization of the high-performance liquid chromatographic separation of a tryptic digest of human growth hormone. *J. Chromatogr.* **594,** 65–73.
7. Hancock, W., Chloupek, R. C., Kirkland, J. J., and Snyder, L. R. (1994) Temperature as a variable in reversed-phase high-performance liquid chromatographic separations of peptide and protein samples. I. Optimizing the separation of a growth hormone tryptic digest. *J. Chromatogr. A* **686,** 31–43.
8. Chloupek, R. C., Hancock, W., Marchylo, B. A., Kirkland, J. J., Boyes, B., and Snyder, L. R. (1994) Temperature as a variable in reversed-phase high-performance liquid chromatographic separations of peptide and protein samples. II. Selectivity effects observed in the separation of several peptide and protein mixtures. *J. Chromatogr. A* **686,** 45–59.

14

Prep/Semiprep Separations of Peptides

Denis B. Scanlon and James Finlayson

1. Introduction

Peptide synthesis has undergone a major transformation in the last three decades, building on the solid-phase synthesis methodology of Bruce Merrifield first published in 1963 *(1)*. During the 1970s, the first automation of peptide synthesis was undertaken using Boc chemistry. In the 1980s, improvements were made in the Boc chemistry automated process and, consequently, the synthesis of more-difficult sequences, as well as longer polypeptides became possible *(2)*. Solid-phase Fmoc synthesis was developed in the early 1980s *(3,4)* and was also applied to automated systems *(5)*. The 1990s saw improvements in both Boc and Fmoc chemistry together with novel modes of activation of the amino acids in both chemistries *(6,7)*. The result was faster cycle times and, hence, reduced synthesis times. The range of protecting groups and resins available today means that sophisticated syntheses utilizing a combination of Boc and Fmoc chemistry are possible *(8)*.

Despite the improvements in chemistry, the possible permutations of the 20 amino acids in a peptide of approx 25–30 amino acids in length results in biomolecules with extraordinary differences in physical properties. Purification of a crude peptide mixture can hold great challenges. Apart from chemical failures resulting in deletions, terminations, or modifications of the desired sequence *(9)*, other problems such as poor solubility or tertiary structure in the presence of C_{18} reversed-phase (RP) silica require a many faceted approach to the purification of the humble peptide. There is a high demand for peptides as research tools and as potential therapeutic agents. The peptide chemist, therefore, has to possess the skill to synthesize and purify from milligram quantities to multigram quantities.

From: *Methods in Molecular Biology, vol. 251, HPLC of Peptides and Proteins: Methods and Protocols*
Edited by: M.-I. Aguilar © Humana Press Inc., Totowa, NJ

This chapter will deal with preparative and semipreparative approaches to peptide purification and will describe purifications in the range of 50–2000 mg of crude material. RP is the most frequent mode of purification and we will present buffer variations and discuss changes in the RP support. We will describe the integration of RP purification into synthesis strategies where isolation and purification of intermediates are integral to the overall success of the synthetic strategy. Whereas detailed mass spectrometric methods will not be discussed in this chapter, the importance of mass spectral data will be stressed in the analysis of crude peptide mixtures, the choice of a high-performance liquid chromatography (HPLC) purification strategy and as a final QC tool for purified peptides.

The purification process is a multistep process and the aim of this chapter will be to describe each step in a simple and logical manner (*see* **Note 1**). Some peptides can be extremely difficult to purify for diverse reasons. It is not the aim of this chapter to make the reader an instant expert. Rather, it is to provide a starting point for developing more complex purification strategies.

2. Materials

2.1. Instruments

2.1.1. Preparative System

1. Waters Delta Prep 3000 Quaternary Pumping System.
2. Waters Model 486 LC Spectrophotometer.
3. Waters 745 Data Module.
4. Isco Foxy 200 fraction collector.

2.1.2. Analytical System

1. Waters 600 Quaternary Pumping System.
2. Waters Model 486 LC Spectrophotometer.
3. Waters 717plus Autosampler.
4. Waters Millenium Software.

2.2. Columns

2.2.1. Preparative System

1. Amicon 75 × 50 mm stainless-steel column packed with Merck Lichrospher 100 RP-18 (12 μm).
2. Waters Prepak Cartridge, PrepNova-Pak HR C_{18}, 6 μm, 40 × 100 mm.
3. Waters Prepak Cartridge, PrepNova-Pak HR C_{18}, 6 μm, 25 × 100 mm.
4. Phenomonex Aqua column, 5 μm, C_{18}, 20 × 60 mm.

2.2.2. Analytical System

1. Merck Lichrocart 125-4 cartridge, C_{18}, 5 µm, 4.6 × 125 mm (crude QC, fraction analysis).
2. Merck Superspher 250-4 cartridge, C_{18}, 5 µm, 4.6 × 250 mm (final QC).

2.3. Reagents

The following is not an exhaustive list of suppliers, but care must be taken in choice of supplier for each reagent. All reagents should be of the highest possible quality.

1. Acetonitrile, Merck, HiPerSolv for HPLC.
2. TFA, Auspep, redistilled.
3. TEA, Auspep, redistilled.
4. Phosphoric acid, BDH Analar (85% aqueous solution).
5. Ammonium acetate, Sigma, SigmaUltra grade.
6. Acetic acid, Glacial, Aldrich >99%.

2.4. Buffers

In all cases Buffer A is the major aqueous component and Buffer B contains a majority of organic modifier.

1. TFA buffer, pH 2.0 ***(10)***: Buffer A: 0.1% aqueous TFA. Buffer B: 90% acetonitrile/water (containing 0.1% TFA).
2. 100 m*M* TEA.Ac, pH 5.5: Buffer A: 100 m*M* TEA titrated to pH 5.5 with glacial acetic. Buffer B: 50% acetonitrile/water containing 100 m*M* TEA titrated to pH 5.5 with glacial acid.
3. 100 m*M* ammonium acetate, pH 5.5: Buffer A: 100 m*M* NH_4OAc titrated to pH 5.5 with glacial acetic acid. Buffer B: 50% acetonitrile/water containing 100 m*M* NH_4OAc titrated to pH 5.5 with glacial acid.
4. Phosphoric acid buffer, pH 2.0: Buffer A: 0.1% aqueous phosphoric acid. Buffer B: 50% acetonitrile/water (containing 0.1% phosphoric acid).
5. TEA.Phosphate buffer system ***(11)***: Buffer A: 100 m*M* TEA, pH titrated in the range 2.5–7.5 with phosphoric acid. Buffer B: 50% acetonitrile/water containing 100 m*M* TEA, pH titrated in the range pH 2.5–7.5 with phosphoric acid.

3. Methods

Figure 1 defines the process. It starts from crude lyophilized peptide from which detailed mass spectral and analytical RP-HPLC data are obtained. Preparative conditions to achieve the separation are then defined. The final part of the process, which is equally as important as all other steps, is final QC of the lyophilized peptide. All the steps in **Fig. 1** will be discussed in turn.

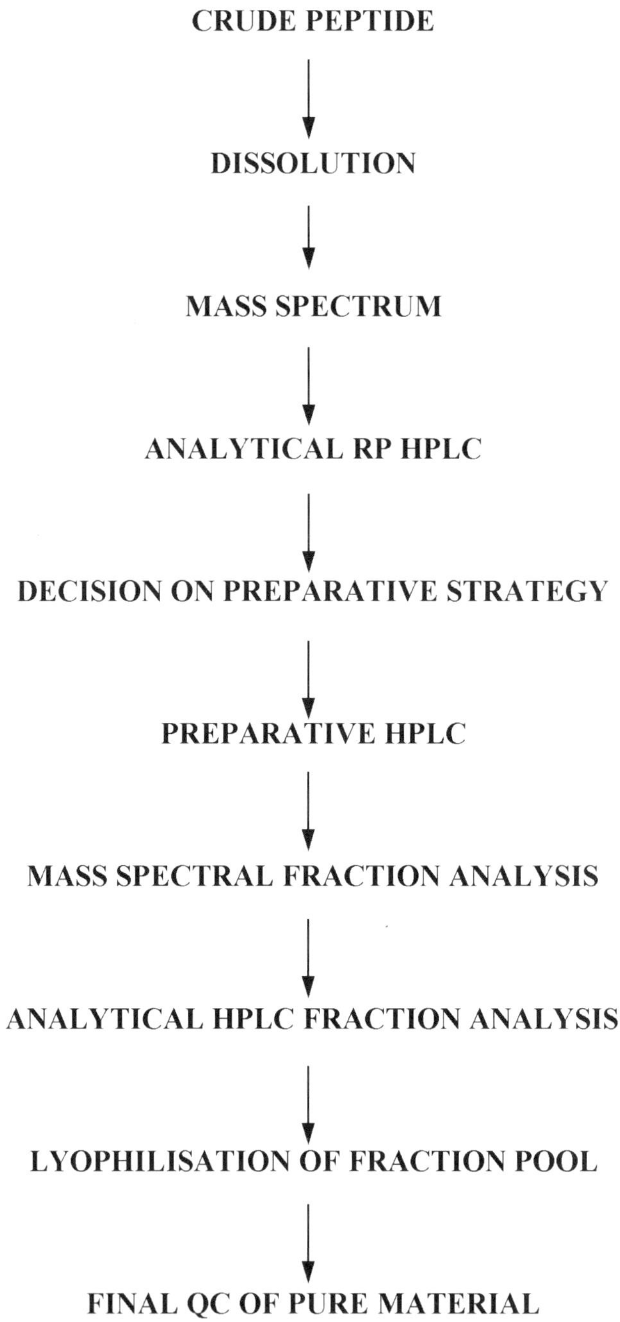

Fig. 1. Flow diagram or the process of peptide purification. Each step in the process will be discussed in turn.

3.1. Crude Sample Preparation

3.1.1. Lyophilization

For the sake of this chapter, it is assumed that the crude peptide is a lyophilized powder and the product of either a Boc or Fmoc solid-phase peptide synthesis. The sample size of crude peptide can range from 50 mg to several grams. Lyophilization removes all volatile organic scavenger from the peptide so that the crude product should be a mixture of the desired sequence together with small amounts of terminated sequences, deletion sequences, or chemically modified sequences from side-reactions during either the synthesis or the cleavage.

3.1.2. Dissolution

Before obtaining analytical data or carrying out any preparative work, the peptide must be dissolved in an aqueous buffer. Dissolution can present special problems. For ease of pumping onto a preparative column, the volume must be kept to a minimum (e.g., 50 to 200 mL). A sample concentration of around 5 mg/mL is desirable (but not crucial). This is more concentrated than the solution required to obtain analytical data (usually 0.5 mg/mL). A simple aqueous buffer is best (e.g., CH_3CN/water) as this will allow the sample to be pumped onto the preparative column without problems such as producing substantial column back pressure (causing the pumps to shut down). Simple is not always possible, however, and **Note 2** lists many options for dissolution of difficultly soluble peptides.

3.1.3. Filtering

Samples must be filtered before pumping onto any column. Failure to do this will block the porous frit end fittings on all analytical and preparative columns. Once blocked, these frits are very difficult to clean. There are many filtration membranes available commercially, but the simplest filtration device is Whatman No. 4 filter paper which acts as a 1-µm filter. Preclarification of the crude peptide solution by centrifugation is useful in cases of poor solubility. A 0.45-µ*M* membrane filter can also be used after the filter paper step but in our experience is seldom necessary.

3.2. Analytical Data

A critical step in the purification process is to obtain detailed analytical data on the crude peptide. The mass spectrum gives vital information regarding the nature of crude peptide; first, whether the crude sample has the correct molecular weight and second, the nature of the impurities in the crude peptide mixture. **Table 1** lists many of the possible deletions or modifications that can occur

Table 1
Amino Acid Modifications

Mass change	Deletion	Mass change	Modification
–186	Trp	+16	Met[O]
–163	Tyr	+22	Na+
–156	Arg	+28	Formyl
–147	Phe	+44	CO_3(fromTrp(Boc))
–137	His	+56	tBu
–131	Met	+71	Acm
–129	Glu	+80	Phosphorylation
–128	Gln,Lys	+80	Sulfation(of Tyr)
–115	Asp	+90	Benzyl
–114	Asn	+100	Boc
–113	Leu,Ile	+104	MeBenzyl
–103	Cys	+114	TFA salt
–101	Thr	+120	MeOBenzyl
–99	Val	+165	DDE
–97	Pro	+222	Fmoc
–87	Ser	+242	Trt
–79	desPhospho	+252	Pbf
–71	Ala,desAcm	+266	Pmc
–57	Gly		
–18	Dehydration(–H_2O),Aspartimide		
–17	Pyroglutamic acid from Gln		
–2	Disulphide bond formation		

[a]Possible modifications or deletions present in crude synthetic peptides. The list is far from comprehensive but attempts to document the commonly observed amino acid deletion masses or protecting group modification masses.

in peptide synthesis. The nature of deletions and modifications will influence the RP-HPLC conditions chosen for the preparative separation (*see* **Note 3**).

If the mass spectrum shows the correct molecular ion, then a C_{18} RP-HPLC trace is run, usually in TFA buffer system with an acetonitrile gradient. A typical gradient will run from 0 to 60% acetonitrile over 30 min. If the peak shape is broad or the resolution is poor then a second analytical trace in TEA.phosphate or ammonium acetate buffer (*see* the buffer list in **Subheading 2.4., item 5**) is run to optimize the separation. Peptides will generally elute between 10 and 25 min under these running conditions. If the peptide is extremely hydrophobic and is retained longer than 25 min then the retention time of the peptide on a C_4 or Cyano(CN) RP-HPLC column (less hydrophobic support) should be investigated. Superior preparative results will be obtained if the peptide elutes in the 10–25-min range analytically. Peptides eluting outside this range generally give poor recoveries from the purification process (*see* **Note 4**).

Hydrophilic peptides eluting before 10 min should be examined analytically with shallow gradients from 0 to 15% acetonitrile over 30 min. Highly end-capped C_{18} RP-HPLC columns (e.g., *see* **Subheading 3.6.4.**) retain hydrophilic peptides in cases where standard columns do not.

3.3. Prep System Setup

3.3.1. Choosing the RP-HPLC Column and Buffer

The choice of the preparative column and buffer is a direct scale-up from the results obtained in the analytical data. If a nonvolatile buffer is selected for the preparative separation, then desalting the purified peptide in volatile TFA buffer, or in some cases, acetic acid buffer (*see* **Note 5**) will be required. The dimensions of the preparative column depends on the quantity of peptide to be purified and the complexity of the crude product. A semipreparative column with the dimensions 25 × 100 mm (e.g., Waters 25 × 100-mm NovaPrepPak cartridge detailed in **Subheading 2.2.1.**) has the capacity to handle 50–200 mg of crude peptide. Larger quantities of peptide (200 mg to multigrams) are best purified on a preparative column with the dimensions 40 × 100 mm or larger (e.g., the NovaPrep-Pak detailed in **Subheading 2.2.1.**). The column capacity increases proportionally with the square of the column radius and in our experience a column with dimensions 75 × 50 mm (short and fat) can purify up to 10 g of crude peptide in a single preparative run (*see* **Note 4**). If a purification is complex, that is the crude chromatogram is complex and the integrated percentage of peptide in the crude trace is low, then results will improve at the lower levels of the theoretical capacity range of the preparative column (i.e., smaller batch amounts purified on a larger dimension column).

3.3.2. Defining the Gradient

A gradient must be set up on the software of a quaternary pumping sytem (with inlet lines A, B, C, D). Starting conditions are usually in the range 0–10% Buffer A and the gradient develops with a slow linear increase in Buffer B concentration (e.g., 0.3–0.5% Buffer B/min). Generally, a linear gradient from 0 to 50% acetonitrile is a good starting point for relatively soluble peptides, and, a gradient from 20 to 80% acetonitrile is appropriate for more hydrophobic peptides (*see* **Section 3.4.2.**). If more detail is known about the elution properties of the peptide compared to the impurities (*see* **Note 3**) then a complex gradient (e.g., nonlinear or with fast step-ups to particular Buffer B concentrations followed by a slow linear gradient) can be designed. The sample is introduced isocratically through line D prior to the commencement of the gradient.

3.3.3. Detection

The product of the gradient elution from the column will flow through a detector and the detection wavelength will be set according to the composition of the peptide. The absorbance of the peptide bond can be monitored at 214–220 nm in analytical mode, but in preparative mode the wavelength should be "tuned out" to make the detection less sensitive. The product should therefore be monitored at 225–232 nm depending on how much sample has been introduced onto the column (*see* **Note 6**).

3.3.4. Fraction Collection

The next set-up task is the programming of the fraction collector. The volume of the fractions collected will depend on the flow rate used, the size of the fraction tubes and the difficulty of the separation to be achieved. These decisions are better made before the introduction of the sample onto the column.

Typically, 0.5-min fractions are a good starting point, but the final volume depends on the difficulty of the separation to be achieved. The fraction collection can be run totally in time mode where fractions are collected over the whole preparative run, or, alternatively, in peak detection mode. Most fraction collectors can be electronically linked with the detector. When a peak is detected (i.e., the absorbance level becomes greater than a predetermined baseline level) the fraction collector switches into collect mode. When the product has eluted the absorbance will fall and the fraction collector will switch off at the same predetermined baseline absorbance level (*see* **Note 7**).

3.3.5. Equilibration of the Column

The final set-up task is equilibration of the system. Some peptides or non-peptidic material can be strongly retained on the column and failure to flush out this material may result in broad "ghost" peaks in subsequent separations. Prior to commencing the preparative run, therefore, a column wash with 100% Buffer B is performed (three column volumes) followed by re-equilibration to Buffer A conditions (five column volumes) (*see* **Note 7**). The sample introduction line D should be washed with water.

3.4. The Prep Run

3.4.1. Pumping On

Because of the volumes involved in the dissolution of peptides in the prep or semi-prep mode, the preparative system must be set up so that the sample can be pumped onto the column. Introduction of the sample through a rheodyne injector is not practical. A quaternary pumping system is required for the task. The filtered buffers are installed in lines A (aqueous) and B (organic) and the introduction of the sample is through line D.

Using the isocratic control section of the software, the peptide solution is introduced on to the column through line D as a mixture with Buffer A. The composition of this mixture will depend on the amount of organic modifier required to solubilize the peptide and the "stickiness" of the peptide on a RP column. Peptides that elute early from an analytical run (20% Buffer B or less) would need to be pumped on at 80% Buffer A/20% line D to ensure the peptide is retained on the column rather than crashing through before the gradient is commenced (*see* **Note 7**). Peptides that are better retained can be pumped on at 50% Buffer A/50%/lined D (*see* **Note 9**). Once the peptide solution is on the column, line D is washed (*see* **Note 10**) to ensure all the sample has been loaded, and, to clean line D ready for the next preparative run.

3.4.2. Starting the Gradient

If a strongly absorbing agent has been used to aid in the dissolution of the peptide (acetic acid, dimethyl sulfoxide [DMSO], or dimethyl formamide [DMF] for instance), the detector will be off-scale during the pumping on stage. After all the peptide solution has been introduced onto the column, Buffer A is pumped isocratically until the absorbance returns to baseline. Once the absorbance has returned to baseline, the gradient is started and the fraction collector set to peak detection mode. Until a peak is detected, the fraction collector diverts the flow to a waste bottle (*see* **Note 7**). Where throughput is desired, however, a robust reproducible method is required which is applicable in almost every case.

3.4.3. Monitoring the Preparative Run/Fraction Collection

Preparative systems with computer-controlled software are capable of continuous monitoring of the preparative trace and have options for running with the product peak spanning full-scale deflection on the monitor or rescaling to view small impurities. Fraction collection parameters are all controlled by the software in such systems and give the chromatographer total flexibility in monitoring the preparative trace (*see* **Note 8**).

Many older preparative systems may only have a chart recorder or data module with the facility to nominate the range of absorbance that will be plotted onto a paper chart. The absorbances suitable for preparative scale chromatography will only be those in the upper end of the range, namely either 0.5, 1.0, or 2.0 absorbance units full scale (Aufs). It is also essential that the fraction collector and the detector be interconnected such that whenever a new fraction is collected a relay is closed that produces an absorbance spike, and, therefore, a mark on the preparative trace. In this manner, the exact position of each fraction can, therefore, be determined on the preparative trace.

3.4.4. Fraction Analysis

Running full analytical RP-HPLC traces for each fraction collected is an incredibly time-consuming process and analytical HPLC time in most laboratories is usually at a premium. The preferred method to initially screen fractions is electrospray mass spectrometry. A complete set of fractions from a preparative run can be screened in less time than an analytical RP-HPLC run of a single fraction. Moreover, the quality of the raw mass spectral data can be an excellent guide to the purity of the fraction being analyzed. One must be cautious, however, as good mass spectral data does not guarantee the RP-HPLC trace will be to the required purity specification. In the case of closely eluting peaks, qualitative mass spectrometry can be used as a guide to locate the peak coresponding to the desired peptide. Based on the mass spectral data, pools of fractions or individual tubes can then be run analytically on RP-HPLC. The tubes exhibiting the required purity by analytical HPLC are pooled and lyophilized (*see* **Note 11**).

3.5. Final QC

After lyophilization, the integrity of the dry-powder product must be confirmed. There are times when a peptide is sensitive to a range of handling conditions and during the lyophilization process the purity by RP-HPLC can decrease by a small amount (or in rare cases substantially) from that assessed in a fraction pool (*see* **Note 12**). To guarantee that the final product is of the desired purity, the pure, dried peptide must be subjected to analytical RP-HPLC, mass spectrometry, and in some cases, amino acid analysis.

3.6. Examples

Analytical and preparative traces obtained in the purification of the examples below will illustrate the points made in the text of the chapter.

3.6.1. Substance P ***(12)***

Figure 2 illustrates a large scale purification from a synthesis of Substance P on a scale of 2 g. Because of the quantitiy of peptide, the Amicon 75 × 50 mm preparative column was used to enable the loading of multigram quantities each preparative run. The analytical data for Substance P (*see* **Fig. 2A**) shows that it runs with good resolution in TFA buffer and the two later eluting impurities

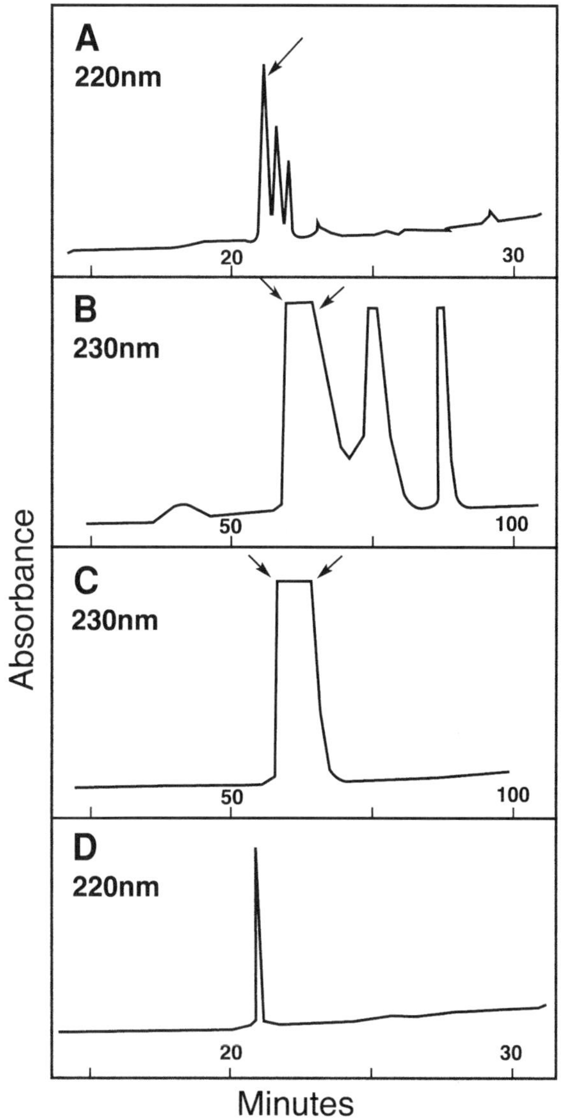

Fig. 2. Substance P: (**A**) Crude analytical trace: TFA buffer, 1.5 mL/min, 0–60%B over 30 min. (**B**) Prep trace, 2 g sample: TFA buffer, 30 mL/min, 0–50%B over 100 min. (**C**) Prep desalt, 360 mg sample: Acetic acid buffer, 30 mL/min, 0–50%B over 100 min. (**D**) Pure analytical trace: TFA buffer, 1.5 mL/min, 0–60%B over 30 min.

are well resolved. The mass spectral data for the crude peptide showed the correct ion for Substance P at m/e 1347 plus peptide impurities with Arg, Lys, and Arg+Lys deletions (data not shown). Owing to the hydrophilic nature of the Arg and Lys residues, these deletion peptides are more retained on C_{18} RP-HPLC and hence elute later in the chromatogram (*see* **Note 2**). The Substance P peak is well resolved from the later eluting impurities. A single preparative run in TFA buffer (*see* **Fig. 2B**) was shown by mass spectrometry to remove the Arg and Lys deletion peptides. The pooled fractions were assessed by analytical RP-HPLC and yielded 360 mg > 80% pure A final preparative run on NovaPrep-Pak 40 × 100 mm column, with acetic acid instead of TFA as the buffer counter-ion (*see* **Fig. 2C**), improved the purity to > 95% and exchanged the peptide counter-ion from trifluoroacetate to acetate (*see* **Note 5**). The final yield of pure Substance P (*see* **Fig. 2D**) was 140 mg > 95% pure.

3.6.2. Conotoxin GVIA (13)

Conotoxin GVI is a polypeptide toxin from the sea snail *Conus geographus* and contains three disulfide bonds. The peptide was synthesised by Fmoc solid-phase peptide synthesis, cleaved and air oxidized in 200 m*M* Tris-HCl, pH 8.3 to form the three disulfide bonds. The crude peptide (650 mg), which by analytical RP-HPLC appeared as a peak sitting on an oxidative hump (*see* **Fig. 3A**), was initially purified in TFA buffer on a PrepNova-Pak, 6 μ*M*, 40 × 100 mm column to remove the majority of impurities and bring the peptide up to 80% purity. Repeated preparative chromatography in this buffer system, from experience in our laboratory, did not improve the purity further. Analytically 25 m*M* ammonium acetate buffer, titrated to pH 5.0 with TEA, was found to shift the oxidative hump away from the main product peak. In PITC amino acid analysis, it has been found that peak shape and resolution are enhanced in ammonium acetate buffer if the pH is titrated to below pH 5.6 with TEA. This proved to be the case for Conotoxin GVIA. When passed through the same column with ammonium acetate buffer, the oxidative hump was shifted to a resolvable later eluting peak (*see* **Fig. 3B**). On lyophilization, ammonium acetate buffer is partially, but not completely, volatile. To complete the purification (*see* **Fig. 3C**), the peptide was finally desalted on the PrepNova-Pak, 6 μ*M*, 25 × 100 mm semiprep column in 0.25% acetic acid buffer to yield 25 mg of Conotoxin GVIA at a purity of >95% (*see* **Fig. 3D**).

3.6.3. Conotoxin MII (14)

Purification can be incorporated as part of a multistep synthesis strategy. This concept was applied to the synthesis and purification of Conotoxin MII, which

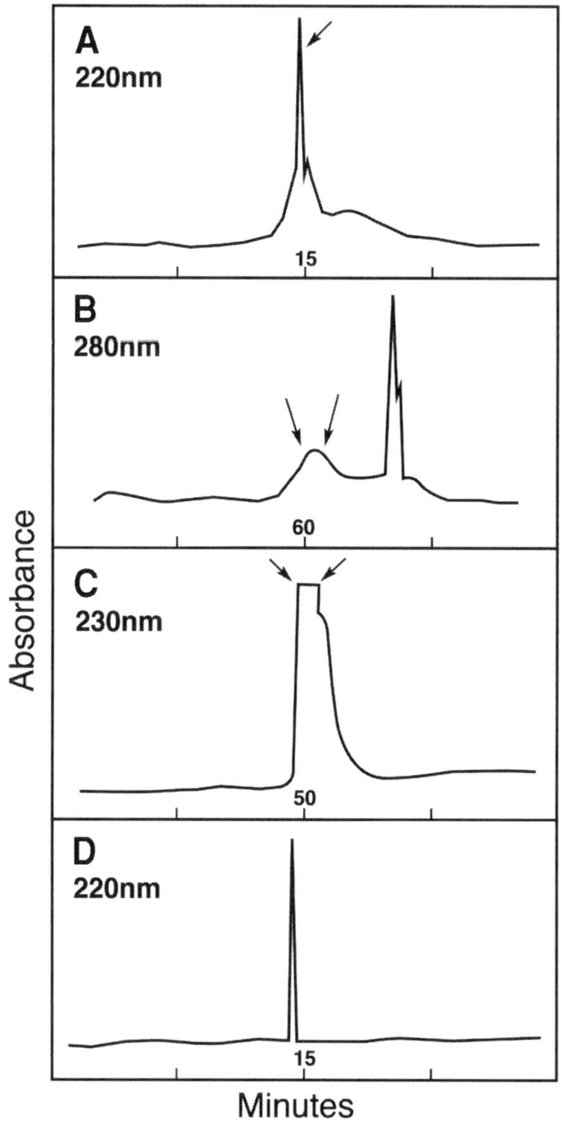

Fig. 3. Conotoxin GVIA: **(A)** Crude analytical trace (air oxidised peptide): TFA buffer, 1.5 mL/min, 0–60%B over 30 min. **(B)** Prep trace, 650 mg sample: TFA buffer, 30 mL/min, 0–15%B over 35 min, 15%B isocratic 35–45 min, 15–45%B 45–135 min, 280 nm detection. **(C)** Prep trace: 25 m*M* ammonium acetate buffer, same gradient as in (B). **(D)** Pure analytical trace: TFA buffer, 1.5 mL/min, 0–60%B over 30 min.

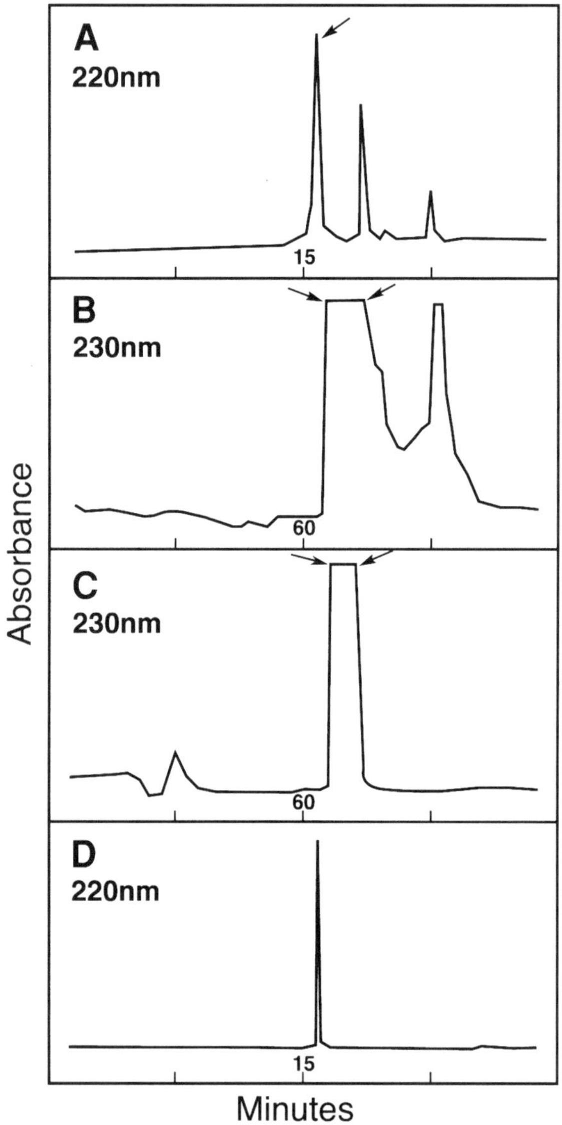

Fig. 4. Conotoxin MII: (**A**) Crude analytical trace (1 disulfide bond, 2× Cys(Acm)): TFA buffer, 1.5 mL/min, 0–60%B over 30 min. (**B**) Prep trace, 1.8 g sample (1 disulfide bond, 2× Cys(Acm)): TFA buffer, 30 mL/min, 0–50% B over 100 min. (**C**) Prep trace (2× disulfide bonds), 180 mg sample: 25 m*M* ammonium acetate buffer, 30 mL/min, 0–12%B over 30 min, 12%B isocratic 30–40 min, 12–40%B 40–100 min. (**D**) Pure analytical trace: TFA buffer, 1.5 mL/min, 0–60%B over 30 min.

is a complex peptide from the venom of the sea snail *Conus magus* and has two disulfide bonds. The disulfide bonds were formed sequentially and selectively by the removal of two pairs of orthogonal Cys protecting groups, namely two Acm groups and two Trityl groups. The Trityl groups were removed during the postsynthesis cleavage of the peptide and the first disulfide bond formed by air oxidation (*see* **Fig. 4A**). The Acm-protected peptide (1.8 g) was purified at this point on a PrepNova-Pak, 6 μ*M*, 40 × 100 mm column in a volatile TFA buffer system to yield 180 mg of monodisulfide-Acm-protected peptide (*see* **Fig. 4B**). This process removed any impurities or excess reagent which could potentially interfere with the oxidative formation of the second disulfide bond.

The second disulfide bond was formed by oxidative removal of the Acm protecting groups. The product of this process was again purified on the PrepNova-Pak, 6 μ*M*, 40 × 100 mm column but with 25 m*M* ammonium acetate buffer pH 5.0, which removed the oxidative impurities in the same manner as that described for Conotoxin GVIA (*see* **Fig. 4C**). To complete the purification, the peptide was finally desalted on a PrepNova-Pak, 6 μ*M*, 20 × 100 mm semiprep column in 0.25% acetic acid buffer to yield 75 mg of Conotoxin MII >95% pure (*see* **Fig. 4D**).

3.6.4. c(RSRNR)

Cyclic(RSRNR) is an extremely hydrophilic cyclic pentapeptide which is poorly retained on C_{18} RP silica HPLC columns. **Figure 5A** shows an analytical trace of c(RSRNR). The early eluting nature of the peptide (9 min) made resolution of the minor back impurity (10 min) extremely difficult. The standard NovaPrep-Pak semiprep column could not retain either the cyclic peptide or the later eluting impurity. The Phenomonex Aqua semiprep column (*see* **Subheading 2.2.1.** for details), however, retained both the product and the impurity on the column for more than 25 min and effected resolution of the two components (*see* **Fig. 5B**). The loading of crude peptide was 60 mg and the yield of the cyclic pentapeptide was 15 mg > 90% pure (*see* **Fig. 5C**).

4. Notes

1. Preparative systems require a "hands-on" approach to ensure that valuable sample is not lost because of equipment malfunction. Computer driven automated systems are available, however, but will not be described as the software and hardware complexities are beyond the scope of this chapter.
2. The first buffer to try when dissolving a crude peptide is 30% CH_3CN water. If this is unsuccessful, then the following is a short list of buffers that may be useful for dissolution of insoluble peptides. As a general rule, these buffers are added to

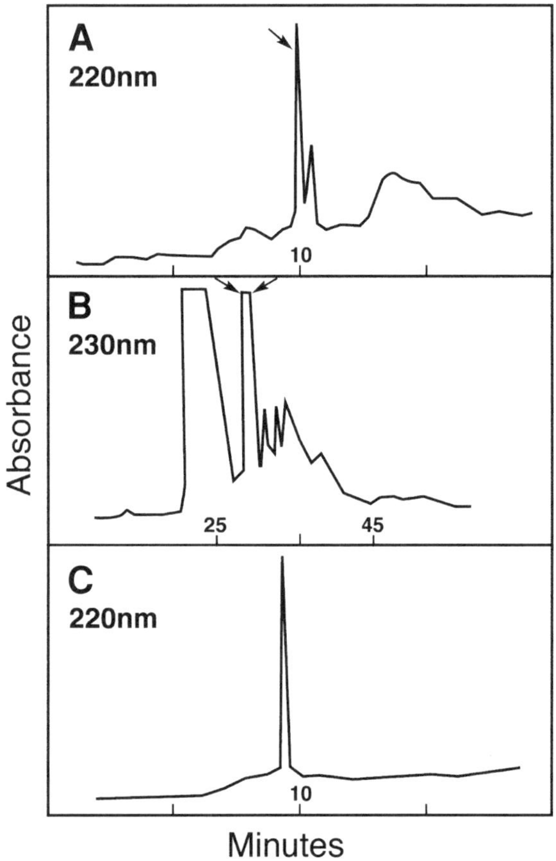

Fig. 5. c(RSRNR): **(A)** Crude analytical trace: TFA buffer, 1.5 mL/min, 0–60%B over 30 min. **(B)** Prep trace, 60 mg sample: Phenomonex Aqua semiprep column, 0–25%B over 50 min. **(C)** Pure analytical trace: TFA buffer, 1.5 mL/min, 0–60%B over

the peptide in a concentrated form and then diluted out with water to the point where the peptide just starts to become cloudy.

a. 50% AcOH/25% water/25% CH_3CN.
b. Formic acid (neat).
c. 200 m*M* NH_4HCO_3 containing 20% CH_3CN, titrated with AcOH to pH 6.0.
d. 500 m*M* Na_2HPO_4 containing 20% CH_3CN, titrated with phosphoric acid to pH 6.0.
e. DMSO or DMF.

3. The mass spectrometer is an extremely sensitive tool for determining the range of impurities in a crude sample. Modifications of the peptide during the synthesis will elute at different times on RP-HPLC. The following are some rules of

thumb for predicting where the modifications detailed in **Table 1** may elute with respect to the desired peptide:

a. Hydrophobic amino acid deletions (e.g., I,V,L,F,M) will elute earlier.
b. Hydrophilic amino acid deletions (e.g., R,K,H) will elute later.
c. Met[O] is a common modifiication which will elute earlier and tBu adducts from the cleavage will elute later.
d. Dehydrations caused by aspartimide formation (e.g., in peptides containing the sequence D-G or D-S) will appear as a closely eluting back peak or shoulder. A number of alternate buffers must be tried to optimize the separation in such cases. Ion exchange is also an option as a negative charge has been lost in the dehydration.
e. Deletions of residues such as G,A,S,T,C,N,Q, which are readily identifiable by mass spectrometry can coelute with the desired peptide on RP-HPLC. A range of buffers and/or column packings should be tried to maximize the resolution of the deletion, but, in cases of major deletions, resysnthesis can be the best option.

4. To maximize yields, purification protocols should be devised so that the minimum number of passes through a preparative column produces the desired result. Care should be taken with extremely hydrophobic peptides, as the higher loading compared to the analytical system can, in some cases, cause the peptide to smear off the column at high acetonitrile concentrations or not come off the column at all.
5. In most cases, TFA buffer will be the initial buffer of choice because of the resolving power and the fact it is volatile on lyophilisation. Where nonvolatile salt buffers are used (e.g., phosphate buffers) a second desalting prep in volatile buffers (e.g., TFA or acetic acid) is necessary prior to lyophilization. Some biological assays, however, are sensitive to TFA being associated with a peptide as a counterion. In these cases TFA must be exchanged by acetic acid. This is most conveniently achieved by a final preparative step with 0.25% AcOH/water as Buffer A instead of 0.1% TFA *(15)*.
6. If Tryptophan is contained in the peptide then 280 nm can be used as a detection wavelength as the indole moiety produces an absorbance maximum at this wavelength. Similarly Tyrosine has an absorbance maximum at 275 nm and the detection can be set at this wavelength if Tyrosine is contained in the peptide. Care must be taken in the choice of detection wavelength as an unexpected low absorbance response in the preparative run can mean that the fraction collector is not triggered and the product flows into the waste container.
7. Collect all eluent from the column! At times, peptides will unexpectedly not be retained (re-equilibration problems), or not elute in the fractions collected but in the "ramp-off" during the re-equilibration of the column. If all eluent from the column is collected then the peptide cannot be lost.
8. Do not expect the preparative trace to model the analytical trace. The preparative trace will be broad and most likely off-scale. Fraction analysis will confirm that resolution is occurring although it is not obvious from the trace.

9. In our experience, it is always better to dilute the sample when pumping on with Buffer A and to be cautious in choosing the percentage of Buffer A as diluent when pumping on.
10. It is often necessary to wash the line initially with a small volume of the solubilizing solvent to prevent peptide precipitation in the line before reaching the column before stepping back to water (or 10% acetonitrile depending on the gradient starting buffer conditions as described in **Subheading 3.4.2.**).
11. Sophisticated preparative systems are available where the column eluent passes through a flow splitter and is continuously monitored by mass spectrometry for a particular peptide molecular weight. Once the peptide molecular weight is detected, the fraction collector is triggered and the product is collected. Analytical RP-HPLC traces are then run to assess the purity of the fractions.
12. During lyophilization, the peptide is susceptible to several oxidative processes. Peptides containing Met can oxidize unexpectedly. Peptides containing Cys can dimerize. Peptides containing multiple Cys can cyclize internally or form disulfide-linked oligomers. The purity should be rechecked following lyophilization.

Acknowledgments

We thank Dr. Philip Mack and Paul Jones for constructive suggestions and careful of editing of the text. We also thank Trevor Smith for invaluable help with the diagrams.

References

1. Merrifield, R. B. (1963) Solid phase synthesis I. The synthesis of a tetrapeptide. *J. Amer. Chem. Soc.* **85,** 2149–2154.
2. Clark-Lewis, I., Aebersold, R., Ziltener, H., Schrader, J. W., Hood, L. E., and Kent, S. B. H. (1986) Automated chemical synthesis of a protein growth factor for hemopoietic cells, Interleukin-3. *Science* **231,** 134–139.
3. Brown, E., Sheppard, R. C., and Williams, B. J. (1983) Peptide synthesis. Part 5. Solid-phase synthesis of [15-Leucine] little gastrin. *J. Chem. Soc. Perkin Trans.* **I,** 1161–1167.
4. Scanlon, D. B., Eefting, M. A., Lloyd, C. J., Burgess, A. W., and Simpson, R. J. (1987) Synthesis of biologically active transforming growth factor-alpha by fluorenylmethoxycarbonyl solid phase peptide chemistry. *J. Chem. Soc. Chem. Commun.* 516–518.
5. Sheppard, R. C. (1986) Modern methods of solid-phase peptide synthesis. *Science Tools* **33,** 9–16.
6. Schnolzer, M., Alewood, P., Jones, A., Alewood, D., and Kent, S. B. H. (1992) *In situ* neutralization in Boc-chemistry solid phase peptide synthesis. *Int. J. Peptide Protein Res.* **40,** 180–193.
7. Angeletti, R. H., Bonewald, L. F., and Fields, G. B. (1997) Six-year study of peptide synthesis. *Methods Enzymol.* **289,** 697–717.
8. Dawson, P. E. and Kent, S. B. H. (1993) Convenient total synthesis of a 4-Helix TASP molecule by chemoselective ligation. *J. Amer. Chem. Soc.* **115,** 7263–7266.

9. Smith, A., Young, J. D., Carr, S. A., Marshak, D. R., Williams, L. C., and Williams, K. R. (1992) State-of-the-art peptide synthesis: comparative characterization of a 16-mer synthesized in 31 different laboratories, in *Techniques in Protein Chemistry III* (Angeletti, R., ed.), Academic, San Diego, CA, pp. 219–229.
10. Bennet, H. P. J., Browne, C. A., and Solomon, S. (1980) The use of perfluorinated carboxylic acids in the reversed-phase HPLC or peptides. *J. Liquid Chromatogr.* **3,** 1353–1365.
11. Rivier, J. E. (1978) Use of trialkyl ammonium phosphate buffers in reverse phase HPLC for resolution and high recovery of peptides and proteins. *J. Liquid Chromatogr.* **1,** 343–366.
12. Tregear, G. W., Niall, H. D., Potts, S. E., Leeman, S. E., and Chang, M. M. (1971) Synthesis of Substance P. *Nature (Lond.) New Biol.* **232,** 86–89.
13. Olivera, B. M., McIntosh, J. M., Cruz, L. J., Luque, F. A., and Gray, W. R. (1984) Purification and sequence of a presynaptic peptide toxin from Conus geographus venom. *Biochemistry* **23,** 5087–5090.
14. Shon, K. J., Koerber, S. C., Rivier, J. E., Olivera, B. M., and McIntosh, J. M. (1997) Three-dimensional solution structure of alpha conotoxin MII, an alpha 3 beta 2 neuronal nicotinic acetylcholine receptor-targetted ligand. *Biochemistry* **36,** 15,693–15,700.
15. Gabriel, T. F. (1987) Simple, rapid method for converting a peptide from one salt form to another. *Int. J. Peptide Protein Res.* **30,** 40–43.

15

Large-Scale Protein Chromatography

Joseph Bertolini, Peter Gomme, and Patrick Thomas

1. Introduction

The scale-up of a chromatographic process from the laboratory involving the processing of milliliters of material, to pilot and manufacturing scale involving the processing of liters or hundred liters of material, requires chromatographic principles to be applied within the boundaries imposed by economic, regulatory, and engineering constraints. In fact, the actual chromatographic process used at large scale is dictated equally by these factors as on the effectiveness of the process for purifying the product of interest.

1.1. Principles of Scale-Up

Chromatography is readily linearly scaleable. The key principles that need to be adhered to are:

1. Column length should be kept constant. Increased capacity is achieved by increasing column diameter.
2. Flow rate, expressed in terms of linear velocity (cm/h) is kept constant.
3. Volume of sample and buffers (elution, equilibration) is kept in proportion to the bed volume.

Resin beads are also scaled-up. Beads are available in various diameters. As a general rule, the resolution of a column is, in theory, approx constant with the number of beads. Therefore, a 1-mL column containing 10-µm beads has the same number of beads as a 1-L column with 100-µm beads, which should give as good resolution using 1000 times the amount of protein. Resolution is a function of the accessible bead surface area and the rate of diffusion of proteins into the bead.

One problem associated with production scale columns is that the weight of the resin may cause distortion to the packing. At laboratory scale, the column

From: *Methods in Molecular Biology, vol. 251, HPLC of Peptides and Proteins: Methods and Protocols*
Edited by: M.-I. Aguilar © Humana Press Inc., Totowa, NJ

walls have a considerable supportive effect that is not present in larger diameter columns. Thus, wider diameter columns have the tendency to become distorted and compacted in the middle, forming a concave surface. This can result in differential flow rates across the resin leading to a reduction in column resolution. For this reason, the bed heights of large columns are maintained at levels generally less than 50 cm. In fact, it is more common with large surface area columns not to exceed 20 cm. If a longer length is required, individual columns can be connected in series.

1.2. The Practical Impact of Chromatographic Parameters on Scale-Up

When developing a chromatographic process, there should be an awareness from the beginning that the process is intended to be scaled-up and that the choice of operating parameters will impact on the practical implementation of the process. This can affect equipment requirements, reagent requirements, and the time and staff required to run the process.

The product yield of the developed process and the number of steps in the process has to be critically assessed. While at laboratory scale, a low-yield process may be tolerated because extra batches can be quickly manufactured, this is usually not the case once a large-scale process is in place. It would represent a drain on time and resources.

The number of chromatographic steps required in a process to effect purification needs to be minimized to limit the expense of column hardware and resin. Furthermore, an increased number of steps can lead to lower recoveries of the desired product. Three steps of 90% efficiency would result in an overall yield of 73%.

It is advantageous to minimize column size as it can have significant cost and operational implications. Contributing to this include: column housing and associated infrastructure, resin requirements, processing times, buffer requirements, and pump requirements.

The functional binding capacity of a resin impacts on the amount of resin required and, in turn, on the size of the column housing. Optimization of resin performance is therefore important. This involves choosing the extent of resin capacity utilization, and operation conditions such as flow rate, loading protein concentration, pH, and conductivity. In some instances, increased capacity may be gained by enabling the protein of interest to pass unimpeded while focusing the chromatographic conditions on removing contaminant proteins present in lower concentrations.

Increased throughput is desirable and can be achieved by utilizing increased flow rate through the process, provided there is no adverse affect on binding and resolution. However, there are other issues that need to be considered. This includes the fact that pressure capabilities of large-scale columns could be exceeded and suitable pumps are required. The choice of pumps can be critical as the shear that is imparted during pumping can affect the integrity of the protein in the solution. A number of different pump types are available including peristaltic, membrane, and lobe pumps which should be evaluated.

The type of resin chosen for a process becomes more critical with scale-up. For example, if the resin is to be used in the manufacture of pharmaceutical product, the resin should have suitable specification, with respect to leachables, and should exhibit acceptable stability, especially in face of repeated treatment with sanitizing and regeneration solutions such as high concentrations of NaOH.

2. Materials

2.1. Resin

The choice of resin is a key determinant of the performance of a chromatographic process. The technical capabilities of a resin can have considerable economic and regulatory implications during the scale-up of a chromatographic process.

The cost of resins can vary considerably. However, the properties of the resins offered should be considered. Thus, information on: allowable flow rate, dynamic binding capacity, cleanability, performance with repeated use, functional group stability, existence of leachables, and regulatory support need to be assessed (*see* **Note 1**).

2.2. Columns

Modern process scale columns are designed to handle the requirements of large-scale chromatographic purification. Today's chromatographic columns are convenient to use, scalable, and reproducible. Columns can vary in dimensions from 280 mm up to 2000 mm and can handle pressures up to 5 bar, thereby allowing the utilization of high flow rates during processing. Construction materials include 316L stainless steel or polyethylene, which are resistant to a wide range of chemicals and temperatures. This allows cleaning-in-place sanitization procedures to be used. The products are usually available with full validation support files which are essential if they are to be used in the manufacture of a registered biopharmaceutical product.

Manufacturers provide a guide to solvent resistance of the materials used in the construction of their columns. Clearly, a column may be exposed to various chemicals in different processes, especially during the course of regeneration and sanitization (*see* **Note 2**).

Preventing batch-to-batch contamination is a major concern in pharmaceutical manufacture and requires the sanitization of resins with respect to bacteria, virus, and prions (*see* **Note 3**).

2.3. Chromatographic Systems

Developers of chromatographic processes at laboratory scale would be well acquainted with chromatographic systems such as AKTA (Pharmacia), Bio-Cad (Perseptive Biosystems), and Biologic (Bio-Rad).

Systems applicable to pilot-scale work are also available. For example, Bioprocess systems (Pharmacia) and K-Prime systems (Millipore). These systems allow computerized and automated operation of chromatographic processes. The software provided in these systems capture how runs were performed with respect to sequence details. This can include flow rates, valves activated, and time that buffers were pumped.

Vendors can provide extensive validation support with respect to the control system (hardware and software), and with respect to the equipment encompassing electronic components and items such as valves, pumps, and tubing. Compliance with GMP is an important requirement if the unit is required for pharmaceutical manufacture (*see* **Note 4**).

3. Methods

3.1. Column Setup

A typical arrangement for a chromatographic procedure is presented in **Fig. 1**. A key feature is the bubble trap to prevent air entering the column, pressure gage to monitor pressure onto the column and a bypass line for loading product. Absorbance detectors are located after the column to allow monitoring of protein binding and elution. Conductivity and pH meters can also be included to confirm buffer changes that occur during equilibration, sample loading, elution and column regeneration. With integrated chromatographic control systems, the monitoring and display of data is achieved by computer based systems.

3.2. Column Preparation and Assessment

One of the key determinants of column performance is the satisfactory packing of the column. Clearly, even packing of the resin within the column and the avoidance of channels and bubbles in the packed resin will contribute to maximal resolution and reproducibility of performance (*see* **Note 5**). Below is described a procedure for packing process columns. Typical of these are the BPG series from Pharmacia which range from 100 to 450 mm in diameter. The procedure for determining the quality of packing through height equivalent to a theoretical plate (HETP) and asymmetry factor calculation is also presented (*see* **Note 6**).

3.2.1. Column Packing

1. Wet the column nets with 20% ethanol to eliminate trapped air. Use the spirit level to check the column is standing horizontally.

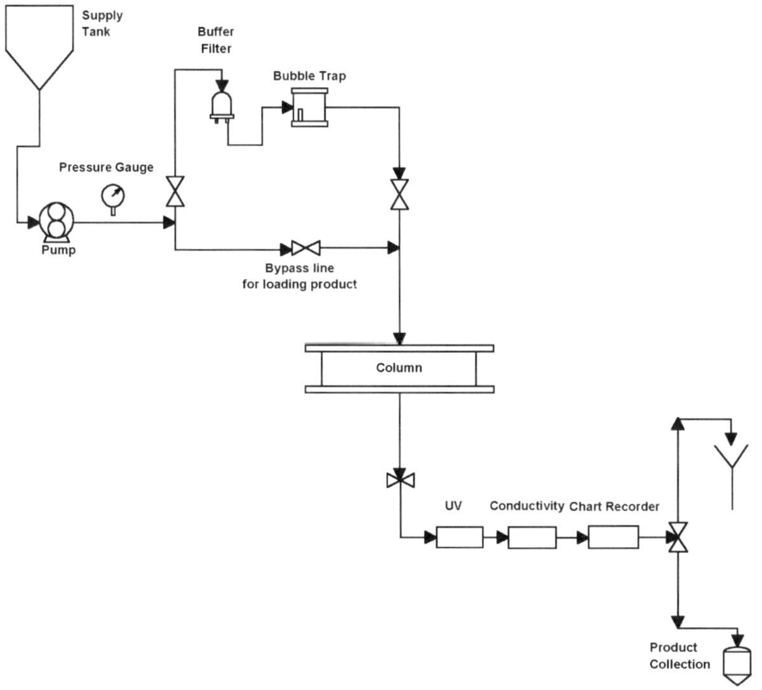

Fig. 1. Chromatographic system setup.

2. Fill the column with 10–15 cm of the packing buffer. Drain the outlet tube to remove all air bubbles. Continue draining until there is only 2–3 cm of buffer remaining in the column.
3. Mix the packing buffer with the media to form a 75% slurry.
4. Ensure that the column outlet is clamped. Carefully pour the homogeneous slurry into the column, making sure there are no air bubbles trapped.
5. Mix the slurry and allow the media to settle.
6. Insert the adaptor and secure the column adaptor in place using the four bolts.
7. Position the adaptor 1–5 mm below the liquid level, using the large handle.
8. Seal the adaptor O-ring by turning the adjuster knob on top of the adaptor clockwise. Do not overtighten.
9. Attach appropriate length tubing to the adaptor.
10. With the O-ring sealed, lower the adaptor and fill it with buffer to expel any air in the adaptor tube. The adaptor tube is full when the buffer creates a meniscus and no further air bubbles appear.
11. Place the pump inlet tubing into the packing buffer. Start the pump at a low flow rate.
12. Remove air bubbles in the pump tubing.

13. While pumping at a low flow rate, connect the tubing to the top of the column. Unclamp the bottom of the column. Slowly increase the pump rate to achieve the desired column packing pressure.
14. Adjust the pump rate if necessary. The back pressure of the bed will increase and the flow rate decrease as the resin packs.
15. Ensure that the pressure does not exceed the column packing pressure.
16. When the bed has settled, note the packed bed height.
17. Stop the pump. Clamp the bottom of the column.
18. Loosen the adaptor O-ring seal by turning the adaptor knob on top of the column counter clockwise.
19. Lower the adaptor quickly by turning the large handle clockwise. The buffer between the sedimented bed and the adaptor passes between the adaptor O-ring and the column wall. During this operation, the bed will begin to rise. Stop lowering the adaptor when it is 0.5–1 cm above the bed.
20. Seal the adaptor O-ring by turning the knob clockwise. Unclamp the bottom of the column and start the pump.
21. Adjust the pump rate to the packing pressure. The bed will be compressed again to its previous bed height.

Note: When there is a maximum of 1 cm of liquid remaining do not open the adaptor O-ring. Instead, unclamp the bottom column tubing and use the large handle to screw the adaptor down the last remaining centimeter. Never lower the adaptor more than a maximum of 5 mm into the bed.

22. Remove any excess buffer or slurry from the top of the adaptor.
23. When not in use store the column in appropriate buffer.

3.2.2. Evaluation of Packing

Before use, the HETP and the asymmetry factor of the packed column must be checked (*see* **Figs. 2** and **3**). The column is tested as follows:

1. Equilibrate column with 0.01 *M* NaOH.
2. Load 0.5 *M* NaOH of volume equivalent to 2.5% of the bed volume of the column.
3. Pump 0.01 *M* NaOH through the column.
4. Monitor conductivity of column eluate with in-line conductivity meter connected to a chart recorder until conductivity peak is observed.
5. Subject peak to analysis described below to calculate HETP and asymmetry factor. An HETP of <0.06 cm is usually acceptable.

3.2.3. Trouble Shooting

Below is a brief guide to some common problems in chromatography.

3.2.3.1. High Back Pressures

1. Valves: Check all valves are fully open between system and collection, are they clean and free of blockages?

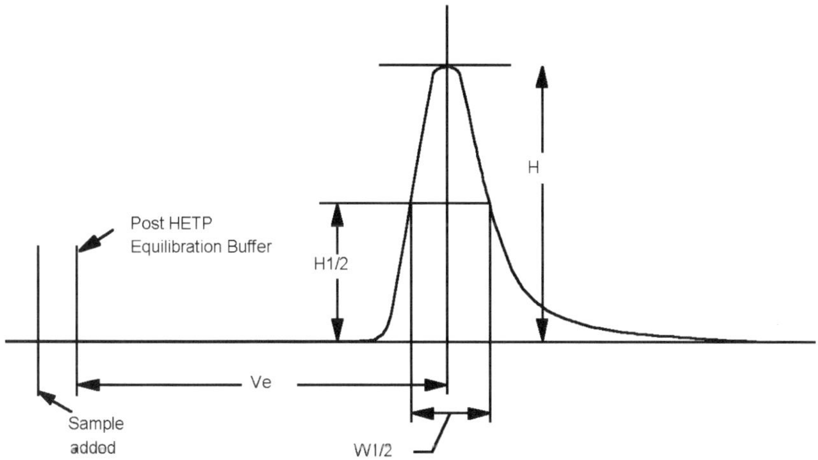

$$\text{HETP} = \frac{L}{N}\ \text{cm}$$

where L = column length in centimetres
N = number of theoretical plates

$$N = 5.54\left(\frac{Ve}{W_{1/2}}\right)^2$$

Method 1

Where Ve = elution volume in liters measured as the volume immediately after the start of post HETP equilibration buffer loading and $W_{1/2}$ = width at half peak height.

Peak height is measured as the vertical distance between the maximum points of the peak to the line tangential to the points of inflection of the trace.

Method 2

Where the equivalent distance of Ve and $W_{1/2}$ in cm can be measured on the chromatogram as they are directly related via chart speed.

Example

Given that: $Ve = 7.330$ cm
$W_{1/2} = 0.845$ cm

Therefore,

$$N = 5.54\left(\frac{Ve}{W_{1/2}}\right)^2$$
$$= 5.54\left(\frac{7.330}{0.845}\right)^2$$
$$= 5.54(8.684)^2$$
$$= 5.54(75.41)$$
$$N = 418$$

Hence, HETP
$$= \frac{L}{N}\ \text{cm}$$
$$= \frac{15}{418}$$
$$= 0.036\ \text{cm}$$

Fig. 2. Calculation of HETP.

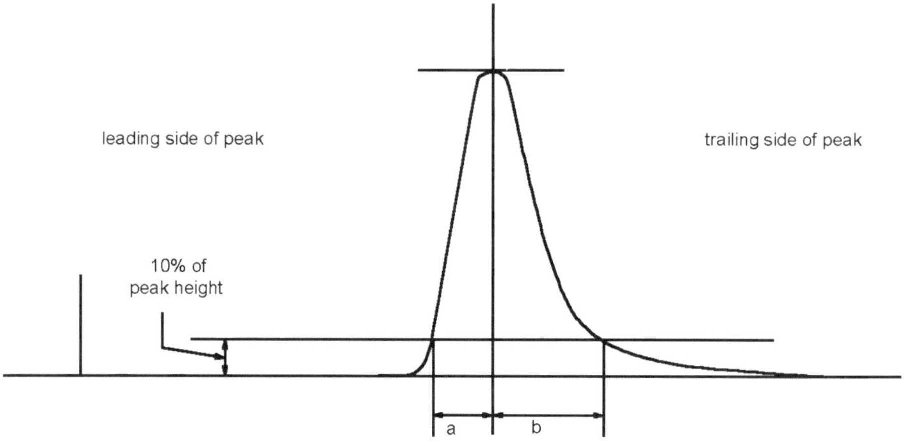

Where *af* = asymmetry factor
a = leading side of the peak at 10% height
b = tailing side of peak at 10% height
h = peak height
af = b/a
af >1 = tailing
af <1 = leading

Fig. 3. Asymmetry calculation. The asymmetry factor should be as close as possible to "1."

2. Blockages: Could there be sources of other blockages, e.g., column net, check for air and that they were wet and washed with 20% ethanol prior to use. Is there any other equipment before or after column that could generate back pressure?
3. Media: Is the media regenerated properly, or is the medium too tightly packed.
4. Internal System Diameters: Are there any changes in internal diameter of tubing, e.g., smaller tubing leading from the column or is the correct sized flow cell being used.

3.2.3.2. Column Leakage

1. Connections: Are end pieces and flanges properly positioned? Are connections from different manufacturers and are they all compatible? Have the correct torque specifications been used?
2. Compatibility: Are all components compatible with solvents, e.g., o-rings, gaskets and tubing?

3.2.3.3. Uneven Chromatograms

1. Flow: Is the linear flow rate as intended (remember cm/h or mL/min).
2. Sample: Has sample handling been altered? Were there any changes to process or any process deviations (filtration, precipitation, incubation times, pH, or temperatures) that may have affected the chromatogram?

3. Buffers: Are buffers properly prepared, consider temperatures, and date of expiry.
4. Dead Volumes: Were all dead volumes filled with the correct buffers or could back mixing have occurred at any stage in the column setup, e.g., air traps?

3.2.3.4. HETP Runs Have Tailing or Leading Peaks

1. Check the column is properly equilibrated in the HETP buffer by checking outlet conductivity.
2. Run a reverse flow HETP. If the peak is not a mirror image of the forward flow HETP, then this indicates a problem due to poorly packed column. Before repacking, try several cycles of reverse flow followed by forward flow at column packing rate to shift any uneven areas in the bed and recheck HETP.
3. If the HETP chromatograms have different symmetries in the forward and reverse directions, then this may indicate a dilution effect prior to loading. Check all connections, especially the valves, for any leakages.

3.2.3.5. Air Trapped in Column

1. Buffers: Are buffers and process at the specified temperature?
2. Connections: Are all connections fitted properly, especially pump where air may leak into the column from the suction side of the pump?
3. Valves: Are all closed valves shut tight?
4. Air-trap: Is the air trap volume sufficient?

4. Notes

1. Resins. One of the major differences between resins is the composition of the base matrices. The rigidity of the matrix determines the flow-rate capabilities of the resin. In addition its stability in the presence of a number of chemical agent can determine the useful life-span of the resin. For example, NaOH is typically used with chromatographic processes to sanitize and clean chromatographic media because it inactivates bacteria, viruses, destroys endotoxin, and solubilizes lipids and hydrophobic proteins. However, not all matrices are equally stable to repeated exposure to NaOH. Thus, if NaOH is to be used, an appropriate choice of resin must be made. Alternatively, a different sanitization regime must be used. Resins are, however, generally resistant to a range of acids, detergents, organic solvents, and chaotropic agents. In fact, the limiting factor can often be resistance of column material itself to various chemical substances. This is discussed in **Note 2**.

   One of the key specifications of a resin is its dynamic binding capacity. This is taken as the loading to the point where absorbance increases by between 5–10%. This indicates a degree of saturation of binding sites resulting in the drop-through of protein. Dynamic binding capacity is often affected by flow rate, with increasing flow rates resulting in reduced dynamic binding capacity. This can be explained by the fact that within the resin diffusion is involved in the transport of protein molecules to binding sites. During the shorter loading period required at the higher flow rates, diffusion does not achieve the mass transfer from the feedstock as would be achieved with the longer loading period associated with slower flow rates.

It is important to examine the performance of the resin across the projected lifetime of the resin. This information is generally not available from resin manufacturers and must be defined for the target process. This involves scaling down the process and subjecting the resins to repeated cycles of loading and regeneration. Total binding capacity is a useful parameter to monitor. This is more convenient to perform than dynamic binding capacity and provides a measure of the total number of functional groups present on the resin beads. Any decrease in total binding capacity would be reflected in a decrease in dynamic binding capacity. Total binding involves loading a model protein (e.g., albumin for anion exchange resins) to saturate the functional groups on the resin. The bound molecules are then eluted and quantitated to derive a measure of the total binding capacity of the resin. Other parameters that can be monitored include: HETP (discussed below), leaching of functional group and matrix components, packability of the resin and evidence of fouling of the resin.

Another important factor in resin selection is availability of the resin over the projected lifetime of the product. Therefore, due consideration should be given to the commercial viability of the manufacturer and its commitment to the product.

Additionally, it is critical that the manufacturer has regulatory support documentation for the resin. This will greatly facilitate the acceptance by the regulatory authorities of a process utilising the chosen resin.

2. Columns. Process scale columns generally are compatible with most cleaning in place solvents, such as NaCl (up to 0.5 *M*) and 2 *M* sodium hydroxide for up to 4 h. Detergents (such as Triton X-100) and chaotrophic agents (up to 8 *M* urea) can be safely used with most bioprocess columns, as can mild concentrations of acids (acetic acid up to 1.7 *M*), trifluoroacetic acid (TFA) up to 0.1%, and nitric acid (up to 0.1 *M*). Alcohols are compatible, such as ethanol (up to 20%) and isopropyl alcohol (up to 30%).

   Some materials are used for particular applications. For example, polyethylene columns are more suitable for conditions of high salt or contact with strong acids (such as hydrochloric acid). Stainless steel is compatible with up to 40% ethanol.

3. Sanitization of Columns. Removal of microbial agents from chromatographic systems is a very important issue in the production of injectable pharmaceuticals. Consequently, the regulatory authorities place heavy emphasis on disinfection, sanitisation of production scale chromatographic columns to ensure that the processes are carried out aseptically.

   The reduction of microbial load is typically achieved in industrial scale chromatographic systems using chemical treatment before, during, and after operation.

   Aqueous mixtures of ethanol in neutral or acidic conditions are often used to inactivate microbes. Ethanol is most effective when used at concentrations of 60–70% v/v, however in production environment 20% solutions are typical. However, at these lower concentrations, ethanol has very little effect on viruses and little or no effect on pyrogens. For these reasons, ethanol is considered only as a bacterial static agent.

Sodium hydroxide is the most widely used agent in the pharmaceutical industry, being used to both sanitize and regenerate chromatographic resins. Concentrations of use (0.1 *M* to 1.0 *M*), incubation times (minutes to days), temperatures, and the frequency of column cleaning vary greatly. Pyrogens are very sensitive to alkaline hydrolysis.

Sodium hypochlorite is the most well known oxidizing agent that is used to sanitize columns. The chemical possesses excellent microbiocidal properties and is harmless to human skin and mucosa. Its use as chromatographic santizing agent has however been limited by its potential to react with the resins resulting in formation of toxic chlorinated byproducts. Peracetic acid used in combination with ethanol (pH 5.5) has been demonstrated to be highly effective antimicrobial agent. Additionally, peracetic acid is an effective antiviral agent. A benefit of using peracetic acid is that it decomposes to acetic acid.

4. Chromatographic Systems. The qualification of chromatographic systems requires documented evidence that it is suitable and in a state of maintenance and calibration consistent with the intended use.

   The qualification process consists of four parts:
   a. design qualification (DQ).
   b. installation qualification (IQ).
   c. operational qualification (OQ).
   d. performance qualification (PQ).

   The DQ describes the requirements and defines the functional and operational specifications of the chromatographic system. For example, what are the pump requirements, what type of detector is required and what capacity is the autosampler required to handle. Other areas that may be included as designing elements include user instructions, maintenance requirements, validation procedures, and training. The DQ is used as the basis for the OQ testing.

   The IQ ensures that the chromatographic system is received as designed and is properly installed. Things to be considered include manufacturer's recommendations, that the equipment compares to the purchase order, all documentation has been included (operating manuals, maintenance instructions, SOP's and validation certificates) and that the system has been installed correctly with all systems able to be powered up. IQ includes running test samples to provide chromatograms that should be compared to reference chromatograms.

   The OQ is used to demonstrate the instrument's ability to perform a designated task in a standard environment and verifies that the chromatographic system is able to comply with the operational and functional requirements set out in the DQ. When performing OQ on chromatographic systems, it is often better to perform the testing in an overall manner rather than isolating individual systems within the instrument. Tests may include: leak testing, baseline drift, injection volume precision, flow rate precision, detector linearity, temperature accuracy, and wavelength accuracy.

The PQ demonstrates that the instrument is able to perform a function in a consistent manner according to a series of preset specifications. Thus, the PQ testing requires a number of cycles to be performed in order to demonstrate instrument consistency. The PQ are application specific and should therefore be performed under conditions that most closely resemble those to be routinely performed. This would include using the same buffers, feedstock, resin, bed height, and elution conditions. Measures would include peak widths, baseline noise, wavelength accuracy, and retention time data.

As each of these qualification tasks is completed, the documentation for DQ, IQ, OQ, and PQ should be completed and finalized.

5. Column Packing. Optimum column packing is essential for a good chromatographic separation. Poorly packed columns can lead to channeling or uneven flows, resulting in zone broadening and reduced resolution.

   The flow rate used for packing is based on the equipment, the column type, and quantity and type of media to be used and can be established empirically by generating a pressure–flow rate curve as described below. Usually, there is extensive relevant information available from the resin and column manufacture.

   Packing involves making a slurry of the resin (approx 75%) before pouring into the column. For short bed heights in wider columns, the stirring is continued when the slurry is in the column to achieve even distribution of resin. Once the top plate is in place and any air is removed from the pump lines, pumping of solvent (e.g., 0.1 *M* NaCl) through the column is commenced at a low flow rate. The flow is progressively increased and the operating pressure recorded. As the media settles, a clear space between the bed and top plate develops. The height of the top plate should not be adjusted. The rate of pumping is increased until a plateau is reached where increased operating pressure is not associated with an increase in flow rate. A plot of pressure verses flow rate allows the determination of the optimal packing flow rate, which is 70–100% of the maximum flow rate. Modern rigid resins may not plateau, therefore, a flow rate up to the allowable pressure limit for the column can be used.

6. Evaluation of Packing. The efficiency of a column is determined by how well it is packed. The method of choice for determination of expressing column efficiency is HETP and the peak asymmetry factor (A_s). Such test methods are simple, quantitative, and do not involve the introduction of colored or contaminating substances. HETP and asymmetry are typically determined by the application of a sample of NaCl (not suitable for ion exchange resins), acetone, or NaOH. Acetone is monitored by UV absorption at 280 nm, whereas NaOH and NaCl is monitored by conductivity. Typically, the sample volume should not be more than be 2.5% of the total bed volume. The sample should be applied as close to the column inlet as possible to avoid dilution. In particular, an air-trap should be bypassed.

   The HETP values should only be compared between columns packed with the same media under identical conditions. Preferably, three test runs should be undertaken before a comparison can be made. Once a certain performance standard has been set, future HETP values will serve as a guide to judge the acceptability of column performance. Acceptable HETP values are usually <0.06 cm.

The eluted peak should be symmetrical. Leading or trailing peaks can be an indication of poor packing or channelling within the resin bed and is the first indication of column deterioration. Peak asymmetry factor should be as close to one as possible.

Increasing asymmetry factor (increased tailing and decreased leading peak) are an indication of a too loosely packed column allowing space between medium and inlet. Alternatively decreasing asymmetry factor (decreased tailing and increased leading peak) indicates a too tightly packed bed or formation of cracks in bed. Double peaks (i.e., extra peaks before or after the main peak) are an indication of cracked bed owing to poor packing or may be caused by pump pulsations. Tailing peaks indicate too loose packing and leading peaks indicate that packing is too hard.

References

1. Janson, J.-C. and Hedman, P. (1982) Large-scale chromatography of proteins. *Adv. Biochem. Eng.* **25,** 43–99.
2. Janson, J.-C. and Ryden, L., eds. (1989) *Protein purification: Principles, high resolution methods, and applications.* VCH, New York.
3. Wheelwright, S. M. (1991) *Protein purification: Design and scale-up of downstream processing.* Hanser, New York.
4. Dasarathy, Y., Ramberg, M., and Andersen, M. (1996) A systematic approach to screening ion-exchange chromatography media for process development. *Biopharm.* **9,** 42–45.
5. Jungbauer, A. and Lettner, H. P. (1994) Chemical disinfection of chromatographic resins, Part I: Preliminary studies and microbial kinetics. *Biopharm.* **7,** 46–56.
6. Huber, L. (1998) Systems qualification for high-performance liquid chromatography systems. *Biopharm.* **11,** 65–66.
7. Dennis, J., Levison, P., and Mumford, C. (1998) Comparative performances of 70 ion-exchange chromatography sorbents for purification of biomolecules. *Biopharm.* **11,** 44–50.

III

Applications

16

Proteolytic Peptide Mapping

Peter Højrup

1. Introduction

Analysis of intact proteins is hampered by the fact that even when using high-resolution techniques, like electrospray mass spectrometry, you will only be able to tell whether the protein has the expected mass. If you find a deviation from the expected you will, in most cases, be left wondering where (and perhaps what) the difference is. These observations are compounded when using low-resolution techniques like gel filtration or sodium dodecyl sulfate (SDS) gel electrophoresis. Although 2D gel electrophoresis is able to show a single charge difference, in the absence of additional information you are still left wondering where and what the differences are.

Performing Edman degradation on an intact protein will, in most cases, yield enough information for you to identify a protein and perhaps enable the construction of oligonucleotide probes, but many proteins are N-terminally blocked, and the need to purify enough protein is often prohibitive. The solution is to perform peptide mapping. This approach is essential for the total characterization of a protein, and the approach also enables quick identification of a protein (if present in a protein database) and to locate differences between proteins.

The steps involved in peptide mapping can be defined as:

1. Purify the protein. Depending on the objective of the peptide mapping, it may be sufficient with a single spot from a 2D gel, or you may have to purify several microgram of material using traditional protein chemical methods.
2. Cleave the protein using a suitable proteolytic enzyme or by chemical means. A software tool like GPMAW (*see* **Note 1**) greatly facilitates the selection of cleavage method and analysis of the resulting peptides.

From: *Methods in Molecular Biology, vol. 251, HPLC of Peptides and Proteins: Methods and Protocols*
Edited by: M.-I. Aguilar © Humana Press Inc., Totowa, NJ

3. Separate the resulting peptides, generally by reversed-phase (RP) high-pressure liquid chromatography (HPLC), but other methods may be applicable depending on the final method of analysis.
4. Analyze the purified peptide using mass spectrometry, Edman degradation, or amino acid analysis.

Peptide mapping is usually carried out for two quite different reasons:

1. **Protein identification.** Currently, the most sensitive identification of an unknown protein is mass spectrometric peptide mapping (peptide mass search, PMS). Using the rapidly expanding protein databases based on nucleotide sequencing, it is possible to identify a protein based on the molecular masses of its constituent (tryptic) peptides if they are determined with sufficient precision. The beauty of the method is that it is not necessary to determine the mass of all peptides; usually six–seven peptides are sufficient for accurate protein identification. As it is not necessary to separate the peptides before analysis, the method is very sensitive, with routine determination of proteins in the low nanogram range (e.g., a single silver stained spot from an SDS gel).

   Two alternative approaches are liquid chromatography coupled to electrospray tandem mass spectrometry (LC–MS/MS) and Edman degradation. Both methods will generate a peptide sequence thus making it possible to do a homology search in case the protein is not present in the database. However, while the sensitivity of LC–MS/MS approaches that of matrix-assisted laser desorption ionization mass spectrometry (MALDI-MS), the sensitivity of Edman degradation is two–three orders of magnitudes lower and further needs a purification step prior to analysis. The topic of LC–MS/MS is beyond the scope of this chapter and will not be treated further, for an introduction please *see* **ref. *1***.
2. **Obtaining the maximum sequence coverage.** Unlike the protein identification where even a limited number of peptides are able to identify the protein, you will try to cover the whole sequence when comparing proteins for identity/similarity or performing complete chemical characterization.

   Comparison of proteins. Comparison of two proteins is done by performing an identical enzymatic digest on each protein followed by identification of each fragment. In some cases, it may be sufficient to perform mass spectrometric mapping of the peptide mixture, but, in most cases, you will not be able to identify sufficient fragments to cover the whole sequence. You will thus have to isolate each fragment by HPLC and perform peptide identification. Owing to the high resolution of RP-HPLC, it will, in many cases, be sufficient to compare the two chromatograms carefully in order to identify differences. When comparing proteins, you should note that choosing an enzyme, which generates long peptide fragments, tends to emphasis similarities whereas short peptide fragments tend to emphasize differences (*see* also **Note 2**).

   Determine the exact chemical structure of a protein. Even though the number of known protein sequences increases exponentially through nucleotide sequence analysis, and you are thus likely to find your protein to be already known, the result of these analyses is the "naked" amino acid sequence of the protein. However, the vast majority of proteins are posttranslationally modified after synthesis, resulting in a structure that deviates from the nucleotide-derived sequence. The preferred strategy to solve the actual protein structure is to perform peptide mapping followed by an appropriate analytical techniques like mass spectrometry, Edman degradation, and/or amino acid analysis. Here, you always aim for total sequence coverage, and in order to obtain this you will in most cases have to perform at least two separate enzymatic digests.

The amount of material needed for maximum sequence coverage analysis is usually much higher than for PMS analysis, usually in the low to middle microgram range. In special cases where you search only for specific modifications using specific extraction of modified peptides (e.g., purification of phosphorylated peptides using IMAC), the sensitivity can approach that of PMS *(2)*.

2. Materials

2.1. Peptide Mapping for Protein Identification

1. Trypsin (modified trypsin, Promega, #V5111) is most easily handled in aliquots of 1 μg trypsin in 10 μL 0.01 *N* HCl per tube at –20°C. Dilute with 70 μL 50 m*M* ammonium bicarbonate before use (final concentration 12.5 ng/μL).
2. 50 m*M* Ammonium bicarbonate. You can store aliquots of 100 m*M* ammonium bicarbonate at –20°C, it is not necessary to adjust the pH. Dilute 1 : 1 for use.
3. Neat acetonitrile (HPLC quality).
4. 10 m*M* Dithiothreitol (DTT). A 1 *M* DTT solution is stable for month at –20°C. Store as 25-μL aliquots and dilute 1 : 100 with 100 m*M* ammonium bicarbonate for use.
5. 55 m*M* Iodoacetamide (Sigma, #I6125) in 100 m*M* ammonium bicarbonate. Must be prepared fresh (iodoacetamide: 10.2 μg/μL).
6. 0.5 mL Polypropylene tubes.
7. Zip-tips (Waters) or homemade micropurification tips based on Poros 50 R2 RP material (Applied BioSystems, #1-1159-05) and Eppendorf GELoader tips (Eppendorf, #0030 001.222) (*see* **Note 3**).
8. Vacuum centrifuge.
9. Concentrated solution of alpha-cyano-4-hydroxycinnamic acid (Aldrich, #14,550-5) in 70% acetonitrile, 0.1% trifluroacetic acid (TFA) in water.
10. MALDI mass spectrometer.

2.2. Alkylation of Protein

1. 8 *M* Urea, 100 m*M* ammonium bicarbonate (or Tris-HCl), pH 8.0.
2. Reduction agent. Typically, DTT is used for reduction, but mercaptoethanol or tris(carboxyethyl)phosphine (TCEP–Sigma, #C-4706) may be used instead. Mercaptoethanol is volatile, whereas TCEP is also active at low pH.
3. Alkylating reagent. Iodoacetate, iodoacetamide, 4-vinylpyridine and N-isopropyliodoacetamide are all commonly used reagents. 4-vinylpyridine has the disadvantage of being prone to polymerization and thus has a limited shelf life.

2.3. Peptide Mapping With Purification of Peptides

1. Suitable enzyme or chemistry (*see* **Table 1**).
2. Buffer for digestion (*see* **Note 4**).
3. HPLC system capable of binary gradient formation (*see* **Note 5**) equipped with an in-line UV detector and a strip-chart recorder or computer integrator.
4. Acetonitrile (HPLC grade, UV cutoff <210 nm). Neat TFA (analytical grade is acceptable, sequencing grade is optimal) (*see* **Note 6**).
5. Automatic or manual fraction collection into polypropylene tubes.
6. Vacuum centrifuge (*see* **Note 7**).

As many of the solvents and reagents are used in very small quantities (e.g., ≤1 mL) it can be advantageous to prepare stock solutions that are stored at –20°C, even of those that are not specifically mentioned in the text.

Most proteolytic enzymes are supplied freeze-dried. When first opened, you may dissolve the enzyme in pure water, make 0.5–2 µg aliquots in small polypropylene tubes, and lyophilize the aliquots. Endo Glu-C protease may be stored frozen in pure water.

3. Methods

3.1. Selection of Proteolytic Enzyme

The selection of an enzyme for digestion is one of the most important considerations when performing peptide mapping, and should be carefully considered in relation to the task you want to perform.

For PMS the most common choice is trypsin. This is one of the most active enzymes, it is well characterized for PMS analysis (*see* **Notes 8** and **15**), and cleaves with high specificity (C-terminal to Lys and Arg). Most tryptic peptides will have molecular masses in the range 800–2500 Da, which is well suited for mass spectrometry. Furthermore, you are certain that all peptides will contain a basic residue (except the C-terminal peptide), which means they will ionize better when analyzed by mass spectrometry in positive mode. Endo-Lys and lysyl endopeptidase are also well suited for PMS, but the peptides generated will, on average, be much larger, which may impede their identification. Other potential enzymes are Endo Asp-N and Endo Glu-C, but the results are usually much more unpredictable.

If the protein sequence of your target protein is known, you should look at the distribution of residues when aiming for the largest sequence coverage before selecting an enzyme. Particularly, the charged residues are important, as most of the specific enzymes will target these residues (*see* **Table 1**). A few computer programs for this type of analysis are available (*see* **Note 1**).

Important aspects for enzyme selection are:

a. *Short peptides.* Very short peptides are difficult to handle, both by MALDI mass spectrometry (<7 residues) and HPLC (<5 residues if they are hydrophilic).

b. *Long peptides.* The mass resolution and sensitivity decreases on the mass spectrometer as you get above 30–35 residues. Likewise large, hydrophobic peptides can be difficult to purify by RP HPLC (>50 residues). If your prime objective is to obtain amino acid sequence by Edman degradation, long peptides are generally to most informative.
c. Hydrophobic peptides tend to stick to the wall of polypropylene tubes, particularly if the sample is dried completely (*see* **Note 7**).
d. If you want to do protein comparison, you should note that small peptides tend to emphasize differences (e.g., making it easier to identify small differences between proteins) whereas large peptides will emphasize similarity (e.g., make it easier to identify similar regions).
e. Enzyme specificity, activity, and buffer compatibility (*see* **Table 1**) should be chosen relative to the intended analysis method (mass spectrometry, HPLC separation, Edman degradation, amino acid analysis, and so on).

For most proteins, you will need at least two different enzymes (digests) in order to get a good coverage (>90%).

3.2. Peptide Mass Mapping

1. Cut out the stained gel band or gel spot using a sharp scalpel or a properly sized puncher (2–4 mm diameter). Cut the spot into smaller pieces (1 mm square) and place them in a small polypropylene tube. Try to cut as close as possible to the actual spot. A small concentrated spot yields better results than a large diffuse one. If you have a 1D gel band, you may not need the whole band, but can reserve part of it for additional experiments.
2. Wash the gel pieces carefully in 50 µL, 50 m*M* ammonium bicarbonate. Discard the supernatant.
3. Add 50 µL neat acetonitrile. Let stand for 5 min until the gel pieces has fully shrunken and become white. Withdraw the acetonitrile and discard. Dry the gel pieces in a vacuum centrifuge for 20 min.
4. Add 50 µL 10 m*M* DTT and incubate for 45 min at 56°C.
5. Cool to room temperature and remove excess solvent.
6. Add 55 m*M* iodoacetamide in 100 m*M* ammonium bicarbonate so the gel pieces are just covered.
7. Let stand for 30 min in the dark before discarding the supernatant.
8. Wash the gel pieces in 20 µL 50% acetonitrile in water.
9. Add 30 µL neat acetonitrile. Mix until the gel pieces are fully shrunken and white. Remove the acetonitrile.
10. Add 20 µLl 100 m*M* ammonium bicarbonate and mix for 5 min.
11. Add additional 30 µL neat acetonitrile and mix until the gel pieces are shrunken and white.
12. Remove the supernatant and dry the gel pieces in a vacuum centrifuge.
13. Add 12.5 ng/µL modified trypsin until the gel pieces are just covered with solvent. If you are using a low sensitivity stain like Coomassie blue, the trypsin con-

Table 1
Proteolytic Enzymes Suitable for Peptide Mapping

Name	EC no.	Size Da	Type	Organism	Accession number	pH optimum (range)	Requirements	Tolerates detergents	Cleavage	Speci-ficity	Manufac-turer	Note
Trypsin	3.4.21.4	23.293 23.463	Serine	Bovine porcine modified	P00760 P00761	8 (5.5–10.5)	Stabilized by small amounts of Ca^{++}	2 *M* Urea >0.1% SDS 50% Acetontrile	Arg. Lys	High	R,W,S A,S R,P,W,S	**16**
Chymotryp-sin AB	3.4.21.1	25.600 25.700	Serine	Bovine	P00766 P00767	8 (7–9)		0.1% SDS 1 *M* Urea 10–30% Acetonitrile	Phe, Trp, Tyr	Medium to high	R,A,W,S	**17**
Endo Glu-C	3.4.21.9	27.000	Serine	*Staphylococcus aureus* V8	P04188	7.8 (4–9)		2 *M* Urea 0.2% SDS 10% Acetonitrile	Glu (Asp) -Glu	High	R,A,P,W,S	**18**
Endo Lys-C	3.4.99.30	30.000	Serine	*Lysobacter enzymogenes*	P15636	8.5–8.8 (7–9)		0.1% SDS 1 *M* urea	Lys	Medium to high	R,S	**19**
Lysyl endo-peptidase	3.4.21.50	46.000	Serine	*Achromabacter lyticus*	P15636	8	Inhibited by ammonium salts	>0.1% SDS 4 *M* Urea 40% Acetonitrile	Lys	HIgh	A	**19**
Endo Arg-C	3.4.21.40	30.000	Serine	Mouse submaxillary		8 (7.5–8.5)		2 *M* Urea 10% Acetonitrile	Arg	High	T,S	**19**
Endo Asp-N	3.4.24.33	27.000	Metallo	*Pseudomonas fragi*		8 (6.0–8.5)	Avoid EDTA	0.1% SDS 1 *M* Urea 10% Acetonitrile	N-terminal to Asp	High	T,R,A,S	**19**

Endo Asn-C		23.000	-SH	Jack Bean		5.0 (4.5–6.5)	1–10 m*M* DTT		Asn	High	T	**19**
Clostripain	3.4.22.8	50.000	-SH	*Clostridium histolyticum*	P09870	7.6	1–10 m*M* DTT	4 *M* Urea 10% Acetonitrile	Arg (Lys)	Medium	R,W,S	**19**
Pepsin	3.4.21.7	34.500	Acid	Pig gastric mucosa	P00791	2 (1–5)			N-terminal to hydrophobic Preferentially if preceded by Phe, Met, Leu, Trp	Low	R,A,W,S	**15**
Thermo-lysin	3.4.24.4	37.500	Metallo	*Bacillus thermoproteolyticus*	P00800	8 (7–9)	Ca^{++} Avoid EDTA	8 *M* Urea	N-terminal to hydrophobic Not if followed by Pro	Low	A,S	**15**
Elastase	3.4.21.36	25.900	Serine	Pig pancreas	P00772	8.5 (7–9)		0.1% SDS 2 *M* Urea	Uncharged (small) aliphatic	Low	R,A,W,S	**15**

Chemical cleavage procedures are not included in the table as experience shows that they are too unspecific for general peptide mapping, however, for those special hard-to-cleave proteins; *see* **refs.** ***8*** and ***9***.

Accession number: numbers refer to the Swiss-Prot database (http://www.ebi.ac.uk/).

Cleavage: All enzymes cleave C-terminal to the indicated residue except for except where noted. Minor, but still significant, secondary cleavages are listed in parenthesis. Cleavage does not take place if the cleavage position is followed by proline or a residue indicated by "-" (e.g., a -Lys-Pro- sequence is not cleaved by trypsin). See also **Notes** for individual enzymes.

Specificity: Please *see* **Note 15** and notes for the individual enzymes.

Detergents: Even very small amounts of SDS is detrimental to mass spectrometric analysis, whereas RP-HPLC is capable of handling small amounts of SDS. When using urea you may dissolve the protein in a small amount of 8 *M* urea you can dilute to the required strength with buffer containing the relevant enzyme.

Manufacturer: R, Roche Biomolecular; S, Sigma Inc.; P, Promega; T, Takara; W, Worthington; A, Wako; C, Calbiochem.

centration should be doubled (*see* **Note 8**). Let stand for 5 min before removing excess trypsin containing solvent.

14. Add 50 m*M* ammonium bicarbonate until the gel pieces are just covered.
15. Let the digestion proceed overnight at 37°C.
16. Extract the excess solvent, which will contain the peptides. Purify the peptides on a Zip tip (Waters) or for higher sensitivity use a home-packed GelLoader tip *(3)* using the recommended procedures (*see* **Note 3**).
17. Run the peptide map on a MALDI mass spectrometer in alpha cyano hydroxy cinnamic acid.
18. Analysis of the resulting mass spectra can easily be carried out using World Wide Web-based tools (*see* **Note 9**), but be careful to check the results (*see* **Note 8**).

3.3. Reduction and Carboxymethylation of Proteins in Solution

1. Dried sample (20–100 µg) is dissolved in 20 µL 8 *M* urea, 0.4 *M* ammoniumbicarbonate buffer (*see* **Note 10**).
2. Add 5 µL 45 m*M* DTT.
3. Stand at 50°C for 15 min. Cool to room temperature.
4. Add 5 µL 100 m*M* iodoacetamide.
5. Stand at room temperature for 15 min in the dark.

Purify the protein by RP-HPLC (a short Poros R1 column, PerSeptive Biosystems, is well suited), gel filtration, or RP solid-phase extraction.

For digestion with trypsin, endo Glu-C or lysyl endoproteinase (*see* **Table 1**) you can proceed as follows:

1. Add 140 µL water to dilute the urea to less than 2 *M*.
2. Add 2–4% w/w of the selected enzyme. For prolonged digestion you can add the enzyme in two equal portions with an interval of 4–6 h.
3. Stand for 16–24 h at 37°C.
4. Freeze the sample or inject directly onto RP-HPLC column. When programming the separation gradient, you should start with 2–5 min of isocratic elution in order to let the urea elute from the column before starting the separation of the peptides.

3.4. Digesting in Solution

If your protein contains disulfide bridges, you should reduce and alkylate the protein prior to digestion, see above.

1. Dissolve the protein in an appropriate buffer (50 m*M* ammonium bicarbonate, *see* **Note 4**) to a concentration of 1–5 µg/µL. If you have less than 20 µg, you should dissolve in 20 µL.
2. Add the enzyme of your choice (*see* **Table 1**) to a final concentration of 2–4% w/w. Hard-to-digest proteins and dilute solutions will need a higher enzyme to

substrate ratio. For limiting the cleavage to a single or a few positions, you can try to lower the concentration to 1% or lower.
3. Most proteins will be digested in 6–8 h at 37°C, but conditions may have to be varied to suit each individual protein. For hard-to-digest proteins, you can use extended digestion times (e.g., 24–48 h or longer). For long digestion times, it is advantageous to add the enzyme in two equal portions separated by 4–8 h.
4. Digestions may be stopped by lowering the pH with the addition a small amount of 2% TFA (raising the pH to above 5.0 in the case of pepsin), by injecting the sample onto the HPLC system or by adding an appropriate enzyme inhibitor.
5. Separate the peptide map by RP-HPLC. Prior to separation, you may want to extract 0.5 μL of your sample for MALDI analysis in order to check the extent of proteolysis.

A control reaction run under identical conditions as the main digest should be run in parallel to show which peptides comes from the sample and which are autodigest products or other contaminants. A control digest of a known protein (e.g., β-lactoglobulin—**Figs. 1** and **2**) can also be run to check the activity of the protease. A special case of peptide mapping is when you want to determine the disulfide bridge structure of a protein.

1. Verify the complete protein structure using standard peptide mapping techniques.
2. Use a software tool to determine the optimal cleavage pattern of the target protein (*see* **Note 1**). Remember that disulfide interchange is minimized at low pH. Enzymes like trypsin, Endo Glu-C, Endo Asp-N, Endo Asn-C, and pepsin are all active below a pH of 7.0 and are thus to be preferred.
3. Digest your protein as described above and separate half the sample by RP-HPLC.
4. Make the other half of the sample 10 m*M* in DTT or TCEP (will reduce you sample even at low pH). Let stand for at least 30 min before separation the sample under conditions identical to the separation in **step 3**.
5. Compare the two chromatograms. Peaks that disappear from the first chromatogram and appear in the second chromatogram must arise from Cys containing peptides and will thus have to be identified by mass spectrometry, Edman degradation, or amino acid analysis (*see* **Note 10**).
6. Closely spaced Cys residues in the primary structure presents a special problem as you will have to cleave between them for a positive identification of the disulfide bridge. In this case you will often have to isolate a large peptide containing multiple disulfide bridges followed by subdigestion using a less specific enzyme (*see* **Table 1**) *(4)*.

3.5. Peptide Map Purification by HPLC

Other chapters in this book deal in detail with the purification of proteins and peptides by HPLC, so the following only emphasis the features relevant to peptide mapping. When purifying a peptide map by HPLC, the objective is usu-

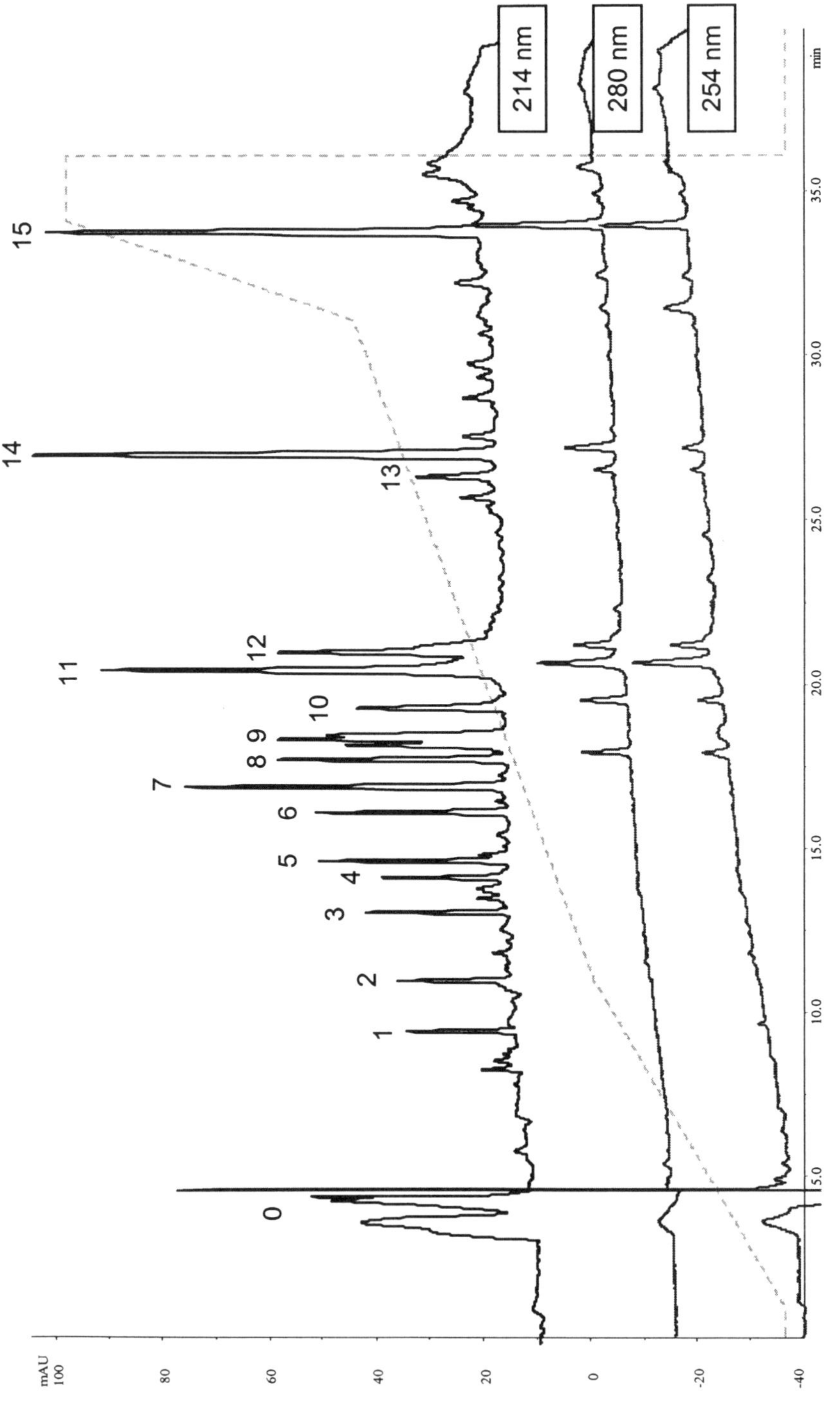

214 nm
280 nm
254 nm
0
1
2
3
4
5
6
7
8
9
10
11
12
13
14
15
mAU
100
80
60
40
20
0
-20
-40
5.0
10.0
15.0
20.0
25.0
30.0
35.0
min

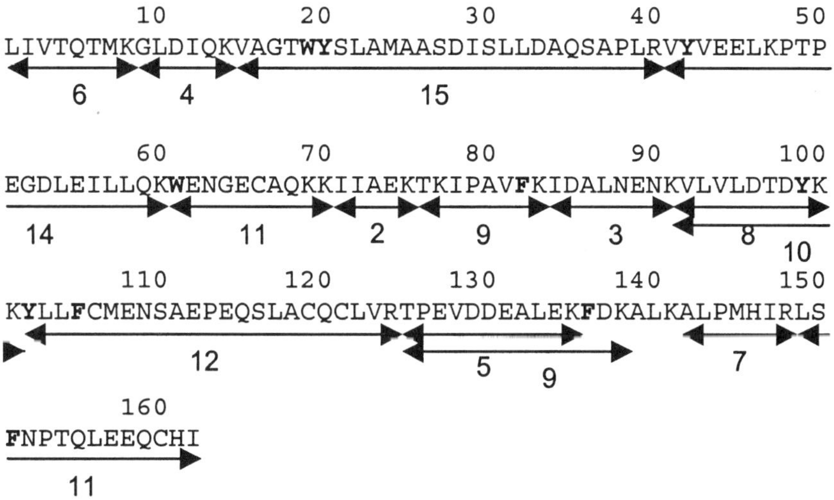

Fig. 2. Identification of the peaks separated in **Fig. 1**. Aromatic residues are shown in bold. The two peptides labeled 11 are linked by a disulphide bond (Cys66–Cys160). Peptide 12 has an internal disulphide bond. The peptides labeled 9 elute very close together (*see* **Fig. 1**).

ally to obtain the highest possible resolution, sensitivity, and yield, but not to handle large amounts. The following description emphasizes this aspect.

For general peptide mapping, use a large pore (300 Å), RP column packed with the smallest possible particle size and the longest length. These parameters are conflicting as backpressure in the system rises with both smaller particle size and longer column length. A compromise is to use a long 250-mm column packed with 5-µm particles or a shorter 150-mm column packed with 3-µm particles. Backpressure in the running system should not exceed 100 bar (increasing the temperature to 45–50°C using a column oven will lower the back pressure in addition to increase the resolution slightly).

Using smaller diameter columns will increase the sensitivity of separation if the HPLC system is optimized for the lower flow at the cost of lower capac-

---

Fig. 1. *(see opposite page)* Tryptic digest of 250 pmol bovine beta lactoglobulin. The traces of 214, 254, and 280 nm are shown. The 254 and 280 nm traces are expanded ×5 compared to the 214-mm trace. The stippled line shows the gradient used. The identification of each peak is shown in **Fig. 2**. Peak 1 and peak 13 were not identified. Peak 13 is likely to be a partial digest product. The peaks at "0" are a mixture of small peptides and the buffer salts from the injection.

ity (*see* **Note 5**). Furthermore, the peak volume from a 2.1-mm column is about 100 μL compared to 400–500 μL from a 4.6-mm column. This makes the volumes more compatible to further downstream processing (e.g., Edman degradation) or speeds up lyophilization (*see* **Note 7**).

An in-line UV detector is essential for separating peptide maps. As we cannot be certain that all peptides contain aromatic residues we have to detect our peptides at a wavelength low enough to observe the absorbance of the peptide bond. However, as solvents like 0.1% TFA and acetonitrile also absorbs at low wavelength we need to find the wavelength with the highest signal to noise ratio. On most systems, this is at 214–215 nm. If you do not operate with the highest purity of solvents, you may have to increase the wavelength to 218–220 nm, but your sensitivity will be greatly reduced.

Multiple wavelength detection (e.g., 254 and 280 nm) can greatly increase the confidence of peptide identification (*see* **Note 11**).

3.6. Separating a Peptide Map

1. Prepare 1 L of A-solvent (5% acetonitrile/0.1% TFA in ultra high-quality water) and 0.5 L of B-solvent (90% acetonitrile in water, 0.85% TFA) (*see* **Note 12**). Degas solvents carefully before use or, better, use in-line degassing, as degassed solvents will reabsorb oxygen from the air within a few hours.
2. Program your HPLC system to a gradient of 1%/min from 0 to 40% B-solvent, followed by 2%/min from 40 to 70% B-solvent. Finally, step to 100% B-solvent for 2 min before equilibrating at 5% for at least 10 column volumes until you have a stable baseline. If your peptide map is not too complex, you may shorten the run time by increasing the gradient to 1.5%/min between 0 and 40% and 2.5–3% /min between 40 and 70%.
3. Make a blind run each morning to equilibrate the column. The first run of a day is usually atypical with a rising or falling baseline that may not be present in subsequent analyses. You can most easily check the condition of your HPLC system and RP phase column by making a separation run using a standard peptide mixture (e.g., a tryptic digest of β-lactoglobulin) (*see* **Note 13**).
4. Dissolve your sample in A-solvent and inject. Keep the injector in the "inject" position for two injection loop volumes before you switch back to "load" position. If your sample is already dissolved in a different solvent (i.e., urea, Tris-HCl, or phosphate) you should allow 2–3 min of isocratic elution before starting the gradient.
5. Collect the isolated peaks in polypropylene tubes. If you run at low flow rates, you may have to take a delay between the detector and the effluent from the capillary into account (*see* **Note 14**).
6. Dry down your collected fractions in a vacuum centrifuge (*see* **Note 7**). For several applications, you may not have to dry the sample completely, just evaporate the major part of the organic modifier.

For problems in peptide map separations, *see* **Note 2**.

4. Notes

1. When the primary structure of the protein being analyzed is known, a couple of software tools are available for planning the peptide map. These tools can predict the peptides generated by various proteolytic enzymes or chemical cleavages. One freeware tool, PAWS, is available for download (http://www.proteometrics.com), although a somewhat more advanced tool, GPMAW (http://welcome.to/gpmaw) *(5)*, will allow you to work with multiple sequences and, in addition to the prediction of peptides, provides additional physical/chemical characteristics of each peptide (e.g., charge, theoretical p*I*, hydrophobicity, and size). GPMAW can even generate a simulated HPLC RP chromatogram. Although this chromatogram is only approximate (separation varies with column, system, and running conditions), it does give valuable indications for problems during separation.
2. If your peptide map contains small hydrophilic peptides, they will elute very early in the chromatogram, perhaps even as part of the solvent peak. Early eluting peaks may also have a bad peak shape, particularly if the column has not been equilibrated sufficiently before the start of the gradient. You may collect these early eluting peaks, dry them sufficiently to get rid of the organic solvent, and then reseparate on a 100–120Å pore size, high carbon load column (e.g., Hypersil ODS2 or equivalent) using 0.1%TFA/90% methanol as a solvent system. The gradient should be a rather shallow 1%/min. gradient from 0 to 30%. If, on the other hand, you have large, hydrophobic peptides, you should use a 300Å, C4 column and elute with either acetonitrile or, for the really hydrophobic peptides, use 2-propanol.
3. The use of reversed phase purification as the last step before analysis of peptide mixtures is necessary to obtain the highest sensitivity and sequence coverage in MALDI peptide mapping. You may use Zip-tips (Waters Inc.), but when analyzing peptide maps isolated from silver stained gel spots, you can obtain considerable higher sensitivity by using homemade micropurification tips based on Eppendorf GELoader tips packed with Poros R2 50 μm RP material (Applied BioSystems, Framingham, MA). For the purification of proteins, you should use Poros R1 material. For details on how to construct these devices, please *see* **ref.** ***3***. The GELoader tip is constricted at the end using the blunt end of a pair of tweezers or a pencil. Column material is dispersed in acetonitrile and sufficient material is applied to make a 3–5 mm length column. The GELoader tip is initially washed with 10–20 μL of neat acetonitrile, then it is equilibrated in 2 × 10 μL of 0.1% TFA. The sample is added in 1–20 μL of solvent followed by washing with 2 × 10 μL of 0.1% TFA. Finally, the sample is eluted with 0.8–1 μL of α-cyano-4-hydroxycinnamic acid in 70% acetonitrile/0.1% TFA (4 mg/mL to concentrated depending on instrument) directly onto the MALDI target.
4. Most buffers without an excessive amount of detergents or salt are suitable for proteolytic digestion. 50 m*M* ammonium bicarbonate is very convenient because it can be removed by evaporation (do not use with lysyl endopeptidase). If you have large quantities or high concentrations of ammonium bicarbonate you may need to add a bit of dilute acetic acid to help completely remove the ammonium bicarbonate. This buffer may be readily substituted for other buffers, e.g., 50 m*M* Tris-

HCl, pH 8.0, or 50 m*M* H_3PO_4, pH 8.0. However, these buffers are not directly compatible with mass spectrometry.

Hard-to-dissolve proteins can often be brought into solution by first dissolving the protein in small amount of 6–8 *M* urea (e.g., 5–10 µL) and then diluting the solution to the appropriate level (e.g., below 2 *M* for trypsin, *see* **Table 1**) with a buffer containing the enzyme for digestion. Other ways of bringing proteins into solution can be the addition of detergents or organic solvents (*see* **Table 1**), however, beware of downstream incompatibilities (e.g., SDS for mass spectrometry and HPLC). Heat denaturation or the addition of organic solvents may denature proteins, thus leading to faster, more efficient digestions.

5. The sensitivity of HPLC separations decreases with the square of the column diameter, thus, a 2-mm column shows four times the sensitivity of a 4-mm column, providing the HPLC system is built to handle the smaller diameters. As the column diameter decreases from 4 to 2 and 1 mm, the optimal flow rate decreases from 1 mL/min to 0.25 mL/min and 0.06 mL/min. Unless you perform flow splitting, you will need special HPLC pumps to be able to perform a gradient in 0.06 mL/min. In addition, the rest of the HPLC equipment has to be optimized for the lower flow rate in order to minimize extracolumn band broadening. The inner diameter of capillaries has to be decreased (*see* **Table 2**). You also need to make connections as short as possible, particularly after the column. The size of the detector flow cell is particularly important, as a standard flow cell with a volume of 10–12 µL will significantly decrease the resolution of a 2-mm column, even though the rest of the system is optimized (*see* **Table 2**). When changing to a smaller diameter column you may also have to replace the dynamic mixer as you may otherwise experience very long delay times.

   For dual pump systems, each pump should have a minimum flow of no more than 5% of the running flow (e.g., a minimum flow of no more than 12.5 µL/min for a set total flow of 250 µL/min), and a minimum step of no more than 1% of the set flow rate. You may manage with slightly lower performance as the dynamic mixer will smooth out irregularities, but particularly at the beginning of the gradient, you may observe irregular separation behavior.
6. TFA deteriorates over time. As the 0.1 % TFA used for buffer A ages, you will see an increase in baseline during gradient runs. You may increase/decrease the content of TFA in the B-solvent to compensate. A ratio of 1:0.85 is usually a good starting point, but varies slightly from detector to detector. If you experience regular baseline noise, you can try to lower the content of TFA in buffer A to 0.05% and the content in buffer B correspondingly.
7. For small amounts of peptide, the drying step can be critical for recovery as the peptides have a tendency to adhere to the walls of the polypropylene tube *(6)*. For many analytical purposes it is not necessary to change solvent or you may only have to evaporate the organic solvent.
8. An accurate calibration of your mass spectra is essential for correct protein identification. The most accurate calibration is to use known peaks in each spectrum

Table 2
Parameters for HPLC System for Binary Gradient Formation

	Column diameter	Typical flow mL/min	Flow cell volume	Capillary diameter	Sensitivity increase
Normal	4–4.6 mm	1.0	<12 µL	0.25 mm	×1
Narrow	2–2.1 mm	0.20–0.25	<8 µL	0.18 mm	×4
Micro	1 mm	0.05–0.07	<2 µL	0.12 mm	×16

for internal calibration. When using Promega modified trypsin (porcine) you will find the following autolytic peptides (monoisotopic *m/z*) that are suited for internal calibration: 842.510, 1045.564, and 2211.105. If these peaks in general are too high or too low, you should adjust the concentration (not the volume) of trypsin added in **Subheading 3.2., step 13**.

When analyzing the results of a search, you should consider the following when differentiating between hits:

How do the deviations between theoretical and experimental masses fit together (i.e., if they lie on a straight line when plotted against the peptide mass, you have a calibration error).

If a peptide contains a methionine, it is likely to have a +16 Da peak. The oxidized methionine seldom occurs alone.

N-terminal glutamine containing peptides are always accompanied by a –15 Da satellite peak (deamidation to pyroglutamic acid).

Multibasic sites often gives rise to lysine/arginine ladders (tryptic cleavage).

Missed cleavages (i.e., peptides with an internal Lys or Arg residue) are much more likely to occur if there is a neighboring acidic residue (tryptic cleavage).

In a peptide mixture, (small) lysine-containing peptides are often suppressed relative to arginine containing peptides due to the more basic nature of arginine.

The GPMAW program (*see* **Note 1**) is also useful for checking the validity of a hit as you can perform detailed analysis of likely/unlikely peptide identifications, as well as search for unusual enzyme cleavage sites.

9. Several web sites cater for the analysis of peptide mass data. The actual number changes over time, but you may want to check out the following:
   a. ProFound (http://65.219.84.5/service/prowl/profound.html).
   b. MsFit (http://prospector.ucsf.edu/ucsfhtml3.4/msfit.htm).
   c. Mascot (http://www.matrixscience.com)
   d. PeptideSearch (http://www.mann.embl-heidelberg.de/GroupPages/pageLink/peptidesearchpage.html)
   e. PepMapper (http://wolf.bms.umist.ac.uk/mapper/)

   You will have to read the accompanying help files for each search engine for detailed instructions.

10. You cannot usually identify reduced cysteine residues by Edman degradation or amino acid analysis, but the surrounding sequence will in most cases be sufficiently unique that positive identification of the peptide is no problem.
11. When separating a peptide map, it is advantageous to use multiple wavelengths. Modern in-line UV detectors can often be set up for three simultaneous wavelengths. You should use 214 nm for separating peptides, as you will get the highest response at this wavelength (you are measuring the absorbance of the peptide bond). Only the aromatic amino acid residues have any significant absorption at higher wavelength. Tryptophan and tyrosine have absorption maxima at 278 and 280 nm, respectively, whereas phenylalanine have maximum absorbance at 254 nm. You should thus set your detector at 214, 254, and 280 nm. Tryptophan is usually recognized by having an absorbance several times that of tyrosine.
12. The inclusion of TFA in the B-solvent is in order to create a stable baseline as both acetonitrile and TFA absorbs at low wavelength. Depending on your system, you may have to vary the concentration of TFA in the B-solvent between 0.75 and 0.85%. If you have a noisy baseline, decreasing the TFA concentration in both A- and B-solvent may help to minimize the problem.

    Mixing a small amount of organic solvent in the A-solvent and some water in the B-solvent helps to give a better mixing of the solvents (mixing water and acetonitrile is quite difficult), gives a better separation (helps to "wet" the hydrophobic surface, it may "collapse" in pure water) and helps to degas the A-solvent. If you have hydrophilic peptides that are eluted in the solvent front by the initial 5% organic content, you can use an A-solvent without organic content, but the first part of the chromatogram tend to be not so reproducible.

    Column manufacturers recommend that you store the column in 50% methanol/water. However, experience have shown that columns are stable for several years when stored in 0.1% TFA/acetonitrile (preferably with a high content of organic solvent).
13. Easy-to-make test mixture of peptides for checking reversed phase columns for peptide mapping: Dissolve 182 µg β-lactoglobulin (Sigma Inc, #L0130) in 100 µL, 50 m*M* ammoniumbicarbonate, pH 8.2. Add 3.6 µg trypsin (Roche Biochemicals, #109 819) and digest for 2 h at 37°C. Stop the digestion with 900 µL 0.1% TFA. Total solution is now 1000 µL at a concentration of 10 pMol/µL. Divide into 55 µL portions and store at –20°C. Use 50 µL for 4-mm columns and 20 µL for 2-mm columns.
14. If you use an automatic fraction collector, you should be careful to calibrate the fraction collector for exact peak collection. In addition, you should be careful that the collector does not add additional peak tubing to the separation system as this will result in delays and mixing of your sample resulting in a loss of resolution. Manual fraction collection is usually more precise but you may (initially) need an assistant to assist in tube handling and annotation. For manual collection, a strip chart recorder is most convenient, as computer acquisition or integrators often have an unacceptable delay.

15. Most proteolytic enzymes do not have an absolute specificity, but are active toward several residues with varying activities. Activity toward a given type of residue is usually also dependent on the local environment (e.g., 3-D structure and neighboring residues). Excessive digestion times may thus lead to secondary cleavages that are not desired and may be difficult to predict.

    Most of the enzymes that have a single residue specificity (Lys, Arg, Glu, Asp, Asn) will show some level of activity towards "like" residues.

    Although the selection of the specific enzymes can be clearly related to peptide predictions based on the primary structure (e.g., using the GPMAW program), the selection of unspecific enzymes like pepsin, thermolysin, and elastase is usually based on other parameters like low solubility of the target protein. They have a difficult to predict, low specificity that is very dependent on the local structure of the substrate. Pepsin is active at pH 2.0, a pH where proteins often are more soluble than at neutral pH. Thermolysin is active at temperatures as high as 80°C (even in the presence of 8 *M* urea) where most proteins are denatured. Elastase targets a group of small hydrophilic residues not targeted by any of the "specific" enzymes. These enzymes are usually used either for an initial cleavage where you hope for few cleavage positions to either make a protein soluble or to cleave a large protein into smaller, more manageable pieces. In these cases you have to experiment with short digestion times and/or very dilute solutions. Alternatively, the enzymes are used as a final cleavage to identify closely spaced disulfide bridges (*see* **ref. *4***).
16. During digestion, trypsin will autolyse and generate a small amount of ψ-trypsin, which has some chymotryptic activity. In certain cases, depending on the sequence or residual structure of the substrate, this may lead to complete cleavage at single or a few residues in a protein. The activity of trypsin is also affected by neighboring residues as acidic groups will slow cleavages and a following proline residue will in most cases completely inhibit cleavage. If you have neighboring basic residues (e.g., -Lys-Lys-) trypsin may cleave at either residue, but as the activity of trypsin toward terminal residues is considerably lower than towards internal residues, you may find all cleavage variants present in the peptide map even after prolonged digestion times.

    Modified trypsin that is considerably more resistant to autolysis is available from a number of manufacturers (*see* **Table 1**).
17. Chymotrypsin has a high activity toward the C-terminal side of aromatic residues, but other residues like Leu and Met can also be cleaved, particularly if they are situated in a hydrophobic environment (mostly -X-X↓Y where X is hydrophobic). Full digestions of intact proteins often give rise to unsuspected peptides, whereas the specificity when subdigesting peptides is usually very high.
18. Endoproteinase Glu-C, also called *S. aureus* protease V8, is a most useful enzyme in protein mapping. Its specificity nicely complements that of trypsin while still generating peptides of a suitable size. The main activity is directed toward the C-terminal of glutamic acid, but a reasonable activity toward aspartic acid means

that you may have to control digestion times carefully in order to limit cleavage to Glu. In a few cases, activity has also been seen toward Gly, but this is sequence dependent. If the following residue is Glu or Pro, cleavage will not take place (e.g., in a -Glu-Glu- sequence cleavage is only observed after the second Glu). Earlier reports that Endo Glu-C was specific toward Glu at pH 4.0 seems to be caused by the lower activity of the enzyme at this pH *(7)*.

19. Lysyl endopeptidase, Endo Lys-C, Endo Asp-N, and Endo Asn-C all tend to generate very clean peptide maps with a low amount of secondary cleavages. However, some of the peptides generated may be quite large and difficult to handle. Endo Asp-N and Endo Asn-C are considerably more expensive than the other enzymes mentioned but a single optimal digest is more likely to succeed than multiple rounds of digestion and separation.

    Endo Arg-C (submaxillary gland proteinase) and clostripain both target the C-terminal side of arginine, but with a significant amount of lysine cleavage. Furthermore, their activity is not very high making them problematic for digesting intact proteins.

References

1. Cole, R. B., ed. (1997) *Electrospray Ionization Mass Spectrometry.* Wiley, New York.
2. Stensballe, A., Andersen, S., and Jensen, O. N. (2001) Characterization of phosphoproteins from electrophoretic gels by nanoscale Fe(III) affinity chromatography with off-line mass spectrometry analysis. *Proteomics* **1,** 207–222.
3. Kussmann, M., Nordhoff, E., Rahbek-Nielsen, H., et al. (1997) Matrix-assisted laser desorption/ionization mass spectrometry: sample preparation techniques designed for various peptide and protein analytes. *J. Mass Spectrom.* **32,** 593–601.
4. Højrup, P., Jensen, M. S., and Petersen, T. E. (1985) Amino acid sequence of bovine protein Z: a vitamin K-dependent serine protease homolog. *FEBS Lett.* **184,** 333–387.
5. Peri, S., Steen, H., and Pandey, A. (2001) GPMAW—a software tool for analyzing proteins and peptides. *Trends in Biochem. Sci.* **26,** 691–693.
6. Speicher, K. D., Kolbas, O., Harper, S., and Speicher, D. W. (2000) Systematic analysis of peptide recoveries from In-Gel digestions for protein identifications in proteome studies. *J. Biomol. Tech.* **11,** 74–86.
7. Sørensen, S. B., Sørensen, T. L., and Breddam, K. (1991) Fragmentation of proteins by *S. Aureus* strain V8 protease. *FEBS Lett.* **294,** 195–197.
8. Allen, G. (1989) *Sequencing of Proteins and Peptides.* Elsevier/North Holland, Amsterdam.
9. Smith, B. J. (1994) Chemical cleavage of proteins, in *Methods in Molecular Biology, Vol. 32: Basic Protein and Peptide Protecols* (Walker, J. M., ed.), Humana, Totowa, NJ.

17

Mass Spectrometric Characterization of Posttranslationally Modified Proteins—Phosphorylation

Martin R. Larsen

1. Introduction

1.1. Overview

In higher organisms, the majority of proteins are posttranslationally modified at some stage, very often resulting in an essential change in the function of the protein. Some modifications change the protein solubility, others are used as molecular switches and thus modify biological activity, whereas others are used to locate proteins to different cell compartments (e.g., **ref.** ***1***). Because a given modification results in a change in the molecular mass of the affected amino acid, mass spectrometry (MS) with its unique sensitivity, high mass accuracy, and its ability to deal with complex mixtures, is the method of choice for characterization of posttranslational modifications *(2–4)*.

Mass spectrometric strategies for the characterization of posttranslational modifications in purified or recombinant proteins include molecular weight determination of the protein before and after chemical or enzymatic removal of the modifying group. This is followed by analysis of peptides derived from the protein by chemical or proteolytic digestion using various chromatographic separation methods *(4)*. Some of those strategies are not applicable to characterization of modified proteins present in low amount (subpicomole) or proteins separated by gel electrophoresis. Modified and optimized procedures are required. Although mass spectrometric analysis of intact proteins from gels after electroblotting *(5,6)*, electroelution *(7)*, or directly from a first dimension gel matrix *(8)* have been reported, these techniques are not routine in most laboratories, and often require custom-built mass spectrometers. Therefore, the

From: *Methods in Molecular Biology, vol. 251, HPLC of Peptides and Proteins: Methods and Protocols*
Edited by: M.-I. Aguilar © Humana Press Inc., Totowa, NJ

analysis must rely on the characterization of peptides derived by enzymatic digestion using different proteases. Here, the use of miniaturized chromatographic MS sample preparation methods specifically designed to either increase the number of detected peptides from a complex mixture or to selectively purify the modified peptides are essential for successful analysis. After identification of a presumably modified peptide, the presence of the modification is verified by chemical or enzymatic removal of the modifying group. The exact location of the modified residue can be assessed by mass spectrometric sequencing or the peptide can be isolated by liquid chromatography (LC) and further characterized if sufficient amounts are available.

1.2. Selected Modifications

The most common posttranslational modifications that take place in eukaryotic cells are glycosylation and phosphorylation. In this chapter, only strategies for the characterization of phosphorylated proteins will be described in detail. Common strategies for characterization of glycosylated proteins are mentioned below or in Chapter 19. However, the strategy described here for optimal sequence coverage can easily be applied to characterization of glycosylated proteins.

1.2.1. Protein Glycosylation

Protein glycosylation is observed in all eukaryotes, and recently it has also been found in prokaryotes ***(9)***. The majority of glycosylated proteins are secreted or membrane associated. The biological role of the glycostructures (glycans) vary from conformational stability, protection against degradation, to essential molecular and cellular recognition in for example development, growth, function, and cellular communication ***(10,11)***. Glycans are often complex branched structures built up of several different carbohydrate residues resulting in high heterogeneity, which often makes characterization difficult. In general, glycoproteins are characterized by measuring the molecular weight of the intact protein by MS, followed by proteolytic digestion and identification of the sites of glycosylation using differential peptide mapping before and after treatment with appropriate endoglycosidases ***(12)***. Characterization of the glycan structure is frequently performed using MS by measuring the molecular weight of the liberated glycan or the glycopeptide before and after sequential exoglycosidase treatment ***(13,14)***. For recent studies covering characterization of glycosylated proteins, the readers are referred to **refs. *15*** and ***16***.

1.2.2. Phosphorylated Proteins

Reversible phosphorylation of different amino acids, mainly serine, threonine, and tyrosine residues, is of key importance for regulation of the protein activity involved in complex cellular mechanisms including signal transduction, metabolism, or apoptosis development (e.g., **ref. *17***). The regulation mechanism is based on the interaction between protein kinases, which phosphorylates specific residues in the proteins, and protein phosphatases, which catalyze the reverse reaction. Because protein phosphorylation results in a mass increase of 80 Da per phosphate group, MS is the obvious method for characterizing this modification. However, several factors influence the success of the analysis of phosphorylated proteins by MS (*see* **Note 1**). For a comprehensive review, describing analysis of phosphorylated proteins and peptides by mass spectrometry, the readers are referred to **ref. *18***.

In this chapter, the strategy we have used in our laboratory for site-specific assignment of phosphorylation is described (*see* **Fig. 1**). It consists of two possible ways of characterizing phosphorylated proteins; **(A)** peptide mass mapping using microcolumns packed with material of increasing hydrophobicity (Poros R2, Poros Oligo R3, and graphite powder) and **(B)** immobilized metal affinity chromatography (IMAC) for selective purification of the phosphorylated peptide. Both methods are evaluated using alkaline phosphatase treatment to remove the phosphate group (*see* **Note 2**). The two strategies are described below.

2. Materials

2.1. Proteins and Enzymes

β-Casein (bovine milk), ovalbumin (chicken), and alkaline phosphatase (lyophilized) were obtained from Sigma (St. Louis, MO). Modified porcine trypsin was obtained from Promega (Madison, WI). Endoproteinase Glu-C was obtained from Calbiochem (La Jolla, CA). Alkaline phosphatase (1 U/μL) was obtained from Boehringer Mannheim/Roche, Indianapolis, IN).

2.2. Solvents, Chemicals, and Buffers

1. Ultrahigh-quality (UHQ) water (Millipore, Bedford, MA).
2. High-performance liquid chromatography (HPLC)-grade acetonitrile (CH_3CN), formic acid (HCOOH), acetic acid (CH_3COOH), and hydrochloric acid (HCl) (Selby Scientific limited, Clayton, Vic. Australia).
3. $FeCl_3$ (Sigma).
4. α-Cyano 4-hydroxycinnamic acid (4HCCA) (Sigma).
5. Cyanogen-bromide (CNBr) activated Sepharose 4 fast flow (Amersham Pharmacia Biotech AB, Uppsala, Sweden).
6. 50 m*M* ammonium bicarbonate (NH_4HCO_3, Sigma), pH 8.3.
7. PBS: 20 m*M* sodium phosphate buffer, 0.15 *M* NaCl, pH 7.5 (1 PBS tablet [ICN Biomedicals Inc., OH] in 100 mL UHQ).
8. 0.1 *M* Tris-HCl, pH 8.0, 5 m*M* $MgCl_2$.

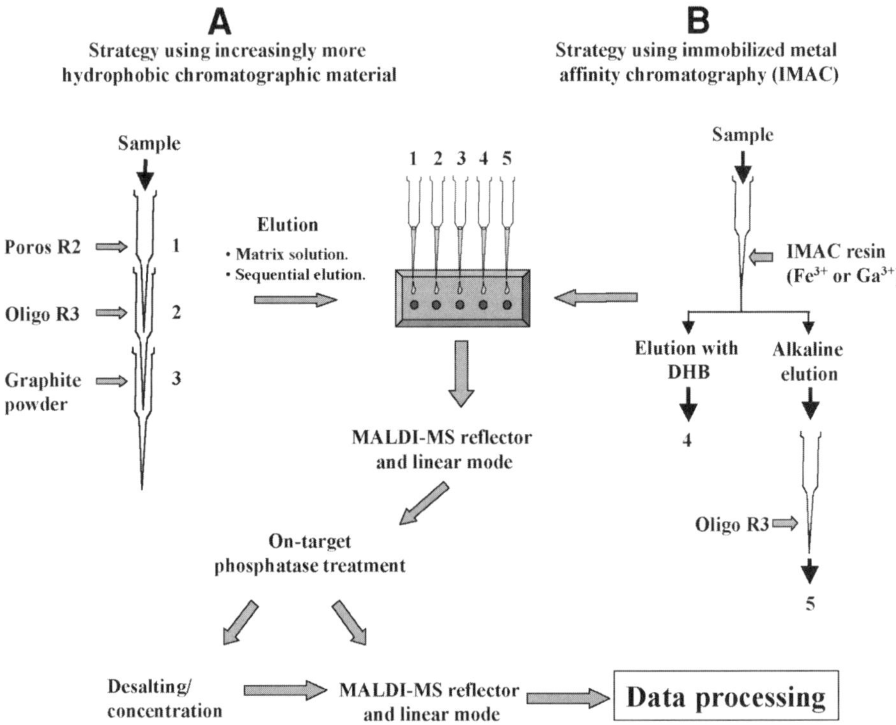

Fig. 1. Strategy for characterization of phosphorylated peptides. **(A)** Desalting and fractionation" of complex peptide mixture using microcolumns packed with increasing hydrophobic material. **(B)** Affinity purification of phosphorylated peptides using immobilized metal affinity chromatography. The numbers 1-5 indicate the different peptide fractions obtained from the two different strategies. All fractions are analyzed by MALDI-MS before and after treatment with alkaline phosphatase.

9. Nitrilotriacetic acid (NTA)-silica (16–24 µm particle size, Qiagen, Valencia, CA).
10. POROS R2 20 and POROS Oligo R3 media (PE Biosystems, Framingham, MA).
11. Graphite powder (activated charcoal C-5510, Sigma).

3. Methods

3.1. Matrix-Assisted Laser Desorption Ionization (MALDI)-Mass Spectrometry

Mass spectra presented in this chapter were obtained on a Voyager™ DE STR Biospectrometry Workstation (PE Biosystems). Spectra were acquired in positive reflector and linear ion mode.

3.2. Optimized MS Sample Preparation Method

Nearly complete sequence coverage, i.e., the fraction of the amino acid sequence that is represented by peptides in the peptide mass map, is essential to ensure detection of all modifications in a protein. The presence of low molecular weight contaminants, e.g., alkali metal ions, often causes reduced signal intensity or complete suppression of the peptide signals resulting in decreased sequence coverage. Desalting the peptide mixture on microcolumns prior to MS reduces this problem ***(19)***. Low molecular mass contaminants are easily removed from the column, resulting in increased signal intensity and signal-to-noise ratio. Consequently, improved sensitivity and significantly higher sequence coverage from peptide mass maps is observed. Combining microcolumns with increasing hydrophobicity can result in a further increase in the number of detected peptides in a mixture ***(20)*** and, therefore, increasing the possibility of detecting the modified peptide(s).

3.2.1. Preparation of GELoader Tip Microcolumns

The microcolumns are prepared from GELoader tips essentially as previously described in ***(19,21)***.

1. A partially constricted GELoader pipette tip (Eppendorf, Hamburg, Germany) is prepared by squeezing the narrow end. The narrow end of a GELoader tip is put flat on a hard surface. Then a 1.5-mL microfuge tube is rolled over the final 1 mm of the tip. Alternatively, the narrow end of the GELoader tip is fixed using a flat-surfaced forcep, and then the GELoader tip is turned once to close the end.
2. Prepare a slurry of 100–200 µL chromatographic material, e.g., POROS R2, POROS Oligo R3 or graphite powder in 70% acetonitrile (approx 1.5 mg/100 µL).
3. An aliquot (20 µL) of 70% acetonitrile is loaded in the top of the constricted GELoader tip and 0.5 µL of the resin slurry is added on top of the acetonitrile. By using a 1-mL syringe, fitted to the diameter of the GELoader tip via a disposable pipette tip cut in both ends, the liquid is gently pressed down to create a small column at the end of the constricted tip. Here, the column should be made dry before performing the next step (no liquid left). The size of the column should be 0.5 cm or less.
4. An aliquot (20 µL) of 2% formic acid (or 0.1% TFA [*see* **Note 3**]) is applied to the top of the column and 10 µL is used to equilibrate it using a gentle air pressure generated by the syringe. The remaining 10 µL 2% formic acid is left on top of the column bed.
5. Apply the analyte sample on top of the remaining 10 µL formic acid.
6. Press the liquid gently through the column by applying an air pressure (syringe). Do not dry the column, leave approx 2 µL on top of the column bed.
7. Wash the column with 20 µL 2% formic acid and leave the column dry.

8. Elute the analytes with 0.5 μL matrix solution directly onto the MALDI-MS target (*see* **Note 4**). The 0.5-μL matrix solution should be spotted in several droplets (5–10) on the target.
9. The POROS R2 and POROS Oligo R3 columns can be reused (up to four times) after washing with 30 μL of 100% acetonitrile, depending on the amount of starting material.

3.3. Differential Peptide Mass Mapping Using Microcolumns Packed With Increasingly Hydrophobic Material (Strategy A, Fig. 1)

In this strategy, an aliquot of the peptide mixture derived from the phosphorylated protein by in-gel digestion using a specific protease (*see* **Note 5**) is consecutively desalted and concentrated using different microcolumns packed with chromatographic material of increasing hydrophobicity. The peptides are initially purified on a Poros R2 GELoader tip microcolumn, which is similar to a C_{8-18} RP column. The flowthrough from this column is applied and concentrated on another microcolumn packed with Poros Oligo R3 material, which is slightly more hydrophobic than R2 material. Finally, the flowthrough from the R3 column is applied and concentrated on a microcolumn packed with graphite powder to purify the hydrophilic and small peptides that are not retained on the previous columns. Alternatively, the Poros R3 column can be omitted if the protein to be characterized is small and few peptides are generated. The bound peptides are eluted with matrix solution (*see* **Note 4**) directly onto the MALDI target. Mass spectra are obtained in both reflector and linear mode for optimal signal intensity of the phospho-peptide (*see* **Note 1**). The analyzed samples are treated with alkaline phosphatase (AP) directly on the target (see procedure below) and reanalyzed. Alternatively, an aliquot of the peptide solution is treated with alkaline phosphatase (*see* **Note 6**) and then purified as described above. The spectra before and after AP treatment are compared, and signals that change 80 Da or in multiples of 80 Da are identified as phosphorylated peptides and assigned to the protein sequence. For site specific assignment of the phosphate group(s) the remaining peptide mixture is used for tandem MS sequencing.

In **Fig. 2**, MALDI-MS analysis of peptides derived from β-casein using endoproteinase Glu-C is shown. β-casein is phosphorylated on five serine residues resulting in a monophosphorylated (KFQSpEEQQQTEDE *m/z* 1705.8) and a tetra-phosphorylated peptide (LNVPGEIVESpLSpSpSpEESITRINKKIE *m/z* 3191.5) observed when digested with endoproteinase Glu-C. An aliquot of the peptide mixture (corresponding to 0.5 pmol) was desalted and concentrated on a Poros R2 column and the flowthrough was further purified on a graphite powder column. The bound peptides were eluted using 4 HCCA in 70% ace-

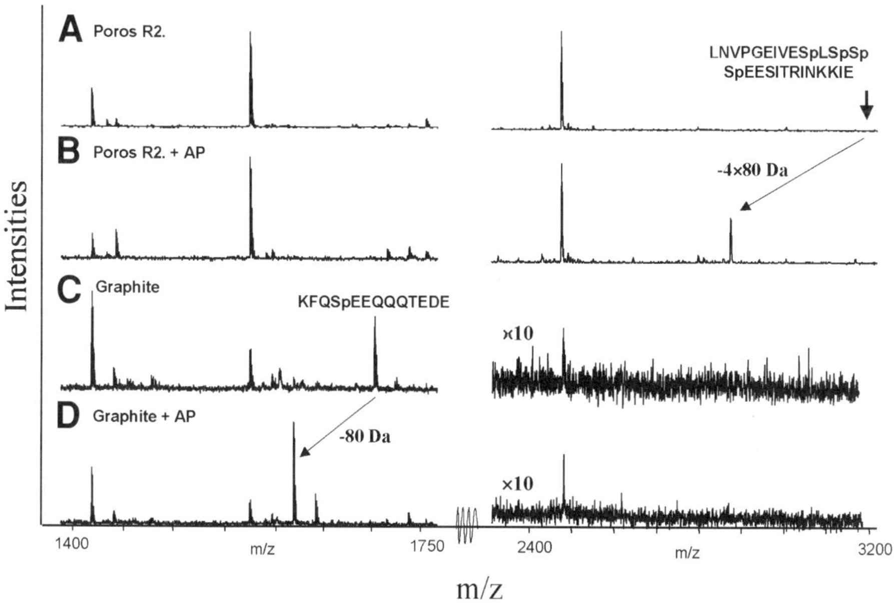

Fig. 2. Analysis of β-casein using microcolumns packed with increasing hydrophobic material. The β-casein was digested with endoproteinase Glu-C generating two phosphorylated peptides (*m/z* 1705.7 KFQSpEEQQQTEDE and *m/z* 3191.2 LNVPGEIVESpL-SpSpSpEESITRINKKIE). Only the mass ranges *m/z* 1400-1750 and *m/z* 2400-3200 are shown. **(A)** MALDI-MS peptide mass map of peptides eluted from the Poros R2 microcolumn. **(B)** Same as in (A) but after on-target alkaline phosphatase treatment. **(C)** MALDI-MS peptide mass map of peptides eluted from the graphite powder microcolumn. **(D)** Same as in (C) but after on-target alkaline phosphatase treatment. The peptides signals that changes 80 Da or multiples of 80 Da or are appearing after alkaline phosphatase treatment are marked. The matrix used in this experiment was α-cyano-4-hydroxy cinnamic acid in 70% acetonitrile.

tonitrile. The Poros R2 MALDI-MS peptide mass map is shown in **Fig. 2A**. The sample was treated with alkaline phosphatase directly on the MALDI target (*see* **Subheading 3.4.**) and reanalyzed (*see* **Fig. 2B**). One signal appeared in the dephosphorylated map, corresponding to the tetraphosphorylated peptide. The absence of the phosphorylated peptide before AP treatment illustrates the suppression effect observed in the MALDI ionization process. The monophosphorylated peptide was not detected in the R2 map. The Graphite powder MALDI-MS peptide mass map is shown in **Fig. 2C**. After dephosphorylation of this sample (*see* **Fig. 2D**) one signal disappeared in (C) and one appeared in (D), corresponding to the monophosphorylated peptide. The monophosphorylated peptide is completely absent in the R2 maps, presumably because it is very hydrophilic and thus not retained on the R2 column.

3.4. On-Target Alkaline Phosphatase Treatment

To increase the overall sensitivity of the analysis alkaline phosphatase treatment was performed directly on the previously analyzed sample *(22)*.

1. The matrix (4CCA or DHB) is redissolved on the target using 1.5 µL 50 m*M* NH_4HCO_3, pH 7.8, containing alkaline phosphatase (0.05 U/µL, Boehringer Mannheim/ Roche).

   We have found that the 4 HCCA crystals can be completely redissolved when using 50 m*M* NH_4HCO_3, pH 7.8, whereas DHB is harder to dissolve.
2. The target is placed in a closed plastic box containing a wet tissue to prevent the samples from drying.
3. The box is incubated at 37°C for 20–30 min.
4. The sample is acidified with 0.5 µL 5% formic acid, and the matrix is allowed to recrystalize.

   In the cases where the previously analyzed sample is desalted and concentrated on microcolumns or where low amount of matrix is apparent from the first analysis, additional matrix (0.2 µL) is added before recrystalization.
5. Prior to MALDI-MS analysis the surface is washed gently with 10 µL 0.1% TFA.

   When very low amounts of peptides are analyzed, the de-phosphorylated peptide sample is transferred to a small column containing Poros R2, Poros Oligo R3 or graphite powder and desalted/concentrated as described above.

3.5. On-Column Alkaline Phosphatase Treatment

For fast removal of phosphate group(s) from proteins and peptides, microcolumns packed with immobilized phosphatase can be used very efficiently. Phosphatase can be immobilized to different kinds of chromatographic material or can be obtained from a variety of manufacturers, e.g., MoBiTec GmbH (Göttingen, Germany, www.mobitec.de).

3.5.1. Immobilization to Sepharose Media

1. Cyanogen-bromide (CNBr) preactivated Sepharose 4B fast flow media (Amersham Pharmacia) (0.1 g) is incubated in 2 mL 10 m*M* HCl for 30 min.
2. The media is washed twice in 10 mL ice-cold 10 m*M* HCl and once in 20 m*M* phosphate buffer, pH 7.6.
3. Alkaline phosphatase (lyophilized) dissolved in 20 m*M* phosphate buffer, pH 7.6 (approx 2 mg/mL) is incubated with the activated media overnight at 4°C.
4. After centrifugation, the supernatant is removed and the sepharose beads are washed in 0.1 *M* Tris-HCl, pH 8, 5 m*M* $MgCl_2$ for 2 h at 4°C.
5. The alkaline phosphatase beads are stored in 50 m*M* NH_4HCO_3, pH 8.3 including 0.02% NaN_3.

3.5.2. AP Microcolumn

The GELoader tip AP microcolumns are made essentially as described above.

1. Prepare a partially constricted GELoader pipette tip (Eppendorf, Hamburg, Germany) by squeezing the narrow end.
2. An aliquot of the immobilized AP material is loaded onto the column in 50 m*M* NH_4HCO_3, pH 8.3 (*see* **Note 7**). Column size should be around 2 cm.
3. Wash the column in 30 μL 50 m*M* NH_4HCO_3, pH 8.3. Do not dry the column.
4. The sample is loaded onto the column in 50 m*M* NH_4HCO_3, pH 8.3.
5. Press the liquid gently and slowly through the column by applying an air pressure using the fitted syringe as above.
6. The dephosphorylated sample can either be analyzed directly by MALDI-MS or desalted/concentrated on RF microcolumns prior to MALDI-MS.

An example of the use of AP GELoader tip microcolumns for dephosphorylation is shown in **Fig. 3**. Aliquots (10% of the total volume) of a tryptic peptide mixture originating from in-gel digestion of 1 pmol ovalbumin were desalted/concentrated on POROS R2 microcolumns before **(A)** and after **(B)** on-column dephosphorylation. Two peptides which changed 80 Da in mass after dephosphorylation could be assigned to the ovalbumin sequences (*m/z* 2088.9 EVVGSpAEAGVDAASVSEEFR, and *m/z* 2525.1 [assuming alkylation with acrylamide of the cysteine residue] LPGFGDSpIEAQCGTSVNV-HSSLR).

3.6. IMAC (Strategy B, Fig. 1B)

Frequently, the signals originating from phosphorylated peptides are suppressed in the presence of nonphosphorylated peptides or because of very low phosphorylation stoichiometry. Therefore, selective isolation of the phosphopeptide(s) using immobilized metal affinity chromatography (e.g., **refs.** ***23–27***) prior to MS analysis may be advantageous. However, some problems are observed in conjunction with this method (*see* **Note 8**). The IMAC column material coated with either Ga^{3+} or Fe^{3+} is packed in constricted GELoader tips and the peptides are applied in a low pH buffer. The phospho-peptides are either eluted with a matrix solution directly onto the MALDI target, with a high pH buffer for further analysis, or alternatively the IMAC beads can be mixed with the matrix directly on the MALDI target *(28)* (*see* **Note 9**).

Various procedures for pre-purifying phosphorylated peptides prior to mass spectrometric analysis using IMAC columns have been used previously *(23–27)*. Here, the protocol adapted from *(27)* is described.

1. A slurry of Fe^{3+}-coated NTA-silica is prepared. The NTA-silica is washed consecutively in UHQ water, 0.1 *M* ethylene diamine tetraacetic acid (EDTA) and

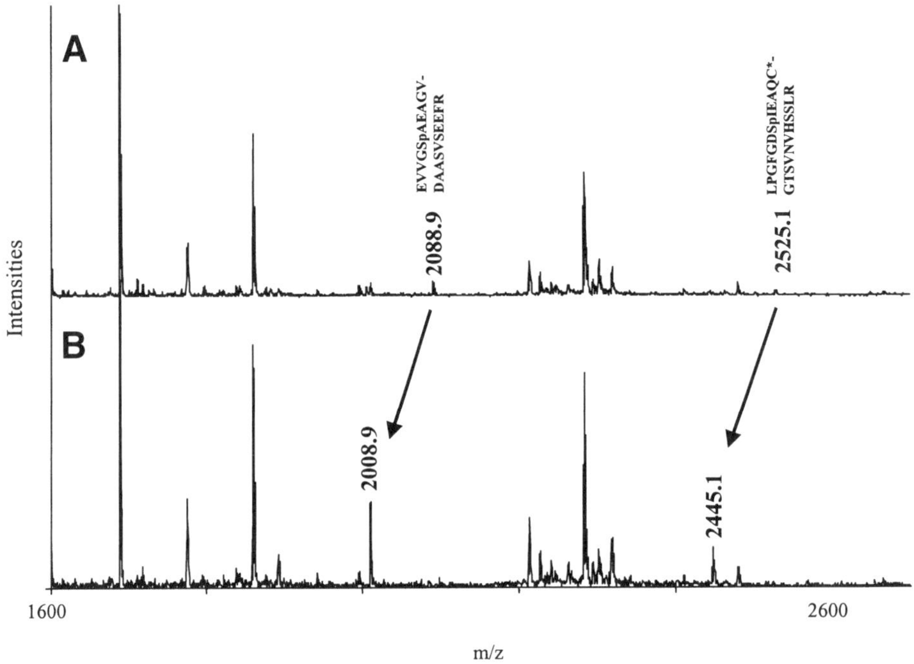

Fig. 3. On-column alkaline phosphatase digestion of tryptic peptides derived from ovalbumin. **(A)** MALDI-MS peptide mass map of peptides after desalting and concentration on a Poros R2 microcolumn. **(B)** MALDI-MS peptide mass map of the desalted and concentrated (A) peptides after on-column alkaline phosphatese treatment. The two phosphorylated peptides are marked in A. *Indicates that the cystein is *S*-acrylamidated.

0.1 *M* acetic acid, followed by incubation for 10 min at RT in 0.1 *M* $FeCl_3$ in 0.1 *M* acetic acid (1 : 1). After incubation, the coated Fe^{3+}-coated NTA-silica is washed in 0.1 *M* acetic acid followed by 0.1 *M* acetic acid and acetonitrile (3 : 1). The Fe^{3+}-coated NTA silica is stored in 0.1 *M* acetic acid.

2. Prepare a partially constricted GELoader pipette tip (Eppendorf) by squeezing the narrow end.
3. Load an aliquot of the Fe^{3+}-coated NTA silica onto the GELoader tip and the column is packed using air-pressure (syringe). The column size should be around 2-cm long.
4. The column is equilibrated with 20 µL 0.1 *M* acetic acid.
5. The peptide mixture is diluted in 0.1 *M* acetic acid and applied to the column using air-pressure (syringe). Do not dry the column.
6. The column is washed once with 0.1 *M* acetic acid and once with 0.1 *M* acetic acid/25% acetonitrile.

7. The phosphorylated peptides are eluted from the column using NH_4OH, pH 10.5/25% acetonitrile or 2.5-dihydroxy benzoic acid (DHB) in 50% acetonitrile. Alternatively, the IMAC beads can be mixed directly with MALDI matrix on the target and analyzed.
8. The eluted peptides in the NH_4OH, pH 10.5/25% acetonitrile fraction are further desalted/concentrated on POROS Oligo R3 or graphite powder microcolumns as described above.

This approach has been illustrated with the analysis of tryptic peptides derived from human calgranulin B isolated by 2D gel electrophoresis ***(29)***. The MALDI-MS peptide mass map of a small aliquot of the peptide mixture derived by in-gel tryptic digestion, and of an aliquot (5%) of the peptide mixture purified on a IMAC microcolumn are shown in **Fig. 4A** and **B**, respectively. Only the phosphopeptide signal together with the signals corresponding to losses of H_3PO_4 and HPO_3 are observed in the IMAC fraction. In addition, very little non-specific binding is apparent in the IMAC peptide mass map. Upon on-target de-phosphorylation with alkaline phosphatase a mass decrease of 80 Da is observed confirming that the signal at *m/z* 2255.85 Da represents a phosphopeptide containing one phosphate group (MHEGDEGPGHHHKPGLGEG$\underline{\text{T}}$P) (*see* **Fig. 4C**).

3.7. Alternative Techniques for Analysis of Phosphorylated Proteins/Peptides

3.7.1. Ion Scanning

An alternative way to identify phosphorylation sites in proteins is by precursor ion scanning ***(30–32)***. This technique allows detection of modified peptides by recording the loss of specific diagnostic fragment ions from the peptide. Different modifications yield different diagnostic low-mass ions during collision-induced dissociation (CID). Thus, phosphorylated peptides yield fragment ions at *m/z* 63 and 79 corresponding to PO_2^- and PO_3^-, respectively, when scanned in negative ion mode. However, for optimal fragmentation of the phosphorylated peptide a change in polarity of the mass spectrometer is required which often eliminates or reduces the signal from the phosphorylated peptide. The technique is mainly used in combination with liquid chromatography (LC) coupled to electrospray ionization mass spectrometry (ESI-MS) ***(32)***. However, it has successfully been applied to the identification of phosphorylation sites by direct analysis of the peptide mixture derived by in-gel digestion of electrophoretically separated proteins using nanoelectrospray MS (e.g., ***31,33***).

A variation of the technique uses the detection of the phospho-amino acid immonium ion ***(34)***. Here, the phospho-tyrosine immonium ion at *m/z* 216 is measured using a high-resolution hybride mass spectrometer (Qq-TOF).

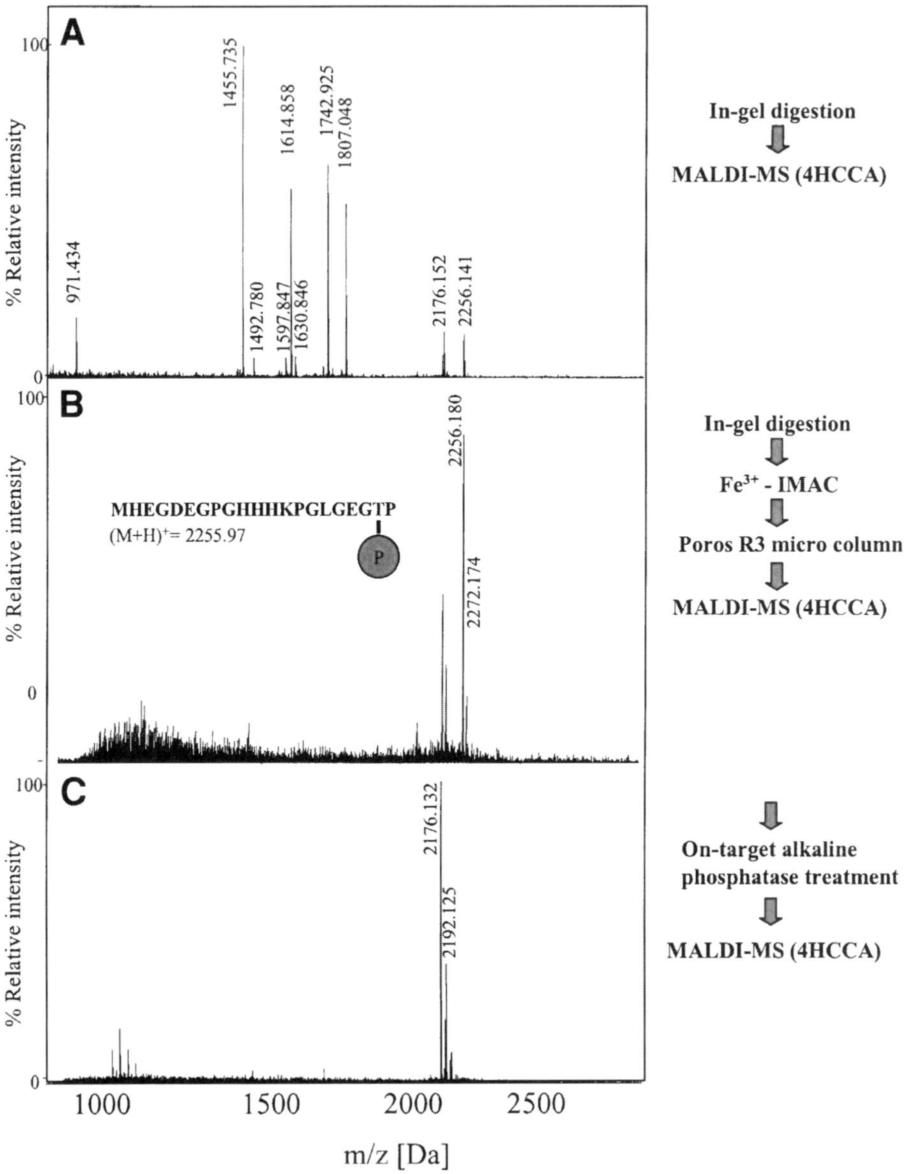

Fig. 4. Localization of a phospho-peptide in a peptide mixture obtained by in-gel tryptic digestion of calgranulin B, using immobilized metal affinity chromatography (IMAC). **(A)** MALDI peptide mass map of the in-gel generated peptide mixture. **(B)** MALDI spectrum of the IMAC purified phosphopeptide. Only the phosphopeptide signal together with the signals corresponding to losses of H_3PO_4 and HPO_3 is observed. **(C)** On-target alkaline phosphatase treatment of the IMAC purified phosphopeptide confirms phosphorylation by a loss of 80 Da. The matrix used in this experiment was α-cyano-4-hydroxy cinnamic acid in 70% acetonitrile. Reproduced with permission from **ref.** ***29*** with permission from Springer-Verlag.

Another ion scanning technique used frequently for analysis of phosphorylated peptides is neutral loss scanning performed using a triple quadrupole analyzer ***(35)***. In this technique, the neutral loss of H_3PO_4 (98 Da) is measured after CID by letting the third quadrupole (Q_3) scan *m/z* 98/n lower than the first quadrupole (Q_1).

3.7.2. Chemical Derivatization

A frequently used method for assignment of phosphorylation sites is based on the conversion of phospho-serine and phospho-threonine residues to *S*-(2-mercaptoethyl)cysteinyl or β-methyl-*S*-(2-mercaptoethyl)cysteinyl residues by beta-elimination/1,2-ethanedithiol addition, followed by reversible biotinylation of the modified protein/peptide (e.g., ***36***). The biotinylated compound can be affinity purified using streptavidin/avidin beads and analyzed, e.g., by MS, for site specific assignment of the modified site.

An alternative chemistry for derivatization of phosphopeptides has been reported ***(37)***, where the phosphopeptides are modified by attachment of cysteamine (1-amino-2-thioethane) to the phosphate group using a carbodiimine condensation reaction. The modified peptides can then be purified using an iodoacetyl chromatographic resin.

The sensitivity of the derivation methods described is limited by the chromatographic separations necessary to remove chemical compounds in between the different derivatization steps and the efficiency of the derivatization.

3.8. Concluding Remarks

Even with the increasing number of tools for phosphoprotein analysis, of which some of them are listed above, characterization of phosphorylated proteins still remains a tremendous challenge in biological MS. The main bottleneck seems to be the inability to detect all components present in a given sample, e.g., all peptides derived from a protein by proteolytic digestion, owing to ion suppression effects and losses during sample preparation. Therefore, we have concentrated on developing new sample preparation techniques for MS in order to specifically increase the number of peptides detected and to recover peptides normally lost during traditional sample preparation techniques.

The methods described in this chapter are focused on the detection of the phosphorylated peptide in a complex peptide mixture by using microcolumns packed with increasing hydrophobic material, which increases the chance for detecting all the components present in a given sample. The strategy shown in **Fig. 1** has, in our hands, been successful in a number of studies for site specific assignment of phosphorylation (e.g., ***22,27,38–40***).

However, successful characterization of phosphorylated proteins ultimately relies on the number of different techniques available in a given laboratory. The more tools available the higher the chance to characterize a phosphorylated protein.

4. Notes

1. Factors affecting phosphoprotein analysis by MS include:
   - Reversible phosphorylation is a regulatory mechanism the cell uses for activation/deactivation of proteins. Therefore, the phosphorylation/-dephosphorylation event must have a fast turnover to allow a fast response. Hence, the phosphorylated species are only present in a very short time period upon different stimuli, and consequently only low amounts of the phosphorylated species is present at a given time. Thus, characterization of phosphorylated proteins requires a fast and efficient way of isolating the phosphorylated from the non-phosphorylated species. Here 2D gel electrophoresis seems to be most advantageous (e.g., ***22***). Immunoprecipitation using anti-phosphotyrosine antibodies have been used to enrich for tyrosine-phosphorylated proteins (e.g., ***41***).
   - The presence of certain modifications reduces the ion yield of the corresponding peptide resulting in suppression of its signal. This is especially the case for acidic modifications, e.g., phosphorylation and glycosylation. Phosphorylated peptides do not ionize as well as nonmodified peptides during the ionization process and consequently yield lower signals, if any, in MS. The peptides can be analyzed in negative ion mode but with less sensitivity. Purification of the phosphorylated peptide, using e.g., IMAC, can overcome this suppression effect.
   - MALDI-TOF/MS is predominantly performed in reflector mode because of the high resolution and mass accuracy. However, phosphopeptides undergo metastable fragmentation, resulting in loss of the phosphate group when operating in reflector mode ***(42)***. Consequently, lower signal intensity is observed. Therefore, it is advantageous to analyze the phosphorylated peptides in both reflector and linear ion mode for optimal detection. In addition, the use of different matrices in MALDI-MS can decrease the metastable fragmentation (e.g., DHB).
2. Alkaline phosphatase is an unspecific phosphatase which cleaves off phosphate groups from a variety of amino acids including serine, threonine and tyrosine. Other phosphatases or a cocktail of different phosphatases can be used.
3. When the microcolumn is used as a cleanup step prior to nano-ESI-MS, the loading/washing solution should be formic acid, as TFA is not compatible with ESI-MS.
4. Several MALDI matrices are available and useful for this sample preparation procedure. However, α-cyano 4-hydroxycinnamic acid (4HCCA) in 70% acetonitrile or DHB in 50% acetonitrile are the most common used. The Poros R2 and Poros Oligo R3 columns can be sequentially eluted using increasing amounts of acetonitrile, and the different fractions are then mixed with the matrix on the MALDI target. This sequential elution is not compatible with the graphite microcolumns since the presence of matrix during peptide elution is essential for the outcome of

the analysis ***(20)***. Hence, the graphite columns are not suitable for desalting of peptides prior to ESI-MS unless picomole amount of sample is available.

5. Proteases for in-gel digestion. Some proteases are more specific when used for in-gel digestion than others. The most common proteases, which we have found to be specific, are trypsin, endoproteinases lys-C and asp-N. However, trypsin is not suitable for very low amount of protein as it generates autoproteolytic fragments, which suppresses signals from peptides derived from the protein in the gel. Other proteases like chymotrypsin and endoproteinase glu-C has not, in our hands been specific enough for assignment of peptides when used for in-gel digestion.
6. Dephosphorylation of the peptide mixture can be performed in one of three ways; in solution, on the target after the first MALDI analysis ***(22)*** or on a GELoader tip microcolumn packed with immobilized alkaline phosphatase.
7. Any kind of buffer can be used for dephosphorylation of protein/peptide mixtures on AP microcolumns, as long as it follows the recommendations from the manufacturers of the AP used. Alkaline phosphatase is a very resistant enzyme and will be active even under hard conditions.
8. The sensitivity of the IMAC technology seems to be dependent on the sequence of the phosphorylated peptide to be purified. The presence of clusters of acidic amino acids (glutamic acid and aspartic acid) seems to enhance the selective purification of a phosphorylated peptide. For example, ovalbumin (chicken) has two serine phosphorylation sites located on two peptides (LPGFGDSpIEAQCGTS VNVHSSLR. *m/z* 2453.1 and EVVGSAEAGVDAASVSpEEFR, *m/z* 2087.9). The peptide at *m/z* 2087.9, which has a small cluster of glutamic acid next to the phosphate group can routinely be purified from 0.5 pmol of starting material, whereas the other peptide require more starting material (Larsen MR, unpublished results). In other cases the phosphopeptide binds with a very high affinity so that it can not come off the column again. Here, the IMAC beads can be analyzed directly by mixing them with matrix on the MALDI target or the phosphopeptides can be eluted with an alkaline phosphatase solution ***(28)***.
9. DHB in 50% acetonitrile is used for elution of peptides off the IMAC column directly onto the MALDI target. In cases where the IMAC beads are mixed with the matrix directly on the MALDI target, 4 HCCA has been used.

References

1. Krishna, R. G. and Wold, F. (1993) Post-translational modification of proteins. *Adv. Enzymol. Relat. Areas Mol. Biol.* **67,** 265–298.
2. Annan, R. S. and Carr, S. A. (1997) The essential role of mass spectrometry in characterizing protein structure: mapping posttranslational modifications. *J. Protein Chem.* **16,** 391–402.
3. Burlingame, A. L. (1996) Characterization of protein glycosylation by mass spectrometry. *Curr. Opin. Biotechnol.* **7,** 4–10.
4. Andersen, J. S., Svensson, B., and Roepstorff, P. (1996) Electrospray ionization and matrix assisted laser desorption/ionization mass spectrometry: powerful analytical tools in recombinant protein chemistry. *Nat. Biotechnol.* **14,** 449–457.

5. Schreiner, M., Strupat, K., Lottspeich, F., and Eckerskorn, C. (1996) Ultraviolet matrix assisted laser desorption ionization-mass spectrometry of electroblotted proteins. *Electrophoresis* **17(5),** 954–961.
6. Eckerskorn, C., Strupat, K., Schleuder, D., et al. (1997) Analysis of proteins by direct-scanning infrared-MALDI mass spectrometry after 2D-PAGE separation and electroblotting. *Anal. Chem.* **69(15),** 2888–2892.
7. Haebel, S., Jensen, C., Andersen, S. O., and Roepstorff, P. (1995) Isoforms of a cuticular protein from larvae of the meal beetle, Tenebrio molitor, studied by mass spectrometry in combination with Edman degradation and two-dimensional polyacrylamide gel electrophoresis. *Protein Sci.* **4,** 394–404.
8. Loo, J. A., Brown, J., Critchley, G., et al. (1999) High sensitivity mass spectrometric methods for obtaining intact molecular weights from gel-separated proteins. *Electrophoresis* **20,** 743–748.
9. Moens, S. and Vanderleyden, J. (1997) Glycoproteins in prokaryotes. *Arch. Microbiol.* **168,** 169–175.
10. Varki, A (1993) Biological roles of oligosaccharides: all of the theories are correct. *Glycobiology* **3,** 97–130.
11. Lis, H. and Sharon, N. (1993) Protein glycosylation. Structural and functional aspects. *Europ. J. Biochem.* **218,** 1–27.
12. Packer, N. H. and Harrison, M. J. (1998) Glycobiology and proteomics: is mass spectrometry the Holy Grail? *Electrophoresis* **19(11),** 1872–1882.
13. Mortz, E., Sareneva, T., Haebel, S., Julkunen, I., and Roepstorff, P. (1996) Mass spectrometric characterization of glycosylated interferon-gamma variants separated by gel electrophoresis. *Electrophoresis* **17,** 925.
14. Kuster, B., Wheeler, S. F., Hunter, A. P., Dwek, R. A., and Harvey, D. J. (1997) Sequencing of N-linked oligosaccharides directly from protein gels: in-gel deglycosylation followed by matrix-assisted laser desorption/ionization mass spectrometry and normal-phase high-performance liquid chromatography. *Anal. Biochem.* **250(1),** 82–101.
15. Burlingame, A. L. (1996) Characterization of protein glycosylation by mass spectrometry. *Curr. Opin. Biotechnol.* **7,** 4–10.
16. Harvey, D. J., Kuster, B., and Naven, T. J. (1998) Perspectives in the glycosciences—matrix-assisted laser desorption/ionization (MALDI) mass spectrometry of carbohydrates. *Glycoconj J.* **15,** 333–338.
17. Graves, J. D. and Krebs, E. G. (1999) Protein phosphorylation and signal transduction. *Pharmacol. Ther.* **82(2–3),** 111–121.
18. McLachlin, D. T. and Chait, B. T. (2001) Analysis of phosphorylated proteins and peptides by mass spectrometry. *Curr. Opin. Chem. Biol.* **5(5),** 591–602.
19. Gobom, J., Nordhoff, E., Mirgorodskaya, E., Ekman, R., and Roepstorff, P. (1999) Sample purification and preparation technique based on nano-scale reversed-phase columns for the sensitive analysis of complex peptide mixtures by matrix-assisted laser desorption/ionization mass spectrometry. *J. Mass Spectrom.* **34,** 105.
20. Larsen, M. R., Cordwell, S. J., and Roepstorff, P. (2002) Graphite powder as an alternative or supplement to reversed-phase material for desalting and concentra-

tion of peptide mixtures prior to MALDI mass spectrometry. *Proteomics* **2,** 1277–1287.

21. Kussmann, M., Nordhoff, E., Nielsen, H. R., et al. (1997) MALDI-MS sample preparation techniques designed for various peptide and protein analytes. *J. Mass Spectrom.* **32,** 593.
22. Larsen, M. R., Sorensen, G. L., Fey, S. J., Larsen, P. M., and Roepstorff, P. (2001) Phospho-proteomics: evaluation of the use of enzymatic de-phosphorylation and differential mass spectrometric peptide mass mapping for site specific phosphorylation assignment in proteins separated by gel electrophoresis. *Proteomics* **1(2),** 223–238.
23. Neville, D. C., Rozanas, C. R., Price, E. M., Gruis, D. B., Verkman, A. S., and Townsend, R. R. (1997) Evidence for phosphorylation of serine 753 in CFTR using a novel metal-ion affinity resin matrix-assisted laser desorption mass spectrometry. *Protein Sci.* **6(11),** 2436–2445.
24. Posewitz, M. C. and Tempst, P. (1999) Immobilized gallium(III) affinity chromatography of phosphopeptides. *Anal. Chem.* **71,** 2883–2892.
25. Li, S. and Dass, C. (1999) Iron(III)-immobilized metal ion affinity chromatography and mass spectrometry for the purification and characterization of synthetic phosphopeptides. *Anal. Biochem.* **270,** 9–14.
26. Figeys, D., Gygi, S. P., Zhang, Y., Watts, J., Gu, M., and Aebersold, R. (1998) Electrophoresis combined with novel mass spectrometry techniques: powerful tools for the analysis of proteins and proteomes. *Electrophoresis* **19,** 1811–1818.
27. Stensballe A, Andersen S, Jensen ON. (2001) Characterization of phosphoproteins from electrophoretic gels by nanoscale Fe(III) affinity chromatography with off-line mass spectrometry analysis. *Proteomics* **1(2),** 207–222.
28. Xhou, W., Merrick, B. A., Khaledi, M. G., and Tomer, K. B. (2000) Detection and sequencing of phosphopeptides affinity bound to immobilized metal ion beads by matrix-assisted laser desorption/ionization mass spectrometry. *J. Am. Soc. Mass Spectrom.* **11(4),** 273–282.
29. Larsen, M. R. and Roepstorff, P. (2000) Mass spectrometric identification of proteins and characterization of their post-translational modifications in proteome analysis. *Fresenius J. Anal. Chem.* **366(6–7),** 677–690.
30. Carr, S. A., Huddleston, M. J., and Annan, R. S. (1996) Selective detection and sequencing of phosphopeptides at the femtomole level by mass spectrometry. *Anal. Biochem.* **239,** 180–192.
31. Wilm, M., Neubauer, G., and Mann, M. (1996) Parent ion scans of unseparated peptide mixtures. *Anal. Chem.* **68(3),** 527–533.
32. Annan, R. S., Huddleston, M. J., Verma, R., Deshaies, R. J., and Carr, S. A. (2001) A multidimensional electrospray MS-based approach to phosphopeptide mapping. *Anal. Chem.* **73(3),** 393–404.
33. Neubauer, G. and Mann, M. (1999) Mapping of phosphorylation sites of gel-isolated proteins by nanoelectrospray tandem mass spectrometry: potentials and limitations. *Anal. Chem.* **71,** 235–242.

34. Steen, H., Kuster, B., Fernandez, M., Pandey, A., and Mann, M. (2001) Detection of tyrosine phosphorylated peptides by precursor ion scanning quadrupole TOF mass spectrometry in positive ion mode. *Anal. Chem.* **73(7),** 1440–1448.
35. Schlosser, A., Pipkorn, R., Bossemeyer, D., and Lehmann, W. D. (2001) Analysis of protein phosphorylation by a combination of elastase digestion and neutral loss tandem mass spectrometry. *Anal Chem.* **73(2),** 170–176.
36. Oda, Y., Nagasu, T., and Chait, B. T. (2001) Enrichment analysis of phosphorylated proteins as a tool for probing the phosphoproteome. *Nat. Biotechnol.* **19(4),** 379–382.
37. Zhou, H., Watts, J. D., and Aebersold, R. (2001) A systematic approach to the analysis of protein phosphorylation. *Nat. Biotechnol.* **19(4),** 375–378.
38. Powell, K. A., Valova, V. A., Malladi, C. S., Jensen, O. N., Larsen, M. R., and Robinson, P. J. (2000) Phosphorylation of dynamin I on Ser-795 by protein kinase C blocks its association with phospholipids. *J. Biol. Chem.* **275(16),** 11,610–11,617.
39. Jellinek, D. A., Chang, A. C., Larsen, M. R., Wang, X., Robinson, P. J., and Reddel, R. R. (2000) Stanniocalcin 1 and 2 are secreted as phosphoproteins from human fibrosarcoma cells. *Biochem. J.* **350 Pt. 2,** 453–461.
40. Jensen, H. H., Hjerrild, M., Guerra, B., Larsen, M. R., Hojrup, P., and Boldyreff, B. (2001) Phosphorylation of the Fas associated factor FAF1 by protein kinase CK2 and identification of serines 289 and 291 as the in vitro phosphorylation sites. *Int. J. Biochem. Cell Biol.* **33(6),** 577–589.
41. Pandey, A., Podtelejnikov, A. V., Blagoev, B., Bustelo, X. R., Mann, M., and Lodish, H. F. (2000) Analysis of receptor signaling pathways by mass spectrometry: identification of vav-2 as a substrate of the epidermal and platelet-derived growth factor receptors. *Proc. Natl. Acad. Sci. USA* **97(1),** 179–184.
42. Annan, R. S. and Carr, S. A. (1996) Phosphopeptide analysis by matrix-assisted laser desorption time-of-flight mass spectrometry. *Anal. Chem.* **68,** 3413–3421.

18

Analyses of Glycopeptides and Glycoproteins by Liquid Chromatography–Mass Spectrometry and Liquid Chromatography–Tandem Mass Spectrometry

Nana Kawasaki, Miyako Ohta, Satsuki Itoh, and Takao Hayakawa

1. Introduction

A variety of recombinant glycoproteins, including erythropoietin (EPO), and tissue plasminogen activator have been developed as medical agents. The carbohydrate moieties are known to be implicated in the biological activity, metabolic fate, stability, and solubility of these compounds, and it is, therefore, important to analyze the structural features of carbohydrate moieties as well as polypeptide chains in glycoprotein products *(1)*.

Liquid chromatography–mass spectrometry (LC–MS) is an effective method for analyses of both polypeptide chains and sugar chains in glycoproteins *(2,3)*. Glycoproteins are digested into peptides and glycopeptides bearing one sugar chain by proteases such as trypsin, endoproteinase Lys-C, Glu-C, and Asp-N, and the digests are then subjected to LC–MS equipped with a reversed-phase column. A mobile phase containing acids such as trifluoroacetic acid (TFA) is generally employed for the elution of both nonglycosylated and glycosylated peptides *(3)*. Amino acid residues, glycosylation sites, and preliminary glycosylation can be characterized from the mass spectra of peaks in this peptide/glycopeptide map. In contrast, the use of ammonium acetate as a mobile phase can preferentially elute the glycopeptides, and the glycopeptides are separated based on the structure of carbohydrates *(4)*. This glycopeptide mapping is useful for the analysis of site-specific carbohydrate heterogeneity in glycoproteins.

Liquid chromatography–tandem mass spectrometry (LC–MS/MS) is valuable for locating the glycopeptides in the peptide/glycopeptide map *(5)*. The

From: *Methods in Molecular Biology, vol. 251, HPLC of Peptides and Proteins: Methods and Protocols*
Edited by: M.-I. Aguilar © Humana Press Inc., Totowa, NJ

precursor-ion scanning, in which marker ions such as *m/z* 204 ($HexNAc^+$) produced from glycoproteins by collision-induced decomposition (CID) are monitored, can trace only glycopeptides in proteolytic-digested glycoproteins.

The present chapter describes the following methods:

1. Proteolytic digestion of a glycoprotein.
2. LC–MS analysis of a proteolytic-digested glycoprotein.
   a. Peptide/glycopeptide mapping.
   b. Glycopeptide mapping.
3. LC–MS/MS analysis of a proteolytic-digested glycoprotein.

Using EPO as an example of a glycoprotein, we also present here applications of LC–MS and LC–MS/MS with glycoproteins.

2. Materials

2.1. Proteolytic Digestion of Glycoproteins

1. Proteases. Choose a protease or proteases that will cleave the glycoprotein into glycopeptides containing one glycosylation site. Enzyme specificity is as follows:
   a. Trypsin: Lys/Arg-↓-X,
   b. Endoproteinase Lys-C (Lys-C): Lys-↓-X,
   c. Endoproteinase Glu-C (Glu-C): Glu(Asp)-↓-X,
   d. Endoproteinase Asp-N (Asp-N): X-↓-Asp(Glu)
2. Protease solution (*see* **Note 1**):
   a. Trypsin: Dissolve TPCK-trypsin (Sigma, St. Louis, MO) in 3 m*M* HCl at a final concentration of 2 μg/μL.
   b. Lys-C: Dissolve Lys-C (Roche Diagnostics, GmbH, Germany) in 100 m*M* ammonium acetate, pH 8.6, at a final concentration of 0.4 μg/μL.
   c. Glu-C: Dissolve Glu-C (Roche Diagnostics, GmbH, Germany) in 100 m*M* ammonium acetate, pH 8.0, at a final concentration of 4 μg/μL.
   d. Asp-N: Dissolve Asp-N (Wako Pure Chemical Industries, Ltd., Osaka, Japan) in 25 m*M* ammonium bicarbonate, pH 8.0, at a final concentration of 0.1 μg/μL.
3. Proteolytic digestion buffer:
   a. Trypsin: 100 m*M* Tris-HCl, pH 8.0.
   b. Lys-C: 100 m*M* ammonium acetate, pH 8.6.
   c. Glu-C: 100 m*M* ammonium acetate, pH 8.0.
   d. Asp-N: 25 m*M* ammonium bicarbonate, pH 8.0.
4. Reduction and carboxymethylation buffer: 0.5 *M* Tris-HCl (pH 8.6) containing 8 *M* guanidine hydrochloride and 5 m*M* EDTA.
5. 2-Mercaptoethanol.
6. Monoiodoacetic acid.
7. Sephadex G-25 column, e.g., PD-10 column (Amersham Biosciences, Uppsala, Sweden).

2.2. LC–MS and LC–MS/MS Analyses of Proteolytic-Digested Glycoproteins

LC–MS/MS analysis (Precursor-ion scanning) generally requires a larger amount of sample than LC–MS analysis (normal scanning). Use of a microbore column (1.0 mm id) is, therefore, recommended for precursor-ion scanning.

1. HPLC equipment:
   a. For LC–MS: A gradient pump equipped with an injector, a column oven, an integrator, and a detector capable of monitoring UV absorbance at 206 nm.
   b. For LC–MS/MS: A gradient pump equipped with an injector, an integrator, and a UV monitor equipped with a microflow cell (0.3 µL), e.g., Magic 2002 (Michrom BioResources, Inc. Auburn, CA).
2. ESI-MS and -MS/MS: Triple-stage quadrupole mass spectrometer equipped with an electrospray ion source, e.g., Finnigan TSQ-7000 (Thermo Finnigan, Inc. San Jose, CA).
3. Column:
   a. For LC–MS: Vydac 218TP52 (250 × 2.1 mm, Vydac, Hesperia, CA), and Pegasil ODS (250 × 2.1 mm, Senshu Science Co. Ltd., Tokyo, Japan).
   b. For LC–MS/MS: Magic C18 (150 × 1.0 mm, Michrom BioResource, Inc. Auburn, CA).
4. HPLC mobile phase (*see* **Note 2**):
   a. For peptide/glycopeptide mapping by LC/MS:
      Solvent A: 0.05% TFA.
      Solvent B: 0.05% TFA/50 %(v/v) acetonitrile.
   b. For glycopeptide mapping by LC–MS:
      Solvent A: 1 m*M* ammonium acetate, pH 6.8.
      Solvent B: 1 m*M* ammonium acetate, pH 6.8/80% (v/v) acetonitrile.
   c. For precursor-ion scanning by LC–MS/MS:
      Solvent A: 0.05% TFA/2%(v/v) acetonitrile.
      Solvent B: 0.05% TFA/80%(v/v) acetonitrile.

3. Methods

3.1. Proteolytic Digestion of Glycoproteins

Glycoproteins that possess multiple disulfide bonds are reduced and derivatized with monoiodoacetic acid for the complete proteolysis as follows **(steps 1–4)**. These steps can be omitted for glycoproteins having only a few disulfide bonds.

1. Dissolve a glycoprotein (360 µg) in 360 µL of reduction and carboxymethylation buffer.
2. Add 2.6 µL of 2-mercaptethanol, and incubate the mixture at room temperature for 2 h.
3. To the reaction mixture, add 7.56 mg of monoiodoacetic acid and incubate the mixture at room temperature for 2 h in the dark.

4. Apply the reaction mixture on a Sephadex G-25 column to remove the reagents, and lyophilize the effluent (*see* **Note 3**).
5. Dissolve the glycoprotein in proteolytic digestion buffer to a final concentration of 1 mg/mL.
6. Add protease solution, and incubate the sample at 37°C for 1–24 h.
   a. Trypsin: Add trypsin solution to a final protein:enzyme ratio of 50:1, and incubate for 1–4 h (*see* **Note 4**).
   b. Lys-C: Add Lys-C solution to a final protein:enzyme ratio of 25:1, and incubate for 24 h.
   c. Glu-C: Add Glu-C solution to a final protein:enzyme ratio of 50:1, and incubate for 20 h (*see* **Note 5**).
   d. Asp-N: Add Asp-N solution to a final protein:enzyme ratio of 50:1, and incubate for 20 h.
7. Store the sample at –20°C before use.

Erythropoietin contains three *N*-glycans at Asn 24, 38, and 83, and one *O*-glycan at Ser-126 ***(6)***. This glycoprotein has two disulfide bonds (*see* **Fig. 1**). The structures of *N*-glycans in EPO are reported to be fucosylated bi-, tri-, and tetraantennary oligosaccharides containing 0–4 sialic acid with 0–3 *N*-acetyllactosamine (*see* **Fig. 2**) ***(3,7–9)***. Glu-C solution (62 µL) was added to EPO (1 mg) dissolved in 1 mL of 100 m*M* ammonium acetate, pH 8.0, and the sample was incubated at 37°C for 20 h. EPO is expected to be digested into four glycopeptides, E5, E6, E10, and E12, and some peptides, as shown in **Fig. 1**.

3.2. LC–MS Analysis of a Proteolytic-Digested Glycoprotein

3.2.1. Peptide/Glycopeptide Mapping

1. Equilibrate a column with the starting solvent (1% of B) at a flow rate of 0.2 mL/min, and monitor the absorbance at 206 nm. Set the column oven to 40°C.
2. Set the ESI voltage and electron multiplier to 4500V and 1200V, respectively. Set the capillary temperature to 225°C. Adjust the pressure of the sheath gas (N_2) and auxiliary gas (N_2) to 70 psi and 10 U, respectively. Set the scan time to 4 s. Perform ionization and acquire the ions at *m/z* 550–2400 in the positive ion mode (*see* **Note 6**).
3. Inject an aliquot (50 µL) of the sample onto the HPLC column.
4. Elute the column using a linear gradient from 1% to 90% of B in 130 min at a flow rate of 0.2 mL/min.

An elution profile of Glu-C-digested EPO is shown in **Fig. 3**. Amino acid residues in nonglycosylated peptides were determined by comparing the experimental masses with the masses predicted from the cDNA-derived amino acid sequence (*see* **Fig. 1**). Mass spectra of glycopeptides E5, E6, E10, and E12 are present in **Fig. 4**. Carbohydrate structures and amino acid residues of the ions A1–A11, B1–B13, and C1–14 were determined on the basis of the experi-

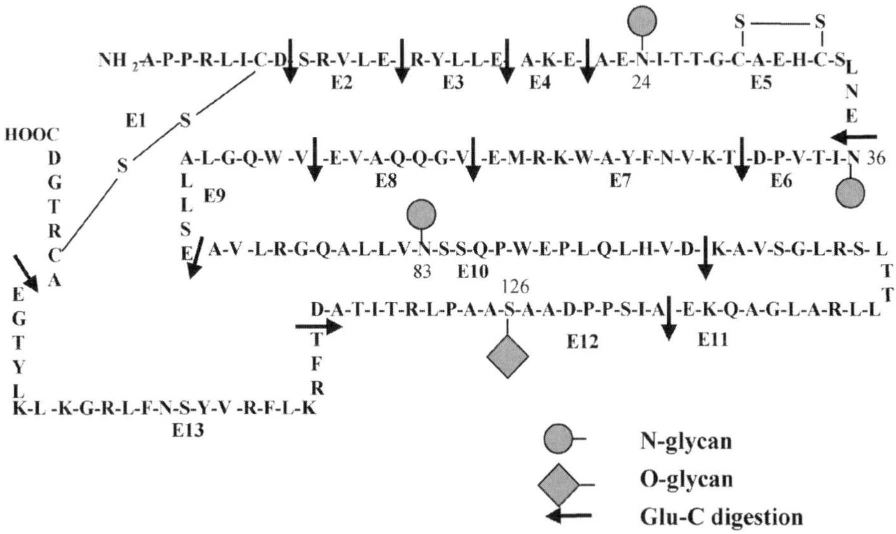

Fig. 1. Amino acid sequence and glycosylation sites of erythropoietin.

mental and theoretical masses of peptides and carbohydrates (*see* **Fig. 2**). The structural assignment of peaks shown in **Fig. 3** is summarized in **Table 1**.

3.2.2. Glycopeptide Mapping

1. Equilibrate a column with the starting solvent (1% of B) at a flow rate of 0.2 mL/min, and monitor the absorbance at 206 nm. Set the column oven to 40°C.
2. Set the ESI voltage, electron multiplier, and capillary temperature to 4500V, 1200V, and 225°C, respectively. Adjust the pressure of the sheath gas (N_2) and auxiliary gas (N_2) to 70 psi and 10 U, respectively. Set the scan time to 3 s. Perform ionization and acquire the ions at *m/z* 1000–2400 in the positive ion mode (*see* **Note 7**).
3. Apply the sample (15 μL) on the HPLC column.
4. Elute the column using a linear gradient from 1% to 6% of B in 60 min followed by a further increase of solvent B up to 36% within 80 min (*see* **Note 8**).

The glycopeptide map of Glu-C-digested EPO is shown in **Fig. 5A**. Four glycopeptides were eluted in the order of E6 (Asn36), E5 (Asn24), E12 (Ser126), and E10 (Asn83), and the glycopeptides were further separated based on the structure of the oligosaccharides attached to each glycosylation site. For example, glycopeptide E6 was separated into 10 peaks (*see* **Fig. 5B**). These glycoforms were eluted in the order of tetra-, tri-, and disialylated glycopeptides, and they were further separated based on the number of *N*-acetyllactosamine repeats.

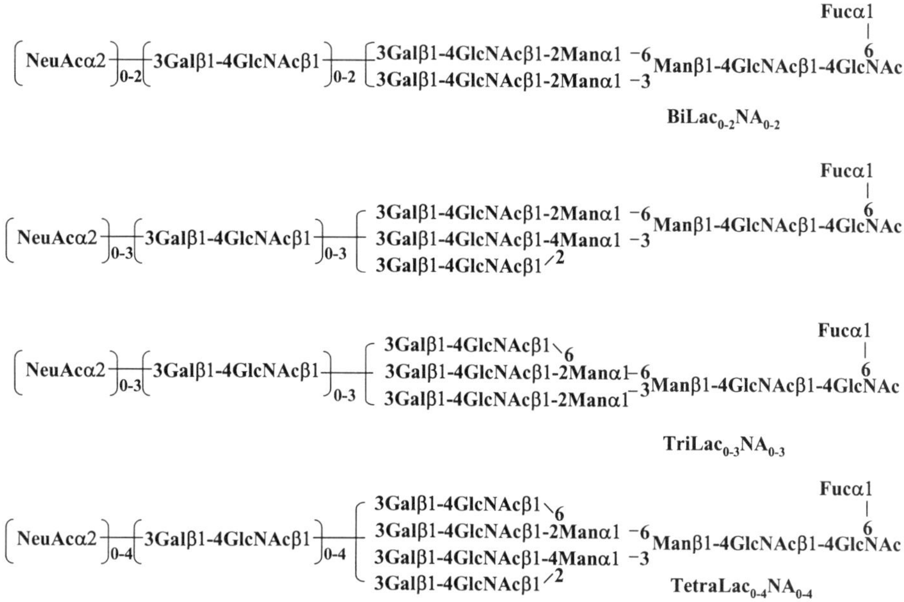

Fig. 2. Carbohydrate structures of erythropoietin. Bi, biantennary; Tri, triantennary; Tetra, tetraantennary; Lac, *N*-acetyllactosamin; NA, NeuAc.

3.3. LC–MS/MS Analysis of a Proteolytic-Digested Glycoprotein

1. Equilibrate a microbore column (1.0 mm id) with the starting solvent (5% of B) at a flow rate of 50 µL/min, and monitor the absorbance at 206 nm.
2. Set the ESI voltage and the electron multiplier to 4500v and 1500 V, respectively. Set the capillary temperature to 225°C. Adjust the pressure of the sheath gas (N_2) and auxiliary gas (N_2) to 70 psi and 10 U, respectively. Set the collision energy and pressure of the collision gas (Ar) to –20 eV and 2.0 mTorr, respectively. Set the scan time to 4 s. Perform ionization in the positive ion mode and acquire the ions at *m/z* 400–2400 with the first quadrupole (Ql), and at *m/z* 204 with the third quadrupole (Q3) in the precursor-ion scan mode (*see* **Note 9**).
3. Inject an aliquot (8 µL) of the sample onto the HPLC column.
4. Elute the column using a linear gradient from 5 to 45% of B in 40 min at a flow rate of 50 µL/min.

Figure 6 shows the TIC chromatograms of Glu-C digested EPO, which was obtained by normal scan (LC–MS) **(A)** and by the precursor-ion scan of *m/z* 204 ($HexNAc^+$) (LC–MS/MS) **(B)**. Nonglycosylated peptides are eliminated by the precursor-ion scan, resulting in a TIC chromatogram showing only glycopeptides ***(10)***.

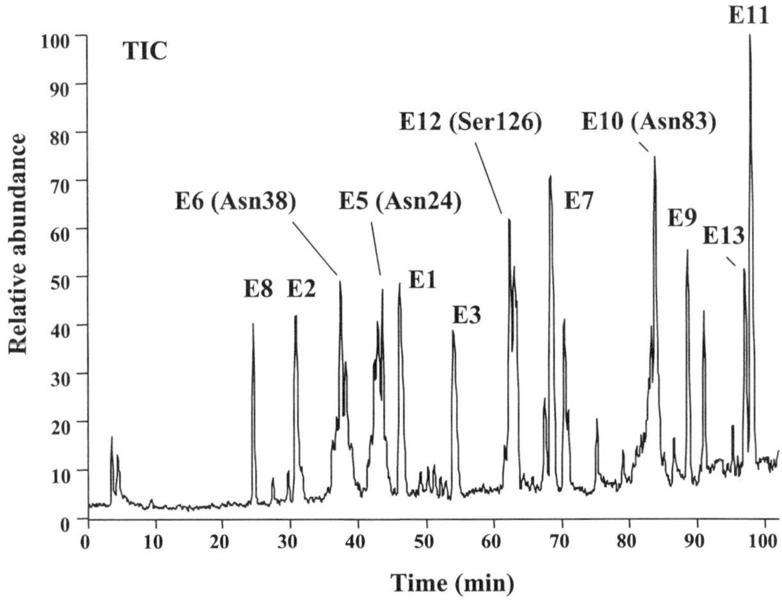

Fig. 3. Glu-C peptide/glycopeptide map of erythropoietin. Column, Vydac 218TP52 (250 × 2.1 mm); eluent A, 0.05% TFA; eluent B, 0.05% TFA/50% (v/v) acetonitrile; gradient, 1 to 90% of B in 130 min; flow rate, 0.2 mL/min; acquired mass, *m/z* 550–2400; ion mode, negative; sample amount, 50 µg erythropoietin.

4. Notes

1. The protease solution can be stored at –20°C for 1 yr.
2. To reduce background, the use of Milli-Q filtrated water is recommended. The mobile phase should be prepared just before use.
3. Instead of Sephadex-G25, reverse-phased (RP) HPLC may be used for desalting; recovery of some glycoproteins, however, is decreased by the derivatization and removal of the denaturant.
4. Peaks that cannot be identified sometimes appear as a result of excess incubation with trypsin. It is worth checking the reaction after 1 h of incubation.
5. Glu-C digests Glu-X without cleavage of Asp-X at a 1:50 enzyme-to-substrate ratio at pH 8.0. Asp-X can be hydrolyzed by Glu-C at a 1:4 enzyme-to-substrate ratio.
6. The positive ion mode is recommended for analyses of peptides and nonsialylated glycoproteins. The negative ion mode is effective for analyses of sialylated glycopeptides.
7. The range of *m/z* 1000–2400 is effective for selective detection of glycopeptides. Most of the nonglycosylated peptides can be eliminated in this range.

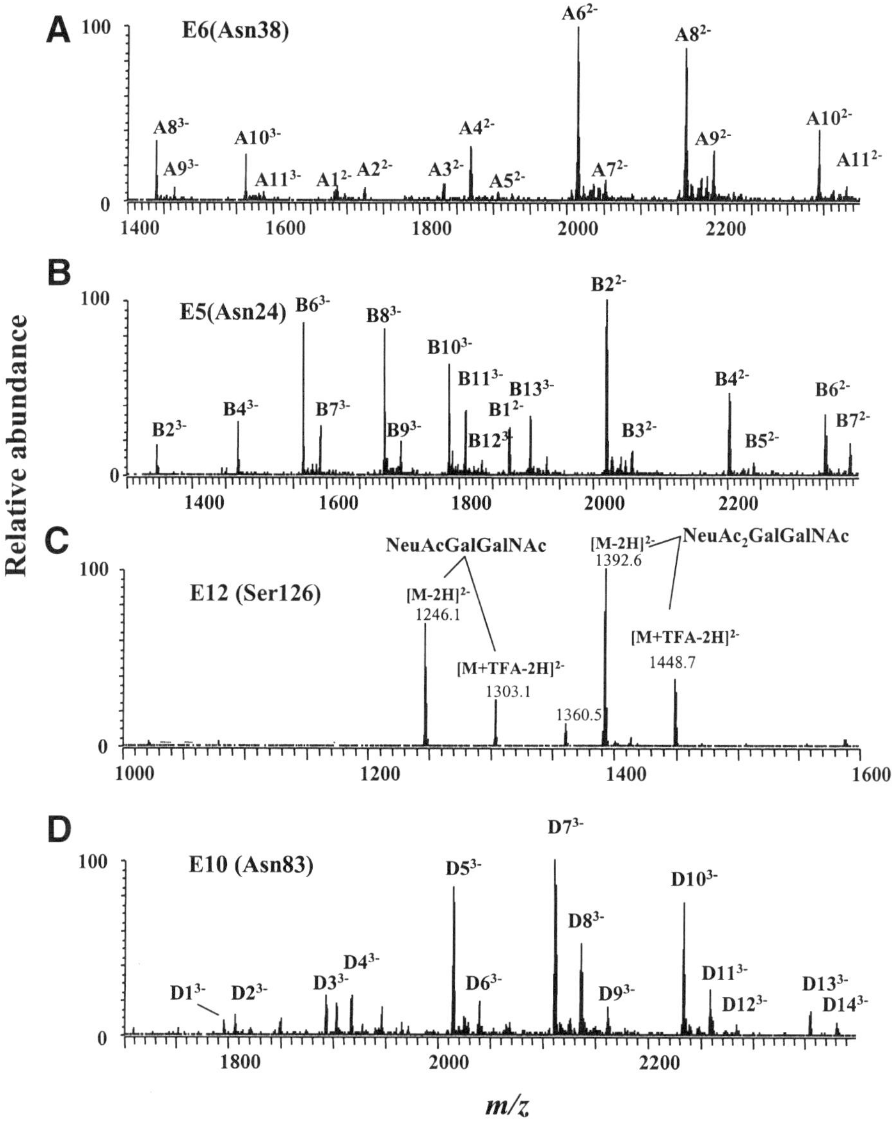

Fig. 4. Mass spectra of glycopeptides E6 (**A**), E5 (**B**), E12 (**C**), and E10 (**D**). Peak assignment is presented in **Table 1**.

8. Since some nonglycosylated peptides are not eluted by ammonium acetate, it is recommended to wash the column with 0.05%TFA/80% acetonitrile for 15 min after the analysis.

Table 1
Structural Assignments of Peaks in Fig. 3 and of Ions in Fig. 4, and Their Theoretical Masses and Observed *m/z* Values

Peptide in **Fig. 3**	Amino acid residues	Ion in **Fig. 4**	Carbohydrate structure[a]	Theoretical mass[b]	Observed *m/z* M^-	M^{2-}	M^{3-}
E8	56–62			729.8	728.4		
E2	9–13			602.7	601.5		
E6	38–43	A1	BiLac$_1$NA$_2$, TriNA$_2$[c]	3375.2		1686.8	
		A2	TriLac$_1$NA$_1$, TetraNA$_1$[d]	3449.3		1724.4	
		A3	TriNA$_3$	3666.5		1831.8	
		A4	BiLac$_2$NA$_2$, TriLac$_1$NA$_2$, TetraNA$_2$[e]	3740.6		1869.1	
		A5	TriLac$_2$NA$_1$, TetraLac$_1$NA$_1$[f]	3814.6		1906.5	
		A6	TriLac$_1$NA$_3$, TetraNA$_3$[g]	4031.8		2014.5	
		A7	TriLac$_2$NA$_2$, TetraLac$_1$NA$_2$[h]	4105.9		2052.8	
		A8	TetraNA$_4$	4323.1		2160.2	1440.1
		A9	TriLac$_2$NA$_3$, TetraLac$_1$NA$_3$[i]	4397.1		2198.0	1465.1
		A10	TetraLac$_1$NA$_4$	4688.4		2343.2	1561.8
		A11	TetraLac$_2$NA$_3$	4762.5		2380.2	1586.5
E5	22–37	B1	BiNA$_1$	3750.7		1875.1	
		B2	BiNA$_2$	4042.0		2020.3	1346.2
		B3	TriNA$_1$	4116.0		2058.5	
		B4	BiLac$_1$NA$_2$, TriNA$_2$[c]	4407.3		2202.7	1468.2
		B5	TriLac$_1$NA$_1$, TetraNA$_1$[d]	4481.4		2239.8	
		B6	TriNA$_3$	4698.6		2348.7	1565.6
		B7	BiLac$_2$NA$_2$, TriLac$_1$NA$_2$, TetraNA$_2$[e]	4772.6		2385.7	1590.7
		B8	TriLac$_1$NA$_3$, TetraNA$_3$[g]	5063.9			1687.3
		B9	TriLac$_2$NA$_2$, TetraLac$_1$NA$_2$[h]	5138.0			1712.2
		B10	TetraNA$_4$	5355.2			1784.5
		B11	TriLac$_2$NA$_3$, TetraLac$_1$NA$_3$[i]	5429.2			1809.6
		B12	TetraLac$_2$NA$_2$	5503.3			1834.0
		B13	TetraLac$_1$NA$_4$	5720.5			1905.7
E1	(1–8)S-S (160–165)			1503.7	1502.1	750.6	
E3	14–18			692.8	691.5		
E12	118–136		NA-Gal-GalNAc	2494.6		1246.1	
			NA$_2$-Gal-GalNAc	2785.9		1392.6	
E7	45–55			1572.9	1571.1		
E10	73–96	D1	BiLac$_1$NA$_2$, TriNA$_2$[c]	5388.5			1795.1
		D2	TriLac$_1$NA$_1$, TetraNA$_1$[d]	5462.6			1820.5
		D3	TriNA$_3$	5679.8			1892.2
		D4	BiLac$_2$NA$_2$, TriLac$_1$NA$_2$, TetraNA$_2$[e]	5753.9			1917.6
		D5	TriLac$_1$NA$_3$, TetraNA$_3$[g]	6045.1			2014.2
		D6	TriLac$_2$NA$_2$, TetraLac$_2$NA$_2$[h]	6119.2			2040.0
		D7	TetraNA$_4$	6336.4			2111.1
		D8	TriLac$_2$NA$_3$, TetraLac$_1$NA$_3$[i]	6410.5			2136.4
		D9	TetraLac$_2$NA$_2$	6484.6			2161.6
		D10	TetraLac$_1$NA$_4$	6701.7			2233.4
		D11	TetraLac$_2$NA$_3$	6775.8			2258.2

(continued)

Table 1 *(continued)*

Peptide in **Fig. 3**	Amino acid residues	Ion in **Fig. 4**	Carbohydrate structure[a]	Theoretical mass[b]	Observed *m/z* M⁻	M²⁻	M³⁻
		D12	$TetraLac_3NA_2$	6849.9			2283.5
		D13	$TetraLac_2NA_4$	7067.1			2355.8
		D14	$TetraLac_3NA_3$	7141.1			2380.8
E9	63–72			1115.3	1113.9		
E13	137–159			2837.4		1418.4	
E11	97–117			2212.6	2211.4		

[a]All N-glycans contain fucosylated core. Bi, biantennary; Tri, triantennary; Tetra, tetraantennary; NA, NeuAc; Lac, *N*-acetyllactosamine.

[b]Average mass value.

[c–i]Isomers.

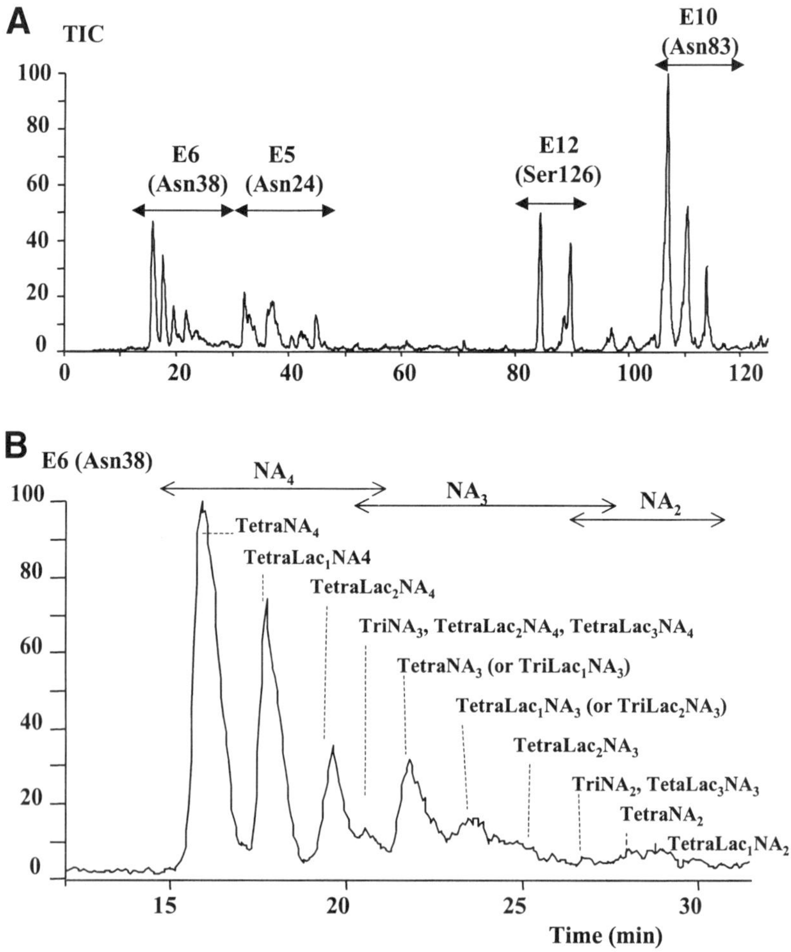

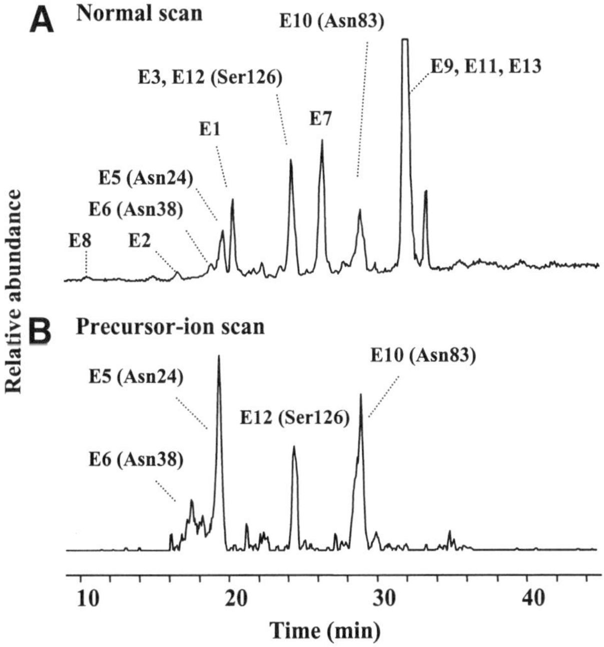

Fig. 6. Normal scan **(A)** and precursor-ion scan **(B)** of Glu-C-digested EPO. Column, Magic C18 (150 × 1 mm); eluent A, 0.05%TFA/2% acetonitrile; eluent B, 0.05% TFA/80% acetonitrile; gradient, 5 to 45% of B in 40 min; flow rate, 50 µL/min; acquired mass (Q1), *m/z* 400–2400. Set mass (Q3), *m/z* 204; ion mode, positive; sample amount, 2 µg (A) and 8 µg (B) of erythropoietin.

9. Monitoring of the oxonium ions *m/z* 204 ($HexNAc^+$) and 366 ($Hex\text{-}HexNAc^+$) is effective for the determination of both *N*-glycosylated and *O*-glycosylated peptides.

References

1. Hayakawa, T. (1991) in *Drug Biotechnology Regulation. Scientific Basis and Practices* (Chiu, Y.-Y. H. and Gueriguian, J. L., eds.), Marcel Dekker, New York, pp. 468–498.

Fig. 5. *(see opposite page)* **(A)** Glu-C glycopeptide map of erythropoietin. Column, Vydac 218TP52 (250 × 2.1 mm); eluent A, 1 m*M* ammonium acetate, pH 6.8. eluent B, 1 m*M* ammonium acetate, pH 6.8/80% (v/v) acetonitrile; gradient, 1 to 6% of B in 60 min followed by 36% of B in 80 min; flow rate, 0.2 mL/min; acquired mass, *m/z* 1000–2400; ion mode, negative; sample amount, 15 µg erythropoietin. **(B)** Expanded view of glycopeptide E6. Bi, biantennary; Tri, triantennary; Tetra, tetraantennary; Lac, *N*-acetyllactosamin; NA, NeuAc.

2. Hirayama, K., Yuji, R., Yamada, N., Kato, K., Arata, Y., and Shimada, I. (1998) Complete and rapid peptide and glycopeptide mapping of mouse monoclonal antibody by LC/MS/MS using ion trap mass spectrometry. *Anal. Chem.* **70,** 2718–2725.
3. Kawasaki, N., Ohta, M., Hyuga, S., Hyuga, M., and Hayakawa, T. (2000) Application of liquid chromatography/mass spectrometry and liquid chromatography with tandem mass spectrometry to the analysis of the site-specific carbohydrate heterogeneity in erythropoietin. *Anal. Biochem.* **285,** 82–91.
4. Ohta, M., Kawasaki, N., Hyuga, S., Hyuga, M., and Hayakawa, T. (2001) Selective glycopeptide mapping of erythropoietin by on-line high-performance liquid chromatography/electrospray ionization mass spectrometry. *J. Chromatogr. A* **910,** 1–11.
5. Huddleston, M. J., Bean, M. F., and Carr, S. A. (1993) Collisional fragmentation of glycopeptides by electrospray ionization LC/MS and LC/MS/MS: Methods for selective detection of glycopeptides in protein digests. *Anal. Chem.* **65,** 877–884.
6. Krantz, S. B. (1991) Erythropoietin. *Blood* **77,** 419–434.
7. Sasaki, H., Bothner, B., Dell, A., and Fukuda, M. (1987) Carbohydrate structure of erythropoietin expressed in Chinese hamster ovary cells by a human erythropoietin cDNA. *J. Biol. Chem.* **262,** 12,059–12,076.
8. Takeuchi, M., Takasaki, S., Miyazaki, H., et al. (1988) Comparative study of the asparagine-linked sugar chains of human erythropoietins purified from urine and the culture medium of recombinant Chinese hamster ovary cells. *J. Biol. Chem.* **263,** 3657–3663.
9. Tsuda, E., Goto, M., Murakami, A., et al. (1988) Comparative structural study of N-linked oligosaccharides of urinary and recombinant erythropoietins. *Biochemistry* **27,** 5646–5654.
10. Itoh, S., Kawasaki, N., Ohta, M., Hyuga, M., Hyuga, S., and Hayakawa, T. (2001) Study on evaluating methods for the quality control of glycoprotein products. (III) Erythropoietin products. Part 3. *Kokuritsu Iyakuhin Shokuhin Eisei Kenkyusho Hokoku* **119,** 65–69.

19

HPLC in the Analysis of Peptide Metabolism

Rebecca A. Lew

1. Introduction

Bioactive peptides are generally synthesized within large precursor molecules, from which the active moiety must be enzymatically excised by one or more specific peptidases *(1)*. For most peptides, this processing occurs intracellularly, within the secretory pathway, but cosecretory or extracellular processing can also occur. Peptidases also play a critical role in the termination of peptide signals, via cleavage to inactive fragments *(2,3)*. The characterization of the peptidases involved in the generation and metabolism of peptides is thus of critical importance to the understanding of the physiology of specific peptide hormones and neurotransmitters.

Given that the metabolism of most peptides involves cleavage at a number of sites, and that the metabolites are often very similar in composition to both the parent peptide and to each other, methods must be devised that will adequately separate, identify, and quantitate all the relevant species. Reversed-phase (RP)-high-pressure liquid chromatography (HPLC) often with the addition of a hydrophobic ion-pair, has been used extensively in the characterization of peptides and their metabolites. Resolution of peptide fragments is usually very high, and a number of complementary methods, such as ultraviolet (UV) absorbance detection (which is fairly standard practice), bioassay, radioimmunoassay, enzyme-linked immunosorbent assay (ELISA), amino acid analysis, and peptide sequencing, may be used for identification and quantification. In recent years, the advent of on-line benchtop mass spectral analysis has simplified and expedited the identification of peptides separated by HPLC, and liquid chromatography–mass spectrometry (LC–MS) has become the preferred method of analysis (*see* **Note 1**).

From: *Methods in Molecular Biology, vol. 251, HPLC of Peptides and Proteins: Methods and Protocols*
Edited by: M.-I. Aguilar © Humana Press Inc., Totowa, NJ

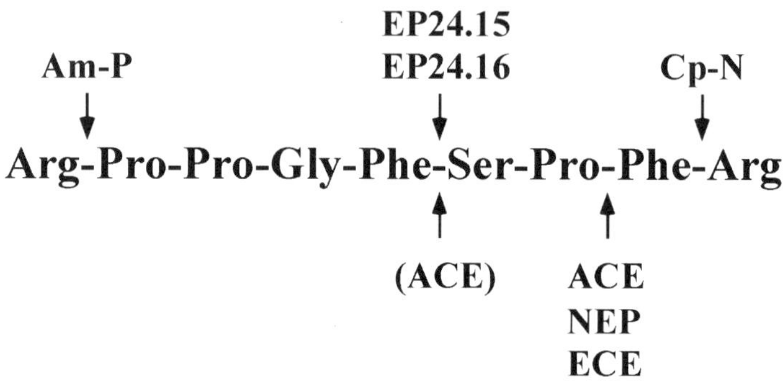

Fig. 1. Major cleavage sites of bradykinin. Shown is the nine amino acid sequence of bradykinin, and the known sites of cleavage by vascular metallopeptidases. Am-P, aminopeptidase P; EP24.15, endopeptidase 3.4.24.15; EP24.16, endopeptidase 3.4.24.16; NEP, neutral endopeptidase; ECE, endothelin-converting enzyme; ACE, angiotensin converting enzyme; (ACE) indicates a secondary cleavage site of ACE; Cp-N, carboxypeptidase N.

A wide range of aspects of peptide degradation can be studied using LC–MS, including: the identification of specific peptidases within distinct cell or tissue types (*see* **Note 2**); purification of peptidases, using substrate cleavage as an assay of activity; characterization (substrate specificity, inhibitor sensitivity, and so on) of isolated or recombinant peptidases (*see* **Note 3**); identification of cleavage sites within a peptide substrate; kinetic analysis of peptide cleavage (*see* **Note 4**); characterization of the metabolism of peptide analogs with agonist or antagonist activities (*see* **Note 5**); development of specific synthetic peptidase inhibitors; and the development and verification of alternative peptidase screening methods, such as fluorescent substrate assays.

In the present chapter, the degradation of the vasodilator peptide bradykinin by vascular endothelial cells will be used as a specific example (*see* **Note 2**). This peptide is rapidly degraded in vivo by a number of peptidases acting at different sites (*see* **Fig. 1**). The focus of this chapter will be on two closely related metalloendopeptidases, EC 3.4.24.15 (EP24.15) and 3.4.24.16 (EP24.16), which cleave bradykinin at the Phe^5-Ser^6 bond. Methods for the preparation of soluble and membrane fractions from cultured cells will be provided, along with a description of the LC–MS analysis of the degradation of bradykinin by such preparations. Furthermore, the identification of the peptidases involved will then be determined by the use of specific inhibitors. Additional aspects of LC–MS analysis of peptidase activity, as outlined above, will be discussed in the **Notes** subheading.

2. Materials

2.1. Preparation of Soluble and Membrane Fractions

1. Cell culture medium: RPMI 1640, 10% fetal calf serum (FCS) (CSL Ltd., Parkville, Victoria, Australia), 2.4 m*M* glutamine, and 2 mL/L of penicillin (10,000 U/mL)/streptomycin (10,000 µg/mL)/fungizone (25 µg/mL) antibiotic mix (CSL Ltd.)—store at 4°C.
2. Collagenase Type II (3 mg/mL (171 U/mg); Worthington Biochemical Corp., Freehold, NJ).
3. Antisera against human von Willebrand factor (Dakopatts, Glostrup, Denmark).
4. Avidin/biotin/peroxidase immunohistochemical kit (Vector Laboratories, Burlingame, CA).
5. Tris-buffered saline (TBS): 25 m*M* Tris-HCl, 125 m*M* NaCl, pH 7.4. Prepare using ultrapure HPLC grade water (*see* **Subheading 2.3.**), in order to minimize possible contamination with bacterial proteases—store at 4°C for up to 1 mo.

2.2. Peptide Digest

1. TBS.
2. Bradykinin: purchased from Auspep (Parkville, Victoria, Australia) or other supplier (e.g., Sigma). The dry peptide is reconstituted in 99% methanol/1% trifluoroacetic acid (TFA), and aliquoted into 20, 50, and 100 µg amounts in Eppendorf tubes. It is then dried on a Speed-Vac centrifugal evaporator, stored at –20°C, and reconstituted in TBS at time of assay.
3. Peptidase inhibitors:
   a. CFP (*N*-[1-**(R,S)**-carboxy-3-phenylpropyl]-Ala-Ala-Tyr-*p*-aminobenzoate) was synthesized at the Baker Institute ***(4)***, and inhibits both EP24.15 (K_i = 19 n*M*) and EP24.16 ($K_i \approx$ 700 n*M*) ***(5)***.
   b. The cFP analog JA2 (*N*-[1-**(R,S)**-carboxy-3-phenylpropyl]-Ala-Aib-Tyr-*p*-aminobenzoate) (K_i[EP24.15] = 23 n*M*, K_i[EP24.16] = 690 n*M*; ***(5)***) was synthesized by Dr. G. Abbenante (Centre for Drug Design and Development, University of Queensland, Australia) as described ***(5)***.
   c. The specific inhibitors Pro-Pheψ(PO_2CH_2)Leu-Pro-NH_2 (PF^{PC}LP-NH_2; K_i[EP24.15] = 66.5 µ*M*, K_i[EP24.16] = 12 n*M*; ***(6)***), Z-(L,D)-Pheψ(PO_2CH_2)-(L,D)-Ala-Arg-Phe (Z-F^{PC}ARF; K_i[EP24.15] = 0.16 n*M*, K_i[EP24.16] = 530 n*M*; ***(7)***), and Z-(L,D)-Pheψ(PO_2CH_2)-(L,D)-Ala-Lys-Met (Z-F^{PC}AKM; K_i[EP24.15] = 0.12 n*M*, K_i[EP24.16] = 230 n*M*; ***(7)***) were synthesized by Prof. Vincent Dive (CEA, Département d'Ingénierie et d'Etudes des Protéines, Gif-sur-Yvette Cedex, France) as previously described ***(6,7)***.
   d. The EP24.16-selective dipeptide Pro-Ile (K_i = 90 µ*M* ***(8)***) was purchased from Auspep (Parkville, Victoria, Australia).
   e. Phosphoramidon was obtained from Sigma (Castle Hill, NSW, Australia).

f. All inhibitors were reconstituted in TBS at ≥10-fold the final assay concentration and stored at –20°C.

2.3. HPLC Analysis

2.3.1. Equipment

1. Hewlett-Packard (now Agilent Technologies) HP1100 series LC system with thermoregulated autosampler, diode array UV detector, and on-line electrospray mass spectrometric detector.
2. ChemStation analysis software.
3. Zorbax Eclipse C18 column, 2.1 × 150 mm, 5 μm (XDB-C18, Hewlett-Packard).

2.3.2. Solvent Reagents

1. HPLC grade acetonitrile: 190 nm cutoff must be used to minimize background UV absorbance when detecting at 214 nm.
2. HPLC grade water: purified by reverse-osmosis (Milli-RO System, Millipore, Bedford, MA), then further purified through Milli-Q/UV Plus ultrapure water system (Millipore) with UV irradiation: critical for reducing contaminants that may interfere with LC–MS analysis.
3. TFA: Peptide Sequencing grade (Applied Biosystems, Norwalk, CT) results in minimal background peaks on LC–MS.
4. Acetic acid: ACS grade.

3. Methods

3.1. Preparation of Soluble and Membrane Fractions

3.1.1. Cell Culture

Detailed methods for cell culture are beyond the scope of this chapter. In brief, sheep aortic endothelial cells are cultured in our laboratory as follows.

1. Under sterile conditions, transect sheep aortae (collected aseptically from an abattoir) longitudinally.
2. Gently pass a sharp sterile scalpel blade over the luminal surface and rinse into 5 mL collagenase in a sterile 15-mL Falcon tube.
3. Incubate cells for 5 min at 37°C to allow dissociation by collagenase.
4. Dilute cells with 10 mL RPMI-1640 medium containing 20% FCS, which inhibits the collagenase.
5. Centrifuge to pellet cells (900*g*, 5 min).
6. Discard supernatant, and resuspend pellet in 10 mL fresh medium.
7. Plate into 90-mm culture plate.
8. Incubate at 37°C in 5% carbon dioxide in air.
9. After several weeks the cells proliferate to form a monolayer. Passage cells when confluent (using 0.1% trypsin/0.02% disodium ethylene diamine tetraacetic acid [EDTA]).

10. Cells are identified as endothelial cells by immunohistochemistry using a rabbit antiserum directed against human von Willebrand factor.

3.1.2. Cell Harvest

1. Wash confluent cells twice in warm (37°C) TBS (approx 5 mL/plate). Typically, a large number of plates (10–15) are processed simultaneously to provide a stock of material for assays.
2. Scrape cells off the culture plates using a rubber policeman into ice-cold TBS (as low a volume as possible, normally about 0.5–1.0 mL/plate).
3. Freeze (–70°C) and thaw the cell suspension three times to break up the cells.
4. Sonicate the cell suspension briefly (3 × 5 s bursts, Sonifier Cell Disrupter, Branson) on ice to further disrupt cell membranes. Care must be taken not to overheat the sample, as this may denature the proteins.
5. Ultracentrifuge at 100,000*g* for 60 min at 4°C.
6. Reserve the cytosolic supernatant.
7. Resuspend the pellet in fresh cold TBS and respin.
8. Discard the second supernatant and resuspend the pellet in fresh cold TBS (as low a volume as possible, usually 0.5 mL) as the crude membrane fraction.
9. Gently pipet the membranes up and down in order to homogenize the pellet as much as possible (*see* **Note 6**). If necessary, briefly sonicate as in **step 4**.
10. Aliquot both soluble and membrane fractions in convenient volumes (e.g., 100 µL) and freeze at –70°C.
11. Determine protein content by standard methods, e.g., that of Lowry ***(9)***.

3.2. Peptide Digest

*3.2.1. Time-Course Experiment (see **Note 7**)*

1. Reconstitute 60 µg bradykinin (in an Eppendorf tube) in 50 µL TBS.
2. Add to 550 µL cytosolic preparation—vortex.
3. Immediately remove 100 µL and place into 400 µL methanol/1% TFA in Eppendorf tube to stop the reaction—this is "time 0" sample.
4. Incubate remaining digest in 37°C water bath.
5. At designated times (e.g., 15, 30, 60, and 120 min), vortex digest and remove 100 µL into 400 µL methanol/TFA.
6. Once all samples have been collected, centrifuge in tabletop microfuge at top speed (approx 14,000 rpm) for 5 min to precipitate any protein (*see* **Note 8**).
7. Carefully transfer supernatant from each sample into new Eppendorf tube.
8. Dry on a centrifugal evaporator (Speed-Vac, Savant) (*see* **Note 9**).

*3.2.2. Inhibitor Experiment (see **Notes 10–12**)*

1. Label Eppendorf tubes for digests in triplicate—for six inhibitors, plus control, total of 21 tubes.
2. Add 50 µL cytosolic preparation to each tube.

3. Add 30 μL TBS to each tube.
4. Add 10 μL of 10X inhibitor (*see* **Subheading 2.** and **Table 1** for concentrations) to appropriate tubes (TBS to control tubes).
5. Vortex and incubate 5 min to allow binding of inhibitor (*see* **Note 13**).
6. Reconstitute 250 μg bradykinin in 250 μL TBS.
7. Add 10 μL bradykinin to each tube—vortex.
8. Incubate in 37°C water bath for 60 min.
9. Add 400 μL methanol/TFA to stop reaction.
10. Spin and dry as in **steps 6–8** in time-course experiment.

3.3. HPLC Analysis (see Note 14)

3.3.1. Solvent Preparation (see **Note 15**)

Solvent A: 0.1% TFA/0.02% acetic acid.

1. Filter 250 mL HPLC water through Durapore 0.45 μm HV filter (Millipore #HVLP04700; be sure that the filter is compatible with the solvent used), using a filter apparatus and vacuum pump.
2. Pour filtered solvent into clean LC flask.
3. Add 0.25 mL TFA and 0.05 mL acetic acid and mix with clean flea on stir-plate. As it is volatile, TFA must be added after filtration step.

Solvent B: 60% acetonitrile/0.1% TFA.

1. Filter 100 mL HPLC water and 150 mL acetonitrile as in **step 1** above.
2. Pour into a second clean flask.
3. Add 0.25 mL TFA and stir as above.

3.3.2. Sample Reconstitution

1. Reconstitute dried samples in solvent A to a final volume of 120 μL.
2. Spin in microfuge, 14,000 rpm for 5 min to remove any precipitate.
3. Transfer sample to 250 μL limited volume insert (Agilent Part #5181-3377).
4. Place limited volume insert into autosampler vials, and load onto autosampler—include a sample of solvent A in position 1, to serve as a blank run.

3.3.3. Gradient Operation

1. Turn on LC–MS as instructed in manufacturer's operations manual.
2. Set diode array detector to store data collected at 214, 254, and 280 nm (reference λ = 350 nm), as well as the entire spectrum from 200–300 nm. Display only 214 nm (*see* **Note 16**).
3. Set column temperature to 50°C.
4. Prime pumps with solvents A and B.
5. Wash column with 100% B at 0.15 mL/min for 15 min.
6. Change to 100% A in a stepwise fashion, then wash column in solvent A (0.15 mL/min) for 15 min, until baseline (absorbance at 214 nm) is stable.

7. Set up linear gradient from 3% B to 100% B over 30 min, followed by a 10-min postrun at 3% B, all at 0.15 mL/min.
8. Set up injection parameters to inject 100 µL from each sample.
9. Set up sample queue parameters for data collection.
10. Initiate autosampler run, with blank as first sample.

3.3.4. Analysis

Following data collection, the chromatograms are analyzed using the ChemStation software package included with the Hewlett-Packard 1100 series LC–MS. As with most HPLC analysis software, retention times, peak heights and peak areas can be calculated and used for peak identification and quantitation. The software also includes deconvolution algorithms for determining masses from the mass spectral data, which can then be used to identify peptide fragments (*see* **Note 17**). Representative chromatograms from the example described are shown in **Figs. 2** (soluble fraction) and **3** (membrane fraction). The soluble fraction of endothelial cells cleaved bradykinin into three major degradation products, BK_{1-5}, BK_{6-9}, and BK_{1-8} **(Fig. 2A)**. The extent of degradation can be calculated by the disappearance of substrate (bradykinin), either by comparison to a time 0 control, or by relation to a standard curve of known amounts of peptide. In our example, approx 28% of the bradykinin was degraded; based on the initial quantity of substrate (10 µg = 9.43 nmoles), the protein in the tissue sample (20 µg), and the time of incubation (1 h), we can calculate the rate of degradation to be 130 nmol/mg protein/h. Addition of the EP24.15/24.16 inhibitor cFP inhibited the production of BK_{1-5}, and BK_{6-9}, suggesting that one or both of these enzymes contribute significantly to cleavage of bradykinin at the 5–6 bond. The degree of inhibition is calculated using the peak area of one or both of the products (*see* **Note 18**), and averaged over the triplicate samples (*see* **Table 1**). Inhibition by more selective inhibitors suggests that EP24.16 is the major effector of bradykinin cleavage in this sample (*see* **Fig. 2C** and **D** and **Table 1**).

Similar analysis of the membrane samples indicated that the predominant cleavage of bradykinin occurred at the 7–8 bond, with generation of both the BK_{1-7} and BK_{8-9} fragments (*see* **Fig. 2A**). This cleavage was significantly inhibited by phosphoramidon, an inhibitor of the membrane-bound metallopeptidases neutral endopeptidase (NEP) and endothelin-converting enzyme (ECE), both of which are known to cleave bradykinin at this site ***(10,11)***. In addition to the BK_{1-7} and BK_{8-9} fragments, BK_{1-5}, BK_{6-9}, and BK_{1-8} were also identified, as seen in the soluble fraction. When examining the effects of EP24.15 and EP24.16 inhibitors, we included phosphoramidon in the reaction to minimize degradation by NEP and ECE (*see* **Note 19**). As with the soluble fraction, most of the cleavage at the 5–6 bond could be attributed to EP24.16

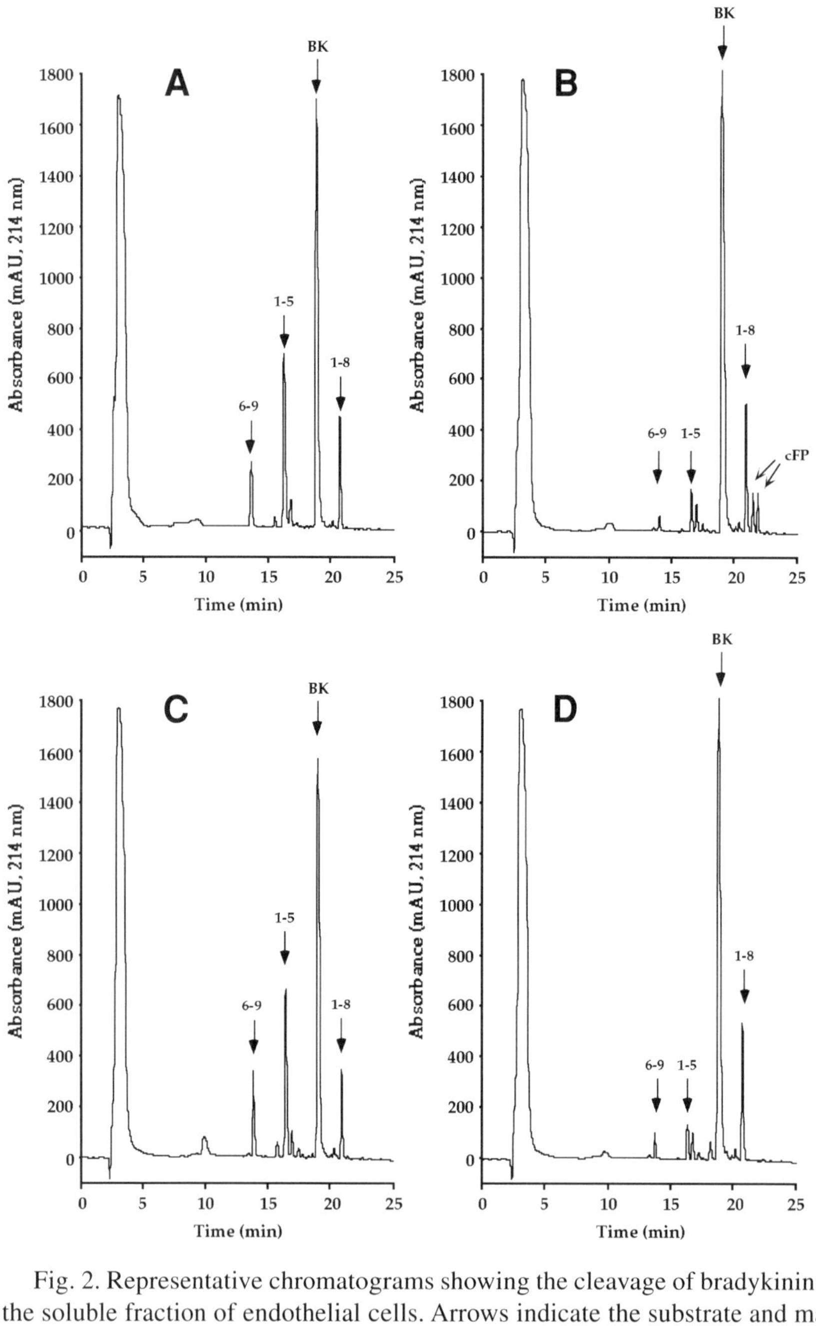

Fig. 2. Representative chromatograms showing the cleavage of bradykinin (BK) by the soluble fraction of endothelial cells. Arrows indicate the substrate and major proteolytic product peaks. Panel **(A)** shows cleavage in the absence of inhibitor, panel **(B)** in the presence of the combined EP24.15/EP24.16 inhibitor cFP (the two stereoisomers of which elute as a doublet), panel **(C)** in the presence of the specific EP24.15 inhibitor Z-F^{PC}ARF, and panel **(D)** in the presence of the specific EP24.16 inhibitor PF^{PC}LP-NH_2.

(*see* **Fig. 3** and **Table 1**), although a slightly greater contribution of EP24.15 was observed.

4. Notes

1. Mass spectral analysis is just one of a range of methods that can be used to detect and identify the products of peptidase cleavage. Prior to obtaining our current LC–MS system, we routinely identified peptides by coelution with standards, if available, and by sequencing or amino acid analysis of collected peaks. This method, of course, requires access to these analytical techniques, which are both costly and time consuming. Typically, product identity would be confirmed during the preliminary stages of a study, and retention times of the resultant peaks, which are highly reproducible, would be used on a more routine basis. We have also used specific radioimmunoassays to detect products of peptide processing ***(12)***; this method is labor-intensive and requires detailed characterization of the cross-reactivity profile of the antisera used, but is extremely sensitive when measuring endogenous peptides. Furthermore, radioimmunoassay can be used to detect N-terminally acetylated peptides, such as various forms of α-*N*-acetylendorphin ***(12)***, which cannot be sequenced by Edman degradation. Indeed, endogenous peptides have been *N*-acetylated in vitro in order to maximize separation on RP-HPLC and to exploit the specificity of N-terminally directed antisera ***(13–15)***.

   Another modification is to follow the degradation of radiolabeled peptides by HPLC separation coupled with counting of collected fractions. This alternative is very sensitive and allows analysis of the metabolism of physiological concentrations of substrate. However, only peptide fragments containing the radiolabel can be detected, and identification can only be accomplished by coelution with standards. Furthermore, it must be verified that the radiolabel does not significantly alter the degradation pattern or kinetics. For us, this technique has been most useful when assessing cleavage in vivo ***(16–19)***, or in cell culture ***(20)***, where the quantities of unlabelled substrate needed for analysis by UV absorbance are prohibitive, and/or may elicit undesirable biological effects.
2. This method for assaying peptidase activity can be applied to other types of biological sample, such as soluble and membrane preparations of whole tissues, subcellular organelles (e.g., secretory vesicles), plasma and other body fluids, and recombinant enzymes. Furthermore, it can be extended to the examination of peptidase activity in live cultured cells ***(20–22)***, although the components in standard culture medium will often interfere with detection of substrate by UV absorption. This can be circumvented by the use of balanced salt solutions in place of complete medium, or by the use of other detection methods, such as using a fluorescent, radioactive or immunoreactive substrate. Similar problems can also occur with other tissue preparations, particularly plasma, but peptidase activity is generally high enough that only small volumes of the source are necessary. Finally, we have used such methods to assess the metabolism of radiolabelled (^{125}I) peptide-based EP24.15 inhibitors in vivo ***(17)***; for this purpose, plasma samples taken at intervals following substrate administration were precipitated with

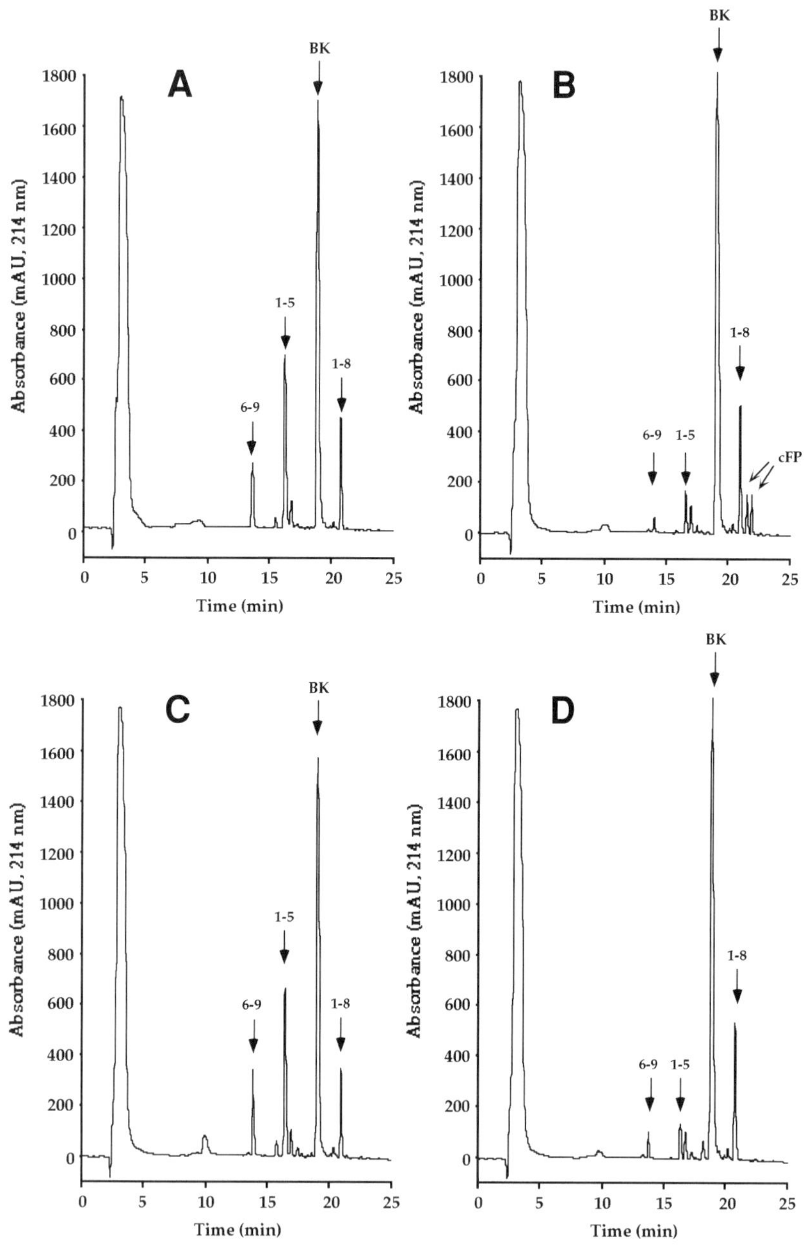

Fig. 3. Representative chromatograms showing the cleavage of bradykinin (BK) by the membrane fraction of endothelial cells. Arrows indicate the substrate and major proteolytic product peaks. Panel **(A)** shows cleavage in the absence of inhibitor, panel **(B)** in the presence of the neutral endopeptidase inhibitor phosphoramidon (marked by "P"), panel **(C)** in the presence of phosphoramidon and the combined EP24.15/EP24.16 inhibitor cFP, and panel **(D)** in the presence of phosphoramidon and the specific EP24.16 inhibitor $PF^{PC}LP\text{-}NH_2$.

Table 1
Inhibition of Bradykinin Cleavage at the Phe5-Ser6 Bond by Specific Inhibitors of Endopeptidases 24.15 and 24.16

Inhibitor	Peptidase	Soluble	Membrane
cFP (10 μ*M*)	EP24.15/EP24.16	81.1 ± 0.3%	71.0 ± 0.8%
JA2 (10 μ*M*)	EP24.15/EP24.16	75.7 ± 1.3%	58.9 ± 3.4%
Z-F^{PC}AKM (50 n*M*)	EP24.15	<5%	6.6 ± 2.1%
Z-F^{PC}ARF (50 n*M*)	EP24.15	<5%	8.0 ± 1.7%
PF^{PC}LP-NH_2 (10 μ*M*)	EP24.16	81.7 ± 1.0%	78.1 ± 1.1%
Pro-Ile (5 m*M*)	EP24.16	61.8 ± 1.7%	53.8 ± 1.8%

Values are mean ± s.e.m. ($n = 9$) derived from three experiments for each preparation. Percent inhibition was determined from the area of the product peaks in the presence or absence of each inhibitor at the concentration indicated.

methanol/TFA, dried, separated by LC, and eluting fractions collected for γ-counting.

3. The described method is ideal for the characterization of purified and recombinant peptidases. The specificity can be determined by using a library of peptide substrates, and the cleavage sites verified by mass spectrometry ***(23)***. If multiple cleavage sites exist, preferred sites can be determined by a time course analysis. The affinity of the enzyme for its substrates can be determined by kinetic analysis (*see* **Note 4**). In addition, the sensitivity of the enzyme to general and specific inhibitors can also be assessed, as is described in this chapter for EP24.15 and EP24.16 inhibitors. We have previously used similar methods (without mass spectral analysis) to characterize a recombinant form of rat EP24.15 ***(24)***.
4. Kinetic parameters of substrates such as affinity constant (K_m) and maximal velocity (V_{max}) can be determined using HPLC analysis methods ***(24)***. Generally, optimal conditions (i.e., amount of enzyme and incubation time) in which less than 10% of substrate is degraded are determined in preliminary studies at a range of substrate concentrations ([S]) spanning two orders of magnitude around the expected K_m. Thc K_m of most peptidases for their endogenous substrates is in the low micromolar range, thus typical concentrations assayed in kinetic analysis are submicromolar to submillimolar. For kinetic analysis, the highest concentrations must be saturating; this is apparent when the observed velocity (moles degraded per unit enzyme per unit time = V) is plotted against [S]. Once an appropriate range of [S] is determined, a minimum of five concentrations are selected for the analysis. Digestion of substrate by enzyme at each [S] is then performed in triplicate, analyzed by HPLC, and the velocity derived and averaged. Because Michaelis–Menten analysis assumes first-order kinetics (generally less than 10% degradation), it is advisable to perform the experiment as a time course; one can then select the most appropriate time point samples to include in the analysis, remembering to adjust the velocity derivation accordingly. Once the data are col-

lated, K_m and V_{max} can be derived from a double-reciprocal (Lineweaver–Burk) plot of $1/V$ vs $1/S$, in which the x-intercept is $-1/K_m$ and the y-intercept is $1/V_{max}$. Similarly, the method can be used to determine K_i values for inhibitors, as well as establish the nature of inhibition (i.e., competitive or noncompetitive) ***(5)***.

5. The characterization of peptidase activity need not be confined to naturally occurring peptide substrates. Many pharmacological tools, including receptor agonists, antagonists and enzyme inhibitors, are peptide-based, and as such, are potentially susceptible to cleavage. We have used HPLC-based techniques similar to those described here to assess the stability of novel EP24.15 inhibitors in vivo ***(17–19)*** and in vitro ***(5,17)***, as well as that of a receptor antagonist in vitro ***(16)***.
6. Membranes may be solubilized at this point by incubating in 1% Triton X-100 (in TBS) for 60 min at 4°C with frequent vortexing. The solubilized membranes are then centrifuged as above; the supernatant contains solubilized protein, and the insoluble pellet is discarded.
7. The simplest way to assess peptidase activity in a new preparation is to set up a time course of digestion. Typically, a convenient volume of cellular material is combined with synthetic substrate, incubated at 37°C, and the digest is sampled at defined time intervals. Each time point requires a readily detectable amount of substrate; in the case of bradykinin (and most peptides), 10 µg is more than sufficient. Thus, the digest would contain at least 50 µg peptide, and usually extra to allow for slight errors in pipetting.
8. Precipitation of any particulate matter is crucial prior to HPLC, as such material may obstruct the column. This is especially important when assaying crude membrane preparations; it is often beneficial to spin such samples twice, to ensure particulates are removed.
9. Recovery of peptides is generally improved by partial, rather than complete drying; therefore, samples are removed once the volume is less than the original sample volume (100 µL), at which point all the organic solvent has evaporated. For most hydrophilic peptides, such as bradykinin, recovery is excellent, even if samples are dried to completion. For more hydrophobic peptides, absolute recoveries should be formally determined; if recovery is low, strategies such as resuspension in a low-percentage organic solvent (below that needed to elute the peptide from the reverse-phase column) can be explored.
10. Once samples have been analyzed by LC–MS, the optimal time of incubation can be determined. For the current purpose, an incubation time in which 25–50% of the substrate has been digested is chosen. This time is then used for further experiments designed to identify the peptidases involved in substrate cleavage. In our example, we found that 50 µL of the soluble fraction of endothelial cells (≈20 µg protein) incubated with 10 µg bradykinin for 60 min in 100 µL final volume TBS resulted in roughly 30% cleavage, based on the area of the bradykinin peak. Thus, in subsequent experiments, these conditions were replicated in the absence or presence of each of the inhibitors.
11. Similarly, bradykinin (10 µg) was also incubated with crude endothelial membranes (10 µL ≈ 40 µg protein) for 120 min in 100 µL TBS in the presence of

10 μ*M* phosphoramidon (to inhibit neutral endopeptidase present in the membranes) with or without each of the EP24.15/16 inhibitors.

12. In general, the experiment should be repeated at least twice with different cell preparations, and the results pooled. In this example, we performed the experiment three times in triplicate.
13. Although most peptidase inhibitors are high affinity with fast on-rates, there are some instances where inhibitors are relatively slow-acting. In these circumstances, or when in doubt, the enzyme source should be preincubated with the inhibitor for a longer time prior to addition of substrate.
14. The exact HPLC conditions (solvents, gradient, column) used must be optimized for the specific substrate and products being studied. Details of the theory and implementation of LC/MS are described in Chapters 1, 2, and 9.
15. Solvents are best if made fresh daily, but can be used for up to three days before significant changes in solvent quality are noticed. Calculate the required volume by multiplying the number of samples by the gradient duration (50 min) and the flow rate (0.15 mL/min), e.g., for 21 samples, plus a blank run, a total of 165 mL is needed. To ensure adequate volumes, prepare 250 mL of both solvent A and B; it is better to waste some solvent than to risk running out. Glassware must be ultraclean, thus, use dedicated measuring cylinders, filter apparatus and flasks, and never wash with detergents. After use, rinse with acetonitrile, followed by HPLC grade water and allow to air-dry. Cover or invert to prevent introduction of dust particles.
16. Although detection of peptides by UV absorption is most commonly monitored at 214 nm (the wavelength at which double bonds, such as the peptide bond, absorb), detection at other wavelengths can be used to confirm the presence of aromatic residues (at 280 nm or 254 nm), or in the case of quenched fluorescent substrates, the presence of the fluorophore in cleavage products. For these purposes, a diode array detector, which can detect absorption over a range of wavelengths simultaneously, is extremely useful.
17. In addition to the speed of on-line mass spectral analysis, the technique allows for virtually unequivocal identification of fragments of known peptide sequences based on mass. However, there are occasionally situations in which mass alone cannot verify a peptide's identity. The cleavage of bradykinin by membrane-bound enzymes is a case in point: as arginine is found at both the N- and C-terminals, the products of aminopeptidase P and carboxypeptidase M activity (BK_{2-9} and BK_{1-8}, respectively) would have the same mass. In this circumstance, it is necessary to confirm peak identity with peptide sequencing and/or coelution of standard.
18. The choice of which peak (substrate, product 1, and product 2) to use in calculating inhibition depends on the cleavage pattern observed. For example, if the substrate is cleaved at only one site, and those products are stable, then any peak can be used. If, however, multiple cleavages occur, then analysis is obviously complicated. In our example, bradykinin is cleaved by the soluble fraction at two major sites: the 5–6 bond and the 8–9 bond. Because we are particularly interested in the former cleavage, we wish to focus on the BK_{1-5} and BK_{6-9} products. As both

products are stable, and not further degraded, we can use either for determining inhibition. A further consideration is the "cleanliness" of the peak; that is, are there any background components (e.g., inhibitors) that coelute with a product and may interfere with accurate quantitation?

19. One limitation of the method becomes evident when examining an array of peptidases present in a biological sample. The extent of cleavage at a single peptide bond by a given peptidase will depend not only on the abundance of that peptidase and its affinity for the substrate, but also on the abundance and affinities of competing enzymes acting both at the same site and at different sites. Judicious use of specific inhibitors alone, and in combination, can elucidate which enzymes are present, and can give some indication as to their relative contribution to substrate degradation. However, in a complex system, absolute numbers (e.g., 65% of substrate A degradation is due to peptidase X, 20% by peptidase Y, and 15% by peptidase Z) cannot be determined, in part because inhibition of one enzyme will increase substrate availability for other peptidases. Thus, when reporting the results from such experiments, it is prudent to use conservative terminology (e.g., substrate A is primarily degraded by peptidase X, but peptidases Y and Z also contribute significantly). However, it is important to report the numerical data (as in **Table 1**), to allow the reader to reach his own conclusions.

Acknowledgments

The author wishes to thank Dr. G. Abbenante and Prof. V. Dive for providing EP24.15 and EP24.16 inhibitors, Shane Reeve for technical assistance and Dr. A. Ian Smith for critical review of the manuscript.

References

1. Docherty, K. and Steiner, D. F. (1982) Post-translational proteolysis in polypeptide hormone biosynthesis. *Ann. Rev. Physiol.* **44,** 625–638.
2. Turner, A. J. and Barnes, K. (1994) Neuropeptidases: candidate enzymes and techniques for study. *Biochem. Soc. Trans.* **22,** 122–127.
3. Schwartz, J.-C., Malfroy, B., and de la Baume, S. (1981) Biological inactivation of enkephalins and the role of enkephalin-dipeptidyl-carboxypeptidase ("enkephalinase") as neuropeptidase. *Life Sci.* **29,** 1715–1740.
4. Chu, T. G. and Orlowski, M. (1984) Active site directed N-carboxymethyl peptide inhibitors of a soluble metalloendopeptidase from rat brain. *Biochemistry* **23,** 3598–3603.
5. Shrimpton, C. N., Abbenante, G., Lew, R. A., and Smith, A. I. (2000) Development and characterisation of novel potent and stable inhibitors of endopeptidase EC 3.4.24.15. *Biochem. J.* **345,** 351–356.
6. Vincent, B., Jiracek, J., Noble, F., et al. (1997) Effect of a novel selective and potent phosphinic peptide inhibitor of endopeptidase 3.4.24.16 on neurotensin-induced analgesia and neuronal inactivation. *Brit. J. Pharmacol.* **121,** 705–710.

7. Jiracek, J., Yiotakis, A., Vincent, B., et al. (1995) Development of highly potent and selective phosphinic peptide inhibitors of zinc endopeptidase 24.15 using combinatorial chemistry. *J. Biol. Chem.* **270,** 21,701–21,706.
8. Dauch, P., Vincent, J.-P., and Checler, F. (1991) Specific inhibition of endopeptidase 24.16 by dipeptides. *Eur. J. Biochem.* **202,** 269–276.
9. Lowry, O. H., Rosebrough, N. J., Farr, A. L., and Randall, R. J. (1951) Protein measurement with the folin phenol reagent. *J. Biol. Chem.* **193,** 265–275.
10. Skidgel, R. A. (1992) Bradykinin-degrading enzymes: structure, function, distribution, and potential roles in cardiovascular pharmacology. *J. Cardiovasc. Pharmacol.* **20 (Suppl. 9),** S4–S9.
11. Hoang, M. V. and Turner, A. J. (1997) Novel activity of endothelin-converting enzyme: hydrolysis of bradykinin. *Biochem. J.* **327,** 23–26.
12. Smith, A. I. and Lew, R. A. (1995) The combination of high performance liquid chromatography and radioimmunoassay for the characterization of peptide processing pathways, in *Peptidases and Neuropeptide Processing* (Smith, A. I., ed.), Academic, San Diego, CA, pp. 125–139.
13. Campbell, D. J., Kladis, A., and Duncan, A.-M. (1993) Nephrectomy, converting enzyme inhibition and angiotensin peptides. *Hypertension* **22,** 513–522.
14. Campbell, D. J., Kladis, A., and Duncan, A.-M. (1993) Bradykinin peptides in kidney, blood, and other tissues of the rat. *Hypertension* **21,** 155–165.
15. Lawrence, A. C., Evin, G., Kladis, A., and Campbell, D. J. (1990) An alternative strategy for the radioimmunoassay of angiotensin peptides using amino-terminal-directed antisera: measurement of eight angiotensin peptides in human plasma. *J. Hypertension* **8,** 715–724.
16. Dean, R., Murone, C., Lew, R. A., et al. (1997) Localization of bradykinin B2 binding sites in rat kidney following chronic ACE inhibitor treatment. *Kidney Int.* **52,** 1261–1270.
17. Lew, R. A., Tomoda, F., Evans, R. G., et al. (1996) Synthetic inhibitors of endopeptidase EC 3.4.24.15: potency and stability *in vitro* and *in vivo*. *Brit. J. Pharmacol.* **118,** 1269–1277.
18. Smith, A. I., Lew, R. A., Shrimpton, C. N., Evans, R. G., and Abbenante, G. (2000) A novel stable inhibitor of endopeptidases EC 3.4.24.15 and 3.4.24.16 potentiates bradykinin-induced hypotension. *Hypertension* **35,** 626–630.
19. Telford, S. E., Smith, A. I., Lew, R. A., Perich, R. B., Madden, A. C., and Evans, R. G. (1995) Role of angiotensin converting enzyme in the vascular effects of an endopeptidase 24.15 inhibitor. *Br. J. Pharmacol.* **114,** 1185–1192.
20. Moeller I., Lew, R. A., Albiston, A. L., Mendelsohn, F. A. O., and Chai, S.-Y. (1999) A globin fragment, LVV-hemorphin-7, induces [3H]thymidine incorporation in a neuronal cell line via the AT_4 receptor. *J. Neurochem.* **73,** 301–308.
21. Lew, R. A., Tetaz, T., and Smith, A. I. (1994) Characterization of a membrane-associated EC 3.4.24.15-like endopeptidase expressed by endothelial cells of the ovine median eminence. *J. Neuroendocrinol.* **6,** 225–232.

22. Norman, M. U., Smith, A. I., and Lew, R. A. (1999) Role of calcium in the release of the bradykinin-degrading peptidase EC 3.4.24.16 from endothelial cells. *Lett. Peptide Sci.* **6,** 349–352.
23. Lew, R. A., Boulos, E., Stewart, K. M., et al. (2000) Bradykinin analogs with β-amino acid substitutions reveal subtle differences in substrate specificity between the endopeptidases EC 3.4.24.15 and EC 3.4.24.16. *J. Peptide Science* **6,** 440–445.
24. Lew, R. A., Hey, N. J., Tetaz, T. J., Glucksman, M. J., Roberts, J. L., and Smith, A. I. (1995) Substrate specificity differences between recombinant rat testes endopeptidase EC 3.4.24.15 and the native brain enzyme. *Biochem. Biophys. Res. Comm.* **209,** 788–795.

20

Isolation and Characterization of Naturally Processed MHC-Bound Peptides From the Surface of Antigen-Presenting Cells

Anthony W. Purcell

1. Introduction

The human major histocompatibility complex (MHC) is located on the short arm of chromosome 6 and encompasses approx 4 Mb, or 0.1%, of the genome. This region is, by far, the most polymorphic of the human genome. More than 220 genes have been identified in this region and at least 10% of these genes have a direct function in immune responses. The human MHC can be divided into three regions that encode the class I, class II, and class III human leukocyte antigen (HLA) gene products. These HLA molecules demonstrate tremendous polymorphism, which reflects the natural evolution of these genes in response to various microbial pathogens in different ethnic populations. HLA class I molecules are expressed on all nucleated cells and associate with short peptides (8–11 amino acids in length) derived from both self and foreign antigens. These peptide ligands are primarily generated in or transported into the cytoplasm and subsequently translocated into the endoplasmic reticulum (ER) where they assemble with nascent MHC class I molecules. These mature, peptide-loaded, complexes are then transported to the cell surface where they are scrutinized by $CD8^+$ cytotoxic T lymphocytes (CTL). Should the peptide ligand be derived from a pathogen and be recognized as foreign in an immunocompetent host, the cell is killed via the cytotoxic armory of the CTL. The expression of HLA class II molecules is confined to a small subset of highly specialized cells called antigen-presenting cells (APCs). The class II molecules associate with longer peptides (9–25 amino acids in length) than class I mole-

From: *Methods in Molecular Biology, vol. 251, HPLC of Peptides and Proteins: Methods and Protocols*
Edited by: M.-I. Aguilar © Humana Press Inc., Totowa, NJ

cules and this association occurs in late endosomal compartments, a distinct and separate cellular compartment to the ER-Golgi route inhabited by assembling MHC class I molecules. Class II molecules are recognized by $CD4^+$ T helper cells and functional recognition of these complexes is intimately involved in both the humoral and cellular immune response. MHC class I and class II molecules form membrane-distal structures that comprise a cleft in which the antigenic peptide ligands reside *(1–3)*. The T-cell receptor (TCR) on $CD4^+$ or $CD8^+$ T cells recognizes MHC molecules in the context of both the class I or class II molecule and the peptide antigen presented in the antigen binding groove of these cell surface molecules *(4)*. Technologies that allow the direct isolation and identification of peptide antigens associated with class I or II molecules have highlighted the ligand specificity of different MHC molecules and allowed direct identification of naturally processed and presented antigens derived from infectious micro-organisms as well as self-peptides associated with autoimmune disorders and cancers.

Several different approaches have been used to isolate MHC-bound peptides from cells, these include analysis of acidified cell lysates *(5–7)*, elution of peptides from the cell surface *(8,9)*, and immunoaffinity purification of the MHC-peptide complexes from detergent solubilized cell lysates *(10,11)*. Each approach has advantages, with the latter providing the best chance of epitope identification owing to the additional specificity of the immunoaffinity chromatography step and subsequent simplification of the range of cellular peptides isolated. However, they all share common features: (i) that upon cell lysis, peptides bound to MHC molecules (and other chaperones/receptors) are protected from intracellular and extracellular proteolysis; and (ii) acid treatment dissociates bound peptides from their MHC complexes. In the first approach, peptides are extracted from whole cell lysates following treatment with trifluoroacetic acid (TFA). The presence of TFA also aids in the precipitation of larger proteins leaving a complex mixture of intracellular and extracellular polypeptides, a proportion of which were bound to and protected from proteolysis by MHC molecules. Typically, these preparations are fractionated by reversed-phase high-pressure liquid chromatography (RP-HPLC) and screened with an immunological readout, such a T-cell assay to confirm the presence of a particular T-cell epitope or to quantitate known T-cell epitopes in different cell types *(5–7)*. An alternative to this method utilizes a nonlytic approach for recovering cell surface-associated peptides. The cells are washed in an isotonic buffer containing citrate at pH 3.3, the acidic nature of this buffer facilitates dissociation of MHC-bound peptides from the cell surface without affecting cell viability *(8)*. The great advantage of this technique is that the same cells may be harvested daily in an iterative approach for obtaining MHC-bound material. Again, although the specificity of this process is a little better for

MHC-bound material than whole cell lysates, some form of biological readout is generally necessary to locate peptides of interest prior to attempts at biochemical characterization.

The use of immunoaffinity chromatography dramatically improves the specificity of the peptide extraction process. The use of appropriate monoclonal antibodies can select a single MHC allele and some antibodies can even select a subpopulation of MHC molecules with defined molecular or functional properties ***(12,13)***. The use of immunoaffinity chromatography to isolate specific MHC molecules provides the most appropriate material for identifying individual peptide ligands restricted by a known MHC allele. Furthermore, in all the approaches discussed, the complexity of the eluates/lysates can be reduced by using cell lines that express reduced numbers of HLA alleles (e.g., homozygous lymphoblastoid or mutant cell lines, such as C1R which express very low levels of endogenous class I molecules, but support high-level expression of transfected class I molecules ***[14]***). This property makes these cells very attractive for examining endogenous peptides presented by individual class I alleles under normal physiological conditions ***(15–18)*** or during infection ***(19,20)***.

This chapter explores methods for the direct isolation of MHC class I or II molecules and focuses on the use of immunoaffinity chromatography and the subsequent separation and identification of the bound peptide antigens.

2. Materials

2.1. Generation of Cell Lysate by Whole Cell Lysate Method

1. A pellet of 1–5 × 10^8 cells (may be stored frozen for up to 6 mo at –70°C). The cells should express high levels of the class I or II molecule of interest (*see* **Note 1**).
2. Lysis buffer: 0.1% TFA (HPLC grade) v/v in Milli-Q H_2O.

2.2. Generation of Cell Lysate by Citrate Shock Method

1. Cells live in culture (either adherent or suspension) 1–5 × 10^8 cells are sufficient.
2. Citrate shock buffer: 0.131 *M* citric acid, 0.066 *M* Na_2HPO_4, 150 m*M* NaCl, pH 3.3.

2.3. Detergent Lysate for Immunoaffinity Chromatography

1. A cell pellet of between 5 × 10^9 and 5 × 10^{10} cells (may be stored frozen for up to 6 mo at –70°C). The nature of the cell type predicates the nature of the immunoaffinity matrix employed in subsequent purification steps and the ease with which individual peptides may be sequenced (*see* **Notes 1** and **2**).
2. Lysis buffer: 0.5% NP-40 (IGEPAL 630 from Sigma (St. Louis, MO) is the equivalent), 50 m*M* Tris-HCl, pH 8.0 (from a 1 *M* stock solution composed of Tris-HCl base (121.1 g/L) with 42 mL conc. HCl), 150 m*M* NaCl, protease inhibitor cocktail (Total Protease inhibitor from Roche or equivalent, should be made up fresh each time), Milli-Q H_2O (make sure this is freshly drawn and is filtered).

2.4. Preparation of Immunoaffinity Column

1. Purified monoclonal antibody at 1–10 mg/mL in Protein A loading buffer. Ideally, the mAb should only recognize the class I or II allele of interest, although affinity and specificity issues frequently require a compromise (*see* **Note 2**).
2. Suitable column (e.g., disposable plastic Econo-Column from Bio-Rad).
3. Protein A loading buffer: 0.05 *M* borate buffer pH 8.0 (composed of 500 mL of a 0.1 *M* boric acid/0.1 *M* KCl stock solution plus 39.7 mL 0.1 *N* NaOH and 460.3 mL Milli-Q H_2O).
4. Protein A wash buffer: 0.2 *M* triethanolamine, pH 8.2 at room temperature (RT) (prepare this solution fresh and pH just prior to use).
5. DMP cross-linker: 40 m*M* DMP-2HCl (Pierce), 0.2 *M* Triethanolamine, pH 8.2 (38 mL), pH to 8.3 with NaOH. Bring to 40 mL with 0.2 *M* triethanolamine pH 8.2 (do not filter).
6. Termination buffer: ice-cold 0.2 *M* Tris-HCl, pH 8.0.

2.5. Immunoaffinity Purification of MHC Class I/Class II Molecules

1. Wash buffer 1: 0.005% NP-40, 50 m*M* Tris-HCl, pH 8.0, 150 m*M* NaCl, 5 m*M* ethylene diamine tetraacetic acid (EDTA), 100 μ*M* phenyl mehtyl sulfanyl fluoride (PMSF), Sigma (0.1 *M* stock in absolute ethanol), 1 μg/mL Pepstatin A, Sigma (1 mg/mL stock in iso-propanol), in Milli-Q H_2O.
2. Wash buffer 2: 50 m*M* Tris-HCL, pH 8.0, 150 m*M* NaCl in Milli-Q H_2O.
3. Wash buffer 3: 50 m*M* Tris-HCl, pH 8.0, 450 m*M* NaCl in Milli-Q H_2O.
4. Wash buffer 4: 50 m*M* Tris-HCl, pH 8.0 in Milli-Q H_2O.
5. Elution buffer: 10% acetic acid in Milli-Q water (use best grade glacial acetic acid, e.g., Sigma ACS grade).

2.6. RP-HPLC Fractionation

1. Buffer A: 0.1% TFA in Milli-Q (filtered with a 0.22-μm filter).
2. Buffer B: 0.09% TFA in neat acetonitrile (HPLC grade, filtered with a 0.22-μm filter).

3. Methods

3.1. Generation of Cell Lysate (Whole Cell Lysis)

1. Cells (5×10^8 to 1×10^9) can be grown in spinner flasks, small bioreactors, or roller bottles to appropriate numbers and harvested by centrifugation (2000*g*, 10 min at 4°C). If necessary, the cells may be harvested by centrifugations and stored at –80°C for up to 6 mo.
2. Wash cells twice in phosphate-buffered saline (PBS).
3. Lyse cells in 0.1%TFA at a cell density of 5×10^7 per mL.
4. Clarify cell lysate by centrifugation and ultrafiltration (e.g., Ultrafree 15 centrifugal filter with a 5-kDa cutoff membrane, Millipore).
5. Concentrate flowthrough (i.e., <5 kDa fraction) by vacuum centrifugation, taking care not to allow the sample to dry completely.

6. Reconstitute in 0.1% TFA (or equivalent and compatible RP-HPLC equilibration buffer) and fractionate by RP-HPLC (*see* **Subheading 3.6.**).

3.2. Generation of Cell Lysate (Citrate Shock)

1. Adherent cells can be treated while still in a tissue-culture flask or attached to microbead/ceramic disk carriers in spinner flask or bioreactor. Suspension cells (5×10^8 to 1×10^9) can be grown in spinner flasks, small bioreactors, or roller bottles to appropriate numbers and harvested by centrifugation (2000*g*, 10 min at 4°C) and resuspended directly in citrate shock buffer.
2. Wash cells twice in PBS.
3. Resuspend cells in citrate shock buffer (at approx 1×10^8 cells/mL density for cells cultured in suspension or enough to just cover adherent cells (e.g., 2–5 mL for T175 flask, 15 mL for rollerbottle with continued rolling).
4. Incubate cells at RT for up to 15 min (less time if you want to retain maximum viability of the cells, typically around 5 min is sufficient for most cell types).
5. Wash cells twice in PBS and resuspend in fresh culture media. They will be ready for reharvesting in 12–24 h, depending on cell type and duration of the previous citrate shock.
6. Spin harvested citrate wash to remove cellular debris, filter with a 0.22-µm filter and then ultrafilter (e.g., Ultrafree 15 centrifugal filter with a 5-kDa cutoff membrane, Millipore).
7. The citrate wash is then desalted by passage over a solid phase extraction column (e.g., C18 Sep-Pak cartridge [Waters]), washed with 0.1% TFA, and bound peptides eluted with 60% acetonitrile/0.09% TFA.
8. The eluate is then dried by vacuum centrifugation, taking care not to totally dry the sample, and reconstituted in 0.1% TFA or equivalent.
9. The reconstituted SPE eluate is then subjected to RP-HPLC fractionation (*see* **Subheading 3.6.**).

3.3. Generation of Cell Lysate (Immunoaffinity Isolation)

1. Cells (5×10^9 to 1×10^{10}) can be grown in spinner flasks, bioreactors or roller bottles to appropriate numbers and harvested by centrifugation (2000*g*, 10 min at 4°C). Cells may be harvested iteratively and the washed cell pellet stored frozen at –80°C for up to 6 mo.
2. Prepare a 2X concentrated Lysis buffer.
3. Determine the amount of lysis buffer required. Cells are lysed at 1.25×10^8 cells per mL of (1X) lysis buffer.
4. Add correct volume of 2X lysis buffer to the frozen cell pellets and thaw the pellets quickly in a bath of tepid (i.e., RT) water. The temperature of the material should remain cold to touch, so do not let the material equilibrate, thaw until small ice clumps are left, and add ice-cold Milli-Q to a final volume so as the lysis buffer is now at 1X strength.
   - **Tip:** Check volume of cell pellet, if the volume of the cell pellet is close or over 50% of the final volume required, you may need to lyse at a lower cell density.

5. Give mixture a brief homogenize (e.g., using a Polytron Disperser) to disperse any left over ice pellets.
6. Mix the lysate by end-over-end mixing at 4°C for 1 h.
7. Centrifuge lysate for 10 min at 2000*g* (4°C). This step removes the nuclei.
8. Centrifuge supernatant for 30 min at 38,000*g* at 4°C—Sorvall high-speed centrifuge, SS34 rotor, 18,000 rpm.
9. Take supernatant from previous step and spin for 1 h 15 m in a Sw28 rotor, Beckman ultracentrifuge 25,000 rpm (100,000*g*) at 4°C. Multiple spins at **steps 7** and **8** may be necessary to fully clarify the lysate.
10. Collect the supernatant. It should be clear. If there is an unclear layer at top of the tubes carefully remove this layer and filter through a 0.8-µm and a 0.45-µm filter.

3.4. Preparation of Cross-Linked Immunoaffinity Column

3.4.1. Swelling Resin and Packing Column

1. Add 3.0 g of protein A-sepharose beads (Pharmacia CL-4B) per 10 mL column bed volume to a 50-mL tube and swell resin in Milli-Q water.
2. Wash twice in Milli-Q and create a 50% slurry.
3. Pour the slurry into column and allow to settle by gravity—check for air bubbles at this stage and agitate the slurry if necessary to remove air bubbles.
4. Equilibrate in Protein A loading buffer.

3.4.2. Adding Antibody to Column

1. Wash column in 10 column volumes (c.v.) of borate buffer.
2. Load antibody onto column at a flow rate of approx 0.5 mL/min.

3.4.3. Washing the Column of Unbound Material

1. When all of the antibody has gone through twice, wash the column with at least 20 c.v. of buffer or until the UV_{280nm} returns to baseline level.
2. Wash column with 15 c.v. of 0.2 *M* triethanolamine, pH 8.2 at RT (prepare this solution fresh and pH just prior to use).

3.4.4. Cross-Linking the Sepharose Protein A to the Fc Region of the Monoclonal Antibody

1. Prepare dimethyl pimelimidate (DMP) cross-linker solution.
2. Save a small aliquot of beads before cross-linking to run on a sodium dodecyl sulfate polyacrylamide gel electrophoresis (SDS-PAGE) gel to check for efficient antibody capture by the protein A Sepharose resin.
3. Pump 5 c.v. of DMP crosslinker through the column at RT leaving a meniscus just over the Sepharose column bed.
4. Seal the bottom of the column and allow to sit at RT for 1 h.

3.4.5. Termination of the Cross-Linking Reaction

1. Add 10 c.v. of ice-cold 0.2 *M* Tris-HCl, pH 8.0.
2. Pump through 10 c.v. of ice cold 0.1 *M* borate buffer, pH 8.0. (It may be convenient to stop here—Wash and store the column in Borate buffer, pH 8.0 supplemented with 0.02% NaN_3.)

3.5. Immunoaffinity Purification of MHC Class I/Class II

1. Load cell lysate onto a 0.5 mL protein A Sepharose precolumn that has been preequilibrated in wash buffer 1.
2. Multiple columns may be required and should be replaced upon clogging. These steps are done either by gravity or using a peristaltic pump in the cold room.
3. Collect precleared lysate and load onto the mAb column slowly—use peristaltic pump because gravity feed is too quick, at least initially.
4. For maximal yield, the lysate should be run through the columns twice (this may or may not be practicable!).
5. Wash the columns in the following order:
   - Wash buffer 1: 20 column volumes.
   - Wash buffer 2: 20 column volumes (to remove detergent).
   - Wash buffer 3: 20 column volumes (to remove nonspecifically bound material).
   - Wash buffer 4: 20 column volumes (removes salt to prevent crystal formation).
6. Take a 25-µL aliquot of column for SDS-PAGE analysis.
7. Elute MHC molecules in 5 c.v. of elution buffer.
8. Dispose of column.
9. As soon as possible, run the eluates over an an centrifugal ultrafiltration device (e.g., Ultrafree 15 centrifugal filter with a 5-kDa cutoff membrane, Millipore) **(the filter should be prewashed with 5 mL of 10% AcOH and the acid flowthrough removed)**.
10. Retain the retentate for SDS-PAGE analysis (–70°C).
11. Freeze flow through which contains the eluted peptides at –70°C (or progress straight on to RP-HPLC fractionation).

3.6. RP-HPLC Fractionation

1. Concentrate flowthrough either by: (i) freeze-drying overnight; or by (ii) centrifugal vacuum centrifugation (this is preferable, but take care not to dry the sample too long using this method). Ideally, the sample volume should be reduced up to 95%, but not to complete dryness, because this results in unacceptable sample loss caused by adsorption to the plasticware.
2. Take concentrated filtrate and fractionate on a HPLC system. A minimum of two dimensions of RP-HPLC will be necessary to achieve sufficient separation for biochemical studies, however, a single RP-HPLC step may be sufficient if using an

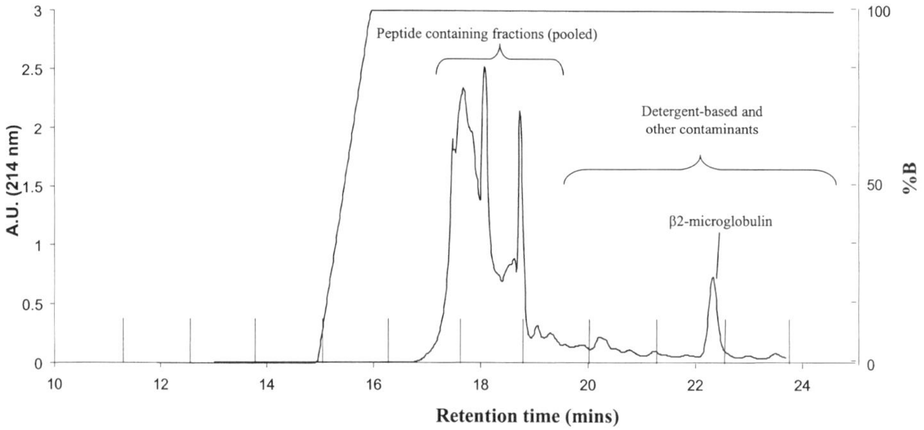

Fig. 1. Rapid isolation of MHC-bound peptides concentrates the eluate and removes contaminants that interfere with MS and cellular-based assays. Purification of HLA molecules from cell lines was performed as described in methods. In this case, human B-LCLs were grown in miniPERM bioreactors (HEREAUS, Hanau, Germany) in EXCEL 610 serum free media (CSL, Melbourne, Australia) supplemented with 1% fetal bovine serum (CSL). 5×10^9 cells were lysed at 4°C in 0.5% Nonidet P-40, 20 m*M* Tris-HCl, and 150 m*M* NaCl (pH 7.4) supplemented with complete protease inhibitor cocktail (Roche). Cell lysates were clarified by two rounds of centrifugation, and the supernatant was filtered and passed over a Tris-blocked Sepharose 4B precolumn. The precleared lysates were then applied to columns containing 2 mL of W6/32 affinity matrix, and the columns were washed in buffer containing 50 m*M* Tris-HCl, 150 m*M* NaCl, and 0.005% Nonidet P-40 (pH 8.0). The columns were subsequently washed extensively with 50 m*M* Tris-HCl and 150 m*M* NaCl (pH 8.0), a high salt buffer, to remove nonspecifically bound material (50 m*M* Tris-HCl and 500 m*M* NaCl [pH 8.0]), and finally, into 50 m*M* Tris-HCl (pH 8.0). Bound HLA-peptide complexes were eluted with 10% acetic acid, which also facilitates dissociation of the peptide ligands. The eluate was then passed through a Centricon 3 membrane (Millipore), and the flowthrough was concentrated by vacuum centrifugation to a final volume of 300 µL This ultrafiltered peptide flowthrough (<3 kDa) was fractionated by RP-HPLC using a SMART system HPLC (Pharmacia Biotech, Uppsala, Sweden). Peptides were separated using a µRPC C2 /C18 (2.1 mm [id] × 10-cm column; Pharmacia Biotech, Uppsala, Sweden) and resolved from contaminating detergent polymers by employing a rapid gradient from 0 to 60% acetonitrile in 0.1% aqueous TFA (12%/min, 200 mL/min). This material is suitable for pool Edman sequencing ***(46)*** and matrix-assisted laser desorption ionization–time-of-flight/mass spectrometry (MALDI-TOF/MS; as shown in **Fig. 3**).

immunological readout to assay fractions. For mass spectrometric-based analysis of fractions, use a rapid gradient of buffer A to B, which results in 2–3 peptide-containing fractions (*see* **Fig. 1**). Using this approach, a single fraction contains greater than 95% of the peptides, whereas the latter fractions contain

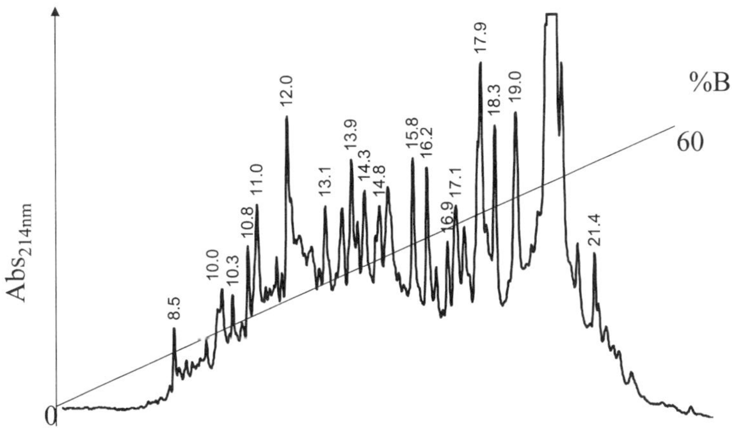

Fig. 2. Optimized RP-HPLC separation of peptide-containing fractions reveals the complexity of HLA-bound peptides and is a portal to their ultimate biochemical and immunological characterization. Peptide-containing fractions from initial RP-HPLC based isolation (*see* **Fig. 1**) were further purified using an optimized gradient based on the RP-HPLC retention behavior of a number of synthetic analogs of known naturally presented peptide ligands. Peptides were separated using a µRPC C2/C18 (2.1 mm (id) × 10-cm column and SMART system HPLC (Pharmacia Biotech) and resolved using linear gradient from buffer A (0.1% TFA) to 40% B (acetonitrile/0.09% TFA; 1.3%/min), then 40% B to 60% B (4%/min) at a flow rate of 150 µL/min. Fractions (75 µL) were retained and analyzed by MALDI-TOF/MS. This material can subsequently be used for MS-based sequence analysis of individual species, screening for T-cell recognition or can be further fractionated using different RP-HPLC modalities and narrow-bore or capillary columns.

NP-40 polymers which hamper MS analysis severely and β2-microglobulin (class I purification only).

3. Further fractionation can be afforded using optimized RP-HPLC protocols. These can vary from allele to allele and are best modeled using synthetic mixtures of known ligands of the class I or II molecule. An example following a general approach is shown in **Fig. 2**. At this stage, further LC–MS or matrix-assisted laser desorption/ionization time-of-flight mass spectometry (MALDI-TOF/MS) should be attempted.
4. Where a specific ligand is targeted, fractions can be screened by ^{51}Cr-release, cytokine assays (e.g., ELISPOT) or proliferative responses using TCR systems. The design and execution of such screening assays requires a great deal of immunological expertise and care must be taken to minimize sample toxicity to T cells. Typically, assays involve T-cell line, clone or T–T hybridoma recognition of fraction-pulsed APC. Several excellent recent reports that use these assays are referred to for more detail ***(21–38)***.

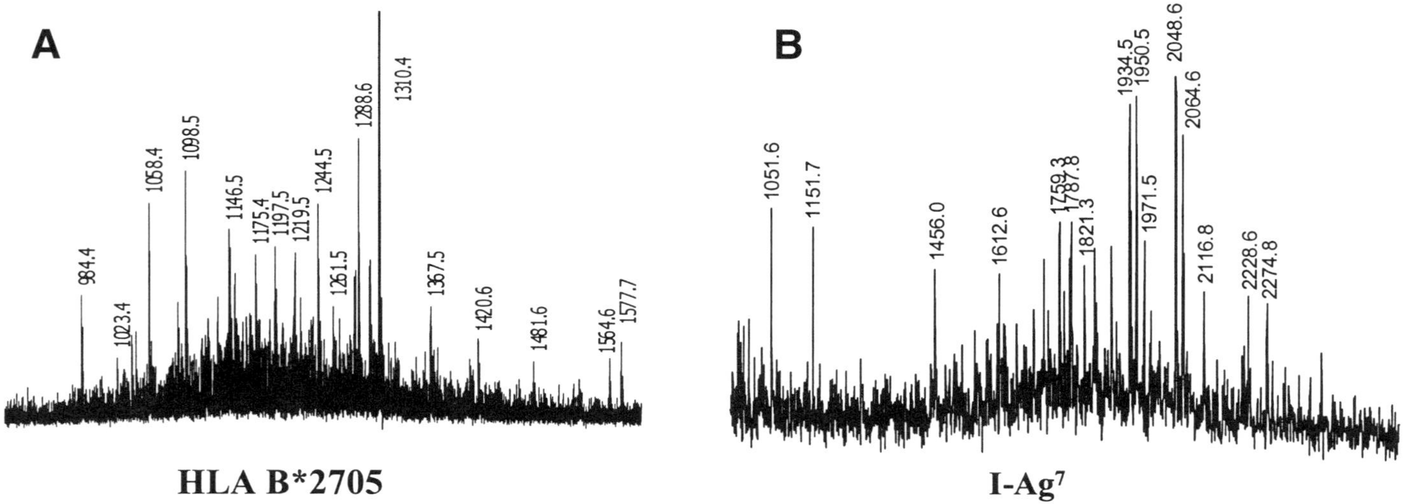

Fig. 3. Examples of biochemical analysis of peptides eluted from MHC class I and II molecules. Peptides eluted from **(A)** human class I (HLA B*2705) and **(B)** murine class II (I-Ag^7) MHC molecules following one round of RP-HPLC separation and visualized by MALDI-TOF/MS.

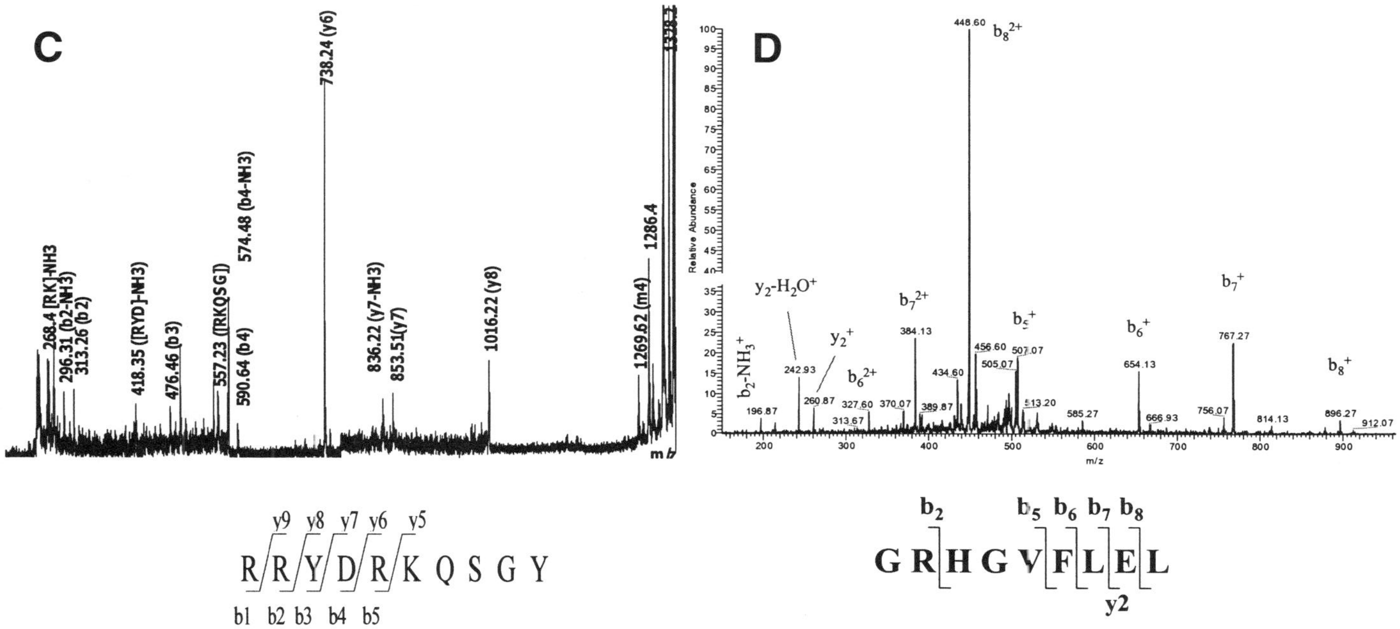

Fig. 3. **(C)** PSD-MALDI and **(D)** ESI-ion trap MS^2 based fragmentation and sequence assignment for a B*2705-restricted self peptides isolated from the surface of human B-lymphoblastoid cells following two dimensions of optimized RP-HPLC separation ***(40,41)***.

5. Additional fractionation can be afforded using either narrow bore or capillary HPLC systems preferably with on-line liquid chromatography–tandem mass spectroscopy (LC-MS/MS) capabilities (*see* **Note 3**).

3.7. Analysis of Fractions Using MS

Although outside the scope of this chapter, most forms of modern MS can be applied directly to fractionated material produced using these protocols. MALDI-TOF/MS can be performed directly by combining 0.5 µL of fraction with 0.5 µL of matrix (e.g., α-cyano-4-hydroxycinnamic acid, 10 mg/mL in ethanol) to give highly informative insights into fraction complexity and ligand repertoire ***(39,40)***. Moreover, for instruments fitted with a reflectron and post-source decay capacity, individual ligands may be sequenced as recently reviewed ***(41)***. The use of electrospray ionization or, more aptly, nanoelectrospray ionization-based instruments can also be applied directly to the analysis of fractionated MHC ligands ***(40)***. Several of these types of instrumentation are also excellent means to obtain sequence information on individual species, including ion-trap, triple quadrupole, and Qq-TOF-based instruments ***(41,42)***. Some examples are shown in **Fig. 3** (pp. 300, 301).

4. Notes

1. Choice of cell type: In order to maximize the yield of MHC class I or II molecules, the cell line used must be given serious consideration. Epstein Barr virus transformed B cell lines that express high levels of HLA A, B, or C class I molecules or HLA DR, DQ, or DP molecules are easily sourced from depositories such as ATCC, and can be used to great effect in biochemical studies of bound ligands. These B lymphoblastoid cell lines (B-LCLs) grow to high density in cell culture and express high levels of HLA gene products. Homozygous cell lines for most common class I or II alleles are well documented and have been used for years by the tissue typing community. The use of B-LCLs dictates the use of a discriminating antibody should a single allele be required to be purified. In the absence of such an antibody, the ideal cell type for these experiments would express a single MHC molecule and have intact antigen processing and presentation pathways. Several mutant cell lines have been generated that approximate such a cell type. The B-lymphoblastoid cell line Hmy2.C1R was generated by gamma irradiation of LICR.LON.Hmy2 ***(43)*** and selected with antibodies against HLA A and HLA B alleles and complement. This resulted in a cell line with no detectable HLA A or B gene products, yet with intact antigen processing and presentation pathways ***(44)***. Thus, these cells are able to support high level expression of individually transfected HLA A, B, or C gene products ***(44)***. Similar cell lines exist for class II elution studies. For example, the murine cell line M12.C3 lacks endogenous Ia and functional I-A^k expression can be restored by introduction of I-A^K α and β chains via transfection ***(45)***. It should be noted, however, that not all cells express class II molecules endogenously, thus restricting the array of antigen pre-

senting cells amenable to mutagenesis for the creation of appropriate cell lines. It is for this reason that B-LCLs homozygous at the DR, DQ, and DP locus are used in studies of human class II ligands whereas splenocytes and lymphomas are typically used in murine studies.

2. Choice of monoclonal antibody: Closely allied to the choice of cell line, and its impact upon the complexity of the total pool of bound peptide ligands, is the specificity and efficacy of the monoclonal antibody/antibodies used in the immunoaffinity isolation of MHC molecules. Monoclonal antibodies with specificity toward classes of MHC molecule, families of MHC molecule, individual alleles of MHC molecules, and even subsets of molecules of an individual allotype have been generated over the years and are hybridomas are readily accessible commercially through bodies such as the ATCC (www.atcc.org).
3. Column and mobile phase choice for multidimensional RP-HPLC are dictated by sample composition, but considerations should include altered ion pair agent, altered mobile phase pH, altered stationary phase ligand or mode of chromatography. A detailed discussion of these considerations can be found in Chapters 1, 2, and 12.

References

1. Scott, C. A., Peterson, P. A., Teyton, L., and Wilson, I. A. (1998) Crystal structures of two I-Ad-peptide complexes reveal that high affinity can be achieved without large anchor residues. *Immunity* **8(3),** 319–329.
2. Madden, D. R. (1995) The three-dimensional structure of peptide-MHC complexes. *Ann. Rev. Immunol.* **15,** 587–622.
3. Garcia, K. C. and Teyton, L. (1998) T-cell receptor peptide-MHC interactions: biological lessons from structural studies. *Curr. Opin. Biotechnol.* **9(4),** 338–343.
4. Garcia, K. C., Teyton, L., and Wilson, I. A. (1999) Structural basis of T cell recognition. *Annu. Rev. Immunol.* **17,** 369–397.
5. Falk, K., Rötzschke, O., Deres, K., Metzger, J., Jung, G., and Rammensee, H.-G. (1991) Identification of naturally processed viral nonapeptides allows their quantification in infected cells and suggests an allele-specific T cell epitope forecast. *J. Exp. Med.* **174,** 425–434.
6. Sijts, A. J., Neisig, A., Neefjes, J., and Pamer, E. G. (1996) Two Listeria monocytogenes CTL epitopes are processed from the same antigen with different efficiencies. *J. Immunol.* **156(2),** 683–692.
7. Rotzschke, O., Falk, K., Wallny, H. J., Faath, S., and Rammensee, H. G. (1990) Isolation and analysis of naturally processed viral peptides as recognized by cytotoxic T cells. *Nature* **348(6298),** 252–254.
8. Storkus, W. J., Zeh, H. J., Salter, R. D., and Lotze, M. T. (1993) Identification of T-cell epitopes: rapid isolation of class I-presented peptides from viable cells by mild acid elution. *J. Immunother.* **14(2),** 94–103.
9. Storkus, W. J., Zeh, H. J., Maeurer, M. J., Salter, R. D., and Lotze, M. T. (1993) Identification of human melanoma peptides recognized by class I restricted tumor infiltrating T lymphocytes. *J. Immunol.* **151(7),** 3719–3727.

10. Falk, K., Rötzschke, O., Stevanovic, S., Jung, G., and Rammensee, H.-G. (1991) Allele-specific motifs revealed by sequencing of self-peptides eluted from MHC molecules. *Nature* **351,** 290–296.
11. Rammensee, H. G., Falk, K., and Rotzschke, O. (1993) Peptides naturally presented by MHC class I molecules. *Annu. Rev. Immunol.* **11,** 213–244.
12. Urban, R. G., Chicz, R. M., Lane, W. S., et al. (1994) A subset of HLA-B27 molecules contains peptides much longer than nonamers. *Proc. Natl. Acad. Sci. USA* **91(4),** 1534–1538.
13. Purcell, A. W., Kelly, A. J., Peh, C. A., Dudek, N. L., and McCluskey, J. (2000) Endogenous and exogenous factors contributing to the surface expression of HLA B27 on mutant antigen presenting cells. *Hum. Immunol.* **61(2),** 120–130.
14. Storkus, W. J., Howell, D. N., Salter, R. D., Dawson, J. R., and Cresswell, P. (1987) NK susceptibility varies inversely with target cell class I HLA antigen expression. *J. Immunol.* **138(6),** 1657–1659.
15. Boisgerault, F., Tieng, V., Stolzenberg, M. C., et al. (1996) Differences in endogenous peptides presented by HLA-B*2705 and B*2703 allelic variants. Implications for susceptibility to spondylarthropathies. *J. Clin. Invest.* **98(12),** 2764–2770.
16. Fruci, D., Butler, R. H., Greco, G., et al. (1995) Differences in peptide-binding specificity of two ankylosing spondylitis-associated HLA-B27 subtypes. *Immunogenetics* **42(2),** 123–128.
17. Paradela, A., Garcia-Peydro, M., Vazquez, J., Rognan, D., and Lopez de Castro, J. A. (1998) The same natural ligand is involved in allorecognition of multiple HLA-B27 subtypes by a single T cell clone: role of peptide and the MHC molecule in alloreactivity. *J. Immunol.* **161(10),** 5481–5490.
18. Skipper, J. C., Kittelsen, D. J., Hendrickson, R. C., et al. (1996) Shared epitopes for HLA-A3-restricted melanoma-reactive human CTL include a naturally processed epitope from Pmel-17/gp100. *J. Immunol.* **157(11),** 5027–5033.
19. Ringrose, J. H., Yard, B. A., Muijsers, A., Boog, C. J., and Feltkamp, T. E. (1996) Comparison of peptides eluted from the groove of HLA-B27 from Salmonella infected and non-infected cells. *Clin. Rheumatol.* **15 Suppl. 1,** 74–78.
20. van Els, C. A., Heberts, C. A., van der Heeft, E., et al. (2000) A single naturally processed measles virus peptide fully dominates the HLA-A*0201-associated peptide display and is mutated at its anchor position in persistent viral strains. *Eur. J. Immunol.* **30(4),** 1172–1181.
21. Gonzalez, J. M., et al. (2000) HLA-A*0201 restricted CD8+ T-lymphocyte responses to malaria: identification of new plasmodium falciparum epitopes by IFN-gamma ELISPOT. *Parasite Immunol.* **22(10),** 501–514.
22. Scheibenbogen, C., Lee, K. H., Maher, S., et al. (1997) A sensitive ELISPOT assay for detection of CD8+ T lymphocytes specific for HLA class I-binding peptide epitopes derived from influenza proteins in the blood of healthy donors and melanoma patients. *Clin. Cancer Res.* **3(2),** 221–226.
23. Matsuo, M., Wada, H., Honda, S., et al. (1999) Expression of multiple unique rejection antigens on murine leukemia BALB/c RLmale symbol1 and the role of dominant Akt antigen for tumor escape. *J. Immunol.* **162(11),** 6420–6425.

24. Nguyen, T., Naziruddin, B., Dintzis, S., Doherty, G. M., and Mohanakumar, T. (1999) Recognition of breast cancer-associated peptides by tumor-reactive, HLA-class I restricted allogeneic cytotoxic T lymphocytes. *Int. J. Cancer* **81(4),** 607–615.
25. Tsomides, T. J., Reilly, E. B., and Eisen, H. N. (1997) Anti-melanoma cytotoxic T lymphocytes (CTL) recognize numerous antigenic peptides having 'self' sequences: autoimmune nature of the anti-melanoma CTL response. *Int. Immunol.* **9(2),** 327–338.
26. Slingluff, C. L., Cox, A. L., Henderson, R. A., Hunt, D. F., and Engelhard, V. H. (1993) Recognition of human melanoma cells by HLA-A2.1-restricted cytotoxic T lymphocytes is mediated by at least six shared peptide epitopes. *J. Immunol.* **150(7),** 2955–2963.
27. Brumeanu, T. D., Kohanski, R., Bona, C. A., and Zaghouani, H. (1993) A sensitive method to detect defined peptide among those eluted from murine MHC class II molecules. *J. Immunol. Methods* **160(1),** 65–71.
28. Purcell, A. W., et al. (1998) Avoidance of self-reactivity results in skewed CTL responses to rare components of synthetic immunogens. *J. Immunol.* **160(3),** 1085–1090.
29. Chen, W., Ede, N. J., Jackson, D. C., McCluskey, J., and Purcell, A. W. (1996) CTL recognition of an altered peptide associated with asparagine bond rearrangement. Implications for immunity and vaccine design. *J. Immunol.* **157(3),** 1000–1005.
30. Crotzer, V. L., et al. (2000) Immunodominance among EBV-derived epitopes restricted by HLA-B27 does not correlate with epitope abundance in EBV-transformed B-lymphoblastoid cell lines. *J. Immunol.* **164(12),** 6120–6129.
31. Wang, W., Man, S., Gulden, P. H., Hunt, D. F., and Engelhard, V. H. (1998) Class I-restricted alloreactive cytotoxic T lymphocytes recognize a complex array of specific MHC-associated peptides. *J. Immunol.* **160(3),** 1091–1097.
32. Kittlesen, D. J., et al. (1998) Human melanoma patients recognize an HLA-A1-restricted CTL epitope from tyrosinase containing two cysteine residues: implications for tumor vaccine development [published erratum appears in J Immunol 1999 Mar 1;162(5):3106]. *J. Immunol.* **160(5),** 2099–2106.
33. Hogan, K. T., Eisenger, D. P., Capp, S. B., 3rd, et al. (1998) The peptide recognized by HLA-A68.2-restricted, squamous cell carcinoma of the lung-specific cytotoxic T lymphocytes is derived from a mutated elongation factor 2 gene. *Cancer Res.* **58(22),** 5144–5150.
34. Wang, W., Eulden, P. H., Pierce, R. A., et al. (1997) A naturally processed peptide presented by HLA-A*0201 is expressed at low abundance and recognized by an alloreactive CD8+ cytotoxic T cell with apparent high affinity [published erratum appears in J Immunol 1999 Mar 1;162(5):3106]. *J. Immunol.* **158(12),** 5797–5804.
35. Dubey, P., Hendrickson, R. C., Meredith, S. C., et al. (1997) The immunodominant antigen of an ultraviolet-induced regressor tumor is generated by a somatic point mutation in the DEAD box helicase p68. *J. Exp. Med.* **185(4),** 695–705.

36. Meadows, L., Wang, W., den Haan, J. M., et al. (1997) The HLA-A*0201-restricted H-Y antigen contains a posttranslationally modified cysteine that significantly affects T cell recognition. *Immunity* **6(3),** 273–281.
37. Fiorillo, M. T., Meadows, L., D'Amato, M., et al. (1997) Susceptibility to ankylosing spondylitis correlates with the C-terminal residue of peptides presented by various HLA-B27 subtypes. *Eur. J. Immunol.* **27(2),** 368–373.
38. Skipper, J. C., Hendrickson, R. C., Gulden, P. H., et al. (1996) An HLA-A2-restricted tyrosinase antigen on melanoma cells results from posttranslational modification and suggests a novel pathway for processing of membrane proteins. *J. Exp. Med.* **183(2),** 527–534.
39. Barnden, M. J., Purcell, A. W., Gorman, J. J., and McCluskey, J. (2000) Tapasin-mediated retention and optimization of peptide ligands during the assembly of class I molecules. *J. Immunol.* **165(1),** 322–330.
40. Purcell, A. W., Gorman, J. J., Garcia-Peydro, M., et al. (2001) Quantitative and qualitative influences of tapasin on the class I peptide repertoire. *J. Immunol.* **166(2),** 1016–1027.
41. Purcell, A. W. and Gorman, J. J. (2001) The use of post source decay in matrix assisted laser desorption-ionisation mass spectrometry to delineate T cell determinants. *J. Immunol. Methods* **249,** 17–31.
42. de Jong, A. (1998) Contribution of mass spectrometry to contemporary immunology. *Mass Spectrom. Rev.* **17(5),** 311–335.
43. Edwards, P. A., Smith, C. M., Neville, A. M., and O'Hare, M. J. (1982) A human-hybridoma system based on a fast-growing mutant of the ARH-77 plasma cell leukemia-derived line. *Eur. J. Immunol.* **12(8),** 641–648.
44. Alexander, J., Payne, J. A., Murray, R., Frelinger, J. A., and Cresswell, P. (1989) Differential transport requirements of HLA and H-2 class I glycoproteins. *Immunogenetics* **29(6),** 380–388.
45. Nelson, C. A., Roof, R. W., McCourt, D. W., and Unanue, E. R. (1992) Identification of the naturally processed form of hen egg white lysozyme bound to the murine major histocompatibility complex class II molecule I-Ak. *Proc. Natl. Acad. Sci. USA* **89(16),** 7380–7383.
46. Rammensee, H.-G., Friede, T., and Stevanovic, S. (1995) MHC ligands and peptide motifs: first listing. *Immunogenetics* **41,** 178–228.

21

Australian Funnel-Web Spider Venom Analyzed With On-Line RP-HPLC Techniques

David Wilson and Paul Alewood

1. Introduction

Venoms have attracted significant study in recent years as a reservoir of complex libraries of natural products possessing a wide range of biological activities. Moreover, venoms contain specific and potent molecules that may be utilized in pharmaceutical development and in the production of environmentally friendly insecticides. The compositions of venoms are typically highly complex and contain a variety of molecules including proteins, peptides, and numerous types of small molecules. This complexity requires highly sensitive techniques to allow separation of these components for study. The techniques should also be able to accommodate large variations in sample size to account for the differences in venom available from different creatures (e.g., some snakes can supply up to 500 mg of crude venom from a single milking, whereas some small insects, such as ants, supply submicrogram amounts *[1]*). These qualities have been found and continue to be advanced in the technique high-performance liquid chromatography (HPLC), in particular reversed-phase HPLC (RP-HPLC). This technique combined with a variety of detection methods can allow the collection of a significant amount of data from very small venom samples.

Early analytical studies of whole venom samples used techniques such as gel electrophoresis, size exclusion, and ion exchange chromatography *(2,3)*. Although these techniques are still useful today, they provide limited information. Gel electrophoresis provides an idea of the overall size distribution of the protein components of the venom that typically have masses greater than 15 kDa. However, gel electrophoresis is limited in the separation it can provide

From: *Methods in Molecular Biology, vol. 251, HPLC of Peptides and Proteins: Methods and Protocols*
Edited by: M.-I. Aguilar © Humana Press Inc., Totowa, NJ

for closely related homologs and components containing noncharged post-translational modifications. Furthermore, the amount of sample required for gel electrophoresis is typically larger than that required for HPLC analysis (at least 5–10-fold; *see* **Note 1**) and recollection of separated components for further analysis is relatively labor intensive. Perhaps the greatest limitation of this technique is the poor accuracy of molecular weight estimation. Masses are estimated from a standard curve generated from a molecular weight marker ladder run alongside the sample and are often very inaccurate. By comparison, size exclusion chromatography has an advantage that it may be performed using small columns, though the inability to provide satisfactory component separation remains.

More recently, advances in HPLC support the high resolution separation of complex mixtures and, thus, provide a desirable technique for venom analysis. Current HPLC systems, in contrast to the early systems, provide great separation of very small quantities of complex mixtures at low flow rates, resulting in improved sample recovery.

1.1. Liquid Chromatography–Mass Spectrometry as a Venom Fingerprinting Technique

The detectors typically used with HPLC are nondestructive UV or diode array, but mass spectrometers can also be used and this technique is referred to as liquid chromatography–mass spectrometry (LC–MS). In fact, if set up correctly, both UV and MS data can be collected from a single run (*see* also Chapter 9).

Because a majority of venom toxins are peptidic or proteinaceous in character, there is a direct link between the toxin component expressed and the genetic coding of the particular specimen. The genetic coding of a group of individual specimens belonging to a particular species should have identical or very similar physicochemical properties with each toxin (peptide/protein) having a defined protein sequence, molecular weight and chromatographic retention time in RP-HPLC analyses. Therefore, the collection and comparison of this data provides a means of establishing a venom component profile for a particular species. The usefulness of this approach of identification through venom profile analysis is illustrated in many recent RP-HPLC/matrix-assisted laser desorption ionization time-of-flight mass spectrometry (RP-HPLC/MALDI-TOF) (*see* **Note 2**) examples including bacteria ***(4–8)***, snake venom ***(9)***, cone snail venom ***(10–12)***, scorpion venom ***(13)***, and tarantula venom studies ***(14–16)***.

The Australian Funnel-web spiders are a group of venomous arachnids comprising more than 35 species from two genera (*Atrax* and *Hadronyche*), which are located predominantly along the southeast coast of Australia. The current taxonomy of the Australian Funnel-web spiders is shrouded in confusion and,

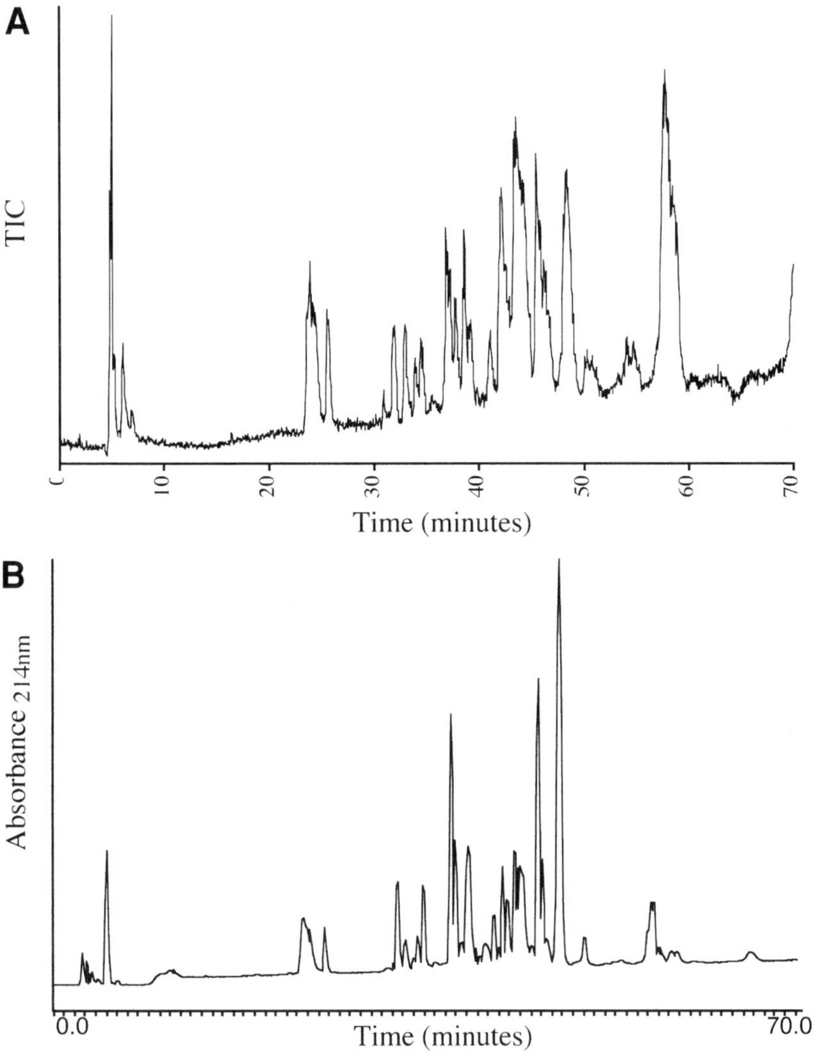

Fig. 1. **(A)** RP-HPLC/ESI-MS chromatogram. **(B)** RP-HPLC/UV chromatogram of venom from an Australian Funnel-web spider illustrating the number of components and complexity.

as a result, species identification based on morphology can prove challenging. The venom of the Australian Funnel-web spiders is complex, containing between 40 and 100 peptide components (*see* **Fig. 1**). Furthermore, the composition of the venom from certain species of these spiders has been shown to differ between the male and female specimens ***(17)***. In this chapter, we investigate the peptidic venoms of selected Australian Funnel-web spiders using

RP-HPLC techniques, including RP-HPLC/electrospray ionization mass spectrometry (ESI-MS) (*see* **Note 3**), in order to identify their complexity, the types of molecules present and the extent to which this approach can identify and characterize different species of the Australian Funnel-web spiders.

2. Materials

2.1. Venom Collection

1. Polyethylene/polypropylene pipet tips and collection tubes (e.g., 1.5–2 mL Eppendorf) (*see* **Note 4**).
2. 10–100 μL Pipet with tip ejection arm (e.g., Gilson P100).

2.2. RP-HPLC/ESI-MS

1. HPLC grade acetonitrile and trifluoroacetic acid (TFA).
2. Solvent A: 0.1% TFA in water.
3. Solvent B: 0.09% TFA in 90% acetonitrile/10% H_2O.
4. RP C_{18} analytical column (Vydac; 2.1 × 250 mm, 5 μm particle size, 300 Å pore size) (*see* **Note 5**).
5. Hewlett Packard Series1100 HPLC pump system (Hewlett Packard, Australia) or similar low flow-rate HPLC pump system such as Applied Biosystems 140B solvent delivery system.
6. Electrospray ionisation mass spectrometer with turbospray facility (e.g., PE-SCIEX triple quadrupole mass spectrometer (PE-SCIEX, Canada) equipped with an Ionspray atmospheric pressure ionization source and Turbospray interface) (*see* **Notes 1**, **2** and **3**).
7. Applied Biosystems 785A programmable absorbance detector.

2.3. RP-HPLC

1. HPLC grade acetonitrile and TFA.
2. Solvent A: 0.1% TFA in water.
3. Solvent B: 0.09% TFA in 90% acetonitrile/10% H_2O.
4. Polyethylene/polypropylene pipet tips and collection tubes (14 mL Falcon or 1.5–2 mL Eppendorf).
5. Waters 600 HPLC controller.
6. Waters 486 tunable absorbance detector.
7. RP C_{18} analytical column (Vydac; 4.6 × 250 mm, 5 μm particle size, 300 Å pore size).

3. Methods

3.1. Venom Collection

1. Collect the venom from live specimens (*see* **Note 6**). Funnel-web spider venom is "milked" from the tips of the fangs of an aggravated specimen using a pipet equipped with a polyethylene/polypropylene tip. Collect, handle and store the venom using polyethylene/polypropylene materials (e.g., pipet tips, Eppendorf

tubes, and so on) (*see* **Note 7**). Wash the "milked" venom from the pipet tip with 800 µL of 0.1% TFA/H_2O and lyophilize in a Savant Speedvac (*see* **Note 8**).

3.2. Online RP-HPLC/ESI-MS

1. Reconstitute the lyophilized venom samples to a concentration of approx 1 mg/mL in approx 30% acetonitrile/water (v/v) (*see* **Note 9**).
2. Spin the venom samples in a bench centrifuge for approx 30 s at maximum speed to pellet all insoluble and particulate material (dirt, cell membrane in the case of dissections) (*see* **Note 10**).
3. Set up the mass spectrometer (i.e., turn mass spectrometer on and set data collection parameters) and turbospray to warm up.
4. Equilibrate the RP-HPLC column by running 100% solvent A through at the flow rate to be used for 20–30 min, but without the column eluent attached to the mass spectrometer (*see* **Note 11**). For the final 5–10 min of equilibration, attach the output of the RP-HPLC column to the mass spectrometer and monitor the mass spectrometer signal to make sure the signal is clean and free of contaminants (*see* **Note 12**).
5. Inject samples (10 µL) onto an on-line RP C_{18} analytical column and simultaneously start both the chromatography gradient and mass spectrometer data collection (*see* **Note 13**). Elute the venom constituents with a flow rate of 130 µL/min using a linear gradient of 100% solvent A: 0% solvent B to 60% solvent B over 60 min. The data analysis and subsequent processing is performed using the Biomultiview (PE-SCIEX, Canada) software package. Electrospray mass spectra are acquired in positive ion mode by direct flow of the column eluent into a PE-SCIEX API III triple quadrupole mass spectrometer (PE-SCIEX, Canada) equipped with an Ionspray atmospheric pressure ionization source and Turbospray interface. The full scan data for each venom sample is acquired at an orifice potential of 80 V over a mass range of 400–2100 Daltons with a step size of 0.2 Dalton (*see* **Note 14**).

3.3. Analytical RP-HPLC

1. Dissolve lyophilized crude venom in 0.1% TFA/H_2O to a concentration of approx 1 mg/mL.
2. Load the venom using a 10–20 µL injection from a Waters 717 Autosampler onto a C_{18} analytical RP-HPLC column (Vydac 4.6 × 250 mm, 300 Å pore size, 5 µm particle size).
3. Elute the venom components with a linear 1% gradient from 0 to 80% Solvent B over 80 min at a flow rate of 1 mL/min governed by a Waters 600 Controller. Monitor the UV absorbance at 214 nm using a Waters 486 tunable absorbance detector.

3.3.1. Venom Profiling of Australian Funnel-Web Spider Venom

The RP-HPLC/UV and RP-HPLC/ESI analysis of venom from juvenile male, mature female, and mature male specimens from two species of the *Atrax* genus of Australian Funnel-web spiders illustrates the complexity of these venoms, the gender differences evident in some species, and how the venom

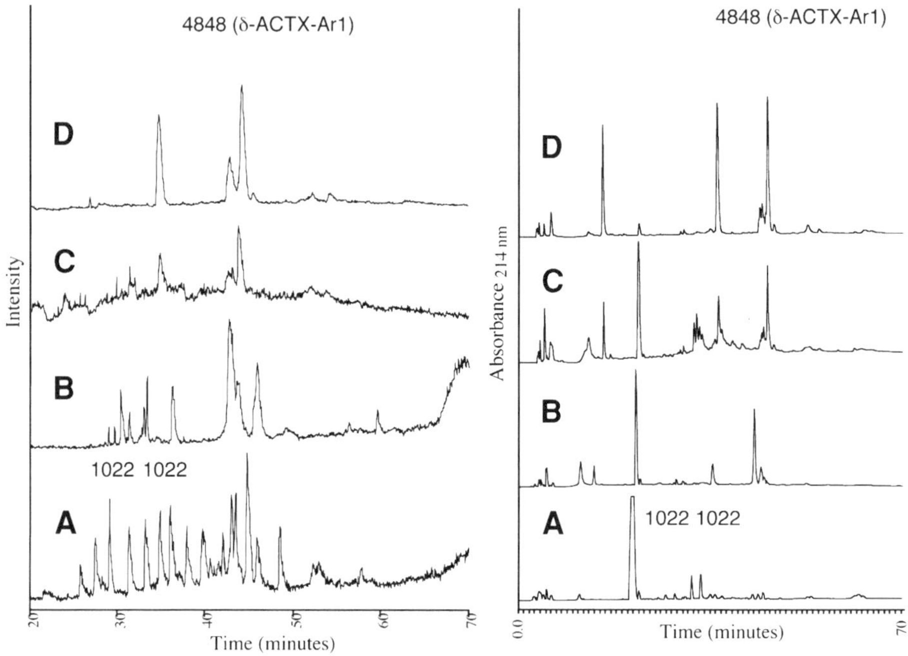

Fig. 2. A comparison of the RP-HPLC/ESI-MS and RP-HPLC/UV chromatograms from the female, and juvenile and mature male venom of *Atrax robustus:* Sydney. RP-HPLC/ESI-MS *(left panel)* and RP-HPLC/UV *(right panel)* chromatogram data obtained for the **(A)** female *Atrax robustus:* Sydney venom sample. **(B)** Mature male *Atrax robustus:* Sydney venom sample. **(C)** Prematuring moult of juvenile male *Atrax robustus:* Sydney venom sample. **(D)** Postmaturing moult mature male *Atrax robustus:* Sydney venom sample. Components of note are highlighted.

profile can be used as a taxonomic tool. Furthermore, the study illustrates the importance of both RP-HPLC/UV and RP-HPLC/ESI data when establishing the venom profile.

3.3.1.1. RP-HPLC-Based Venom Mass Profile Analysis of the Sydney *Atrax robustus:* Sydney Specimens

The venom of the Sydney Funnel-web spider (*Atrax robustus:* Sydney) has been extensively studied and thus was used primarily as a "control" to establish the viability of the methodology developed in these studies. As a result of the commonly known gender difference between the venom composition of this species ***(17)***, venom samples of juvenile male, mature female, and mature male specimens were investigated. The analysis of the venom samples by RP-HPLC/UV and RP-HPLC/ESI (*see* **Fig. 2**) showed a number of obvious differences between the chromatograms, and further confirmed the gender-related venom composition differences. Closer examination of the mass data combined with the elution time data (*see* **Table 1**) revealed a number of components common to both genders, but also further highlighted the differences in venom composition. The most obvious difference is the dominating presence of the molecule δ-ACTX-Ar1 eluting at approx 42 min in the RP-HPLC/UV and RP-HPLC/ESI chromatograms of the male specimen venom samples. In the juvenile male venom sample this component is evident, but at lower levels. Likewise, the female venom sample does not contain the molecule δ-ACTX-Ar1, but possesses two components with *Mr* = 1022 Daltons not observed in the male venom profile. Several other differences are also evident upon examination of the mass data and are illustrated in Table 1. The obvious differences between the corresponding peak intensities of the RP-HPLC/UV and RP-HPLC/ESI-MS chromatograms illustrate the importance of data collection from both detectors. As the intensity of the component peaks in the RP-HPLC/ESI-MS chromatogram are related to the component's capacity to ionize in the mass spectrometer, it does not provide a adequate representation of the relative abundances of each component. In contrast, the RP-HPLC/UV data are directly related to the component concentration and, thus, provide a better profile of the relative abundances of the components in the venom profile. Furthermore, the RP-HPLC/UV data provide evidence of molecules not detected by the mass spectrometer owing to masses outside the scanned mass range or their inability to ionize (*see* **Note 15**).

3.3.1.2. RP-HPLC-Based Venom Mass Profile Analysis of the *Atrax sp.:* Illawarra Specimens

Analysis of the venom of a second species of *Atrax* Funnel-web spider (*Atrax sp.:* Illawarra) provided an excellent example of the differences in venom composition observed between specimens belonging to a distinct species. RP-HPLC/UV and RP-HPLC/ESI analysis of the venom of mature male, mature female, and juvenile male (which matured in captivity and provided pre- and postmaturing moult samples) specimens revealed a very different venom composition to the *Atrax robustus:* Sydney specimens, and possessed some degree of gender-related venom composition differences.

The RP-HPLC/ESI-MS analysis (*see* **Fig. 3**) revealed that the venom composition of the male and female *Atrax sp.:* Illawarra specimens is very similar. Close inspection of the RP-HPLC/ESI-MS mass data revealed some subtle gender-related differences such as the presence of the molecules *Mr* = 5666 Daltons, *Mr* = 3820 Daltons, *Mr* = 4519 Daltons, *Mr* = 4911 Daltons, and *Mr* = 4181 Daltons in the venom of the male specimens only (*see* **Table 2**). Likewise, the components *Mr* = 4410 Daltons and *Mr* = 4361 Daltons were only evident in the venom of the female specimens (*see* **Table 2**). The RP-HPLC/UV

Table 1
Venom Mass Profile "Descriptor"[a] Molecules Determined for Male and Female Specimens of *Atrax robustus:* Sydney

Elution time (min)	Female components (*Mr* = Da)	Male components (*Mr* = Da)
Approx 26	*4004*	*4004*
Approx 28	4395	
Approx 29	*3879*	*3879*
Approx 2.5	*4379*	*4379*
Approx 31	1022	
Approx 33	1022	
Approx 34		7155
Approx 35	*7139*	*7139*
Approx 36	4615	
Approx 38.5	4498	
Approx 40	4469	
Approx 40.5	7946	
Approx 41	4811	
Approx 41.5	7915	
Approx 43	4120	4848
Approx 43.5	*7136*	*7138*
Approx 44	*7286*	*7287*
Approx 44.5	4547	7235
Approx 45	4223	
Approx 46	*7213*	*7213*
Approx 46.5		3791
Approx 47	*4527*	*4526*
Approx 48	*4205*	*4206*
Approx 49	4050	
Approx 56		3510
Approx 58		3991
Approx 58.5	6793	
Approx 60		4022

[a]The "descriptor" molecules are components exclusive to the venom of a particular group (e.g., species group, species, or species gender). In general, "descriptor" molecules are components present in the majority of venom samples collected from a particular group and giving rise to a definitive mass spectral peak. However, in cases where only a single venom sample is obtained, the "descriptor" molecules are based solely on that single sample. Owing to the relationship of some of the Australian Funnel-web spiders, certain "descriptor" molecules are common to the specimens belonging to the species group, whereas other "descriptor" molecules are specific for a particular species variant or gender.

[b]Components common to the venom (i.e., possessing identical mass and retention time information [*Mr* ± 3 Daltons]) of both male and female specimens are highlighted in italics.

Table 2
Venom Mass Profile "Descriptor" Molecules Determined for Male and Female Specimens of *Atrax sp.:* Illawarra

Elution time (min)	Components (*Mr* = Da)
Approx 30	3944
Approx 38	4472
Approx 39	4579
Approx 40	7600
Approx 40.5	7105
Approx 41.5	7443
Approx 42.5	4850
Approx 44	4102
Approx 44.5	5666 (male only)
Approx 46	4221
Approx 47	3820 (male only + juvenile male)
Approx 48	4519 (male only)
Approx 49	3948
Approx 49.5	4911 (male only)
Approx 50	4181 (male only)
Approx 50.5	3851
Approx 56	4410 (female only)
Approx 57	4278
Approx 58	4361 (female only)
Approx 63	5355
Approx 68	5338

analysis (*see* **Fig. 3**) was only performed on the female and juvenile male venom samples owing to sample limitations of mature male venom.

The RP-HPLC/UV and RP-HPLC/ESI-MS analyses of the venom from *Atrax sp.:* Illawarra specimens illustrated that thc composition of these venom samples is substantially different to the venom of the *Atrax robustus:* Sydney specimens. Furthermore, the venom of the juvenile male, mature male, and female *Atrax sp.:* Illawarra specimens is very similar in composition overall, but also contains a number of distinct differences.

The development of a venom mass profiling technique using RP-HPLC/UV and RP-HPLC/ESI-MS data, as illustrated above, has proven to be a useful tool for taxonomic distinction of different species variants of Australian Funnel-web spider belonging to the *Atrax* genus. The method also provides a method to distinguish between genders, and can distinguish between two very closely related specimens.

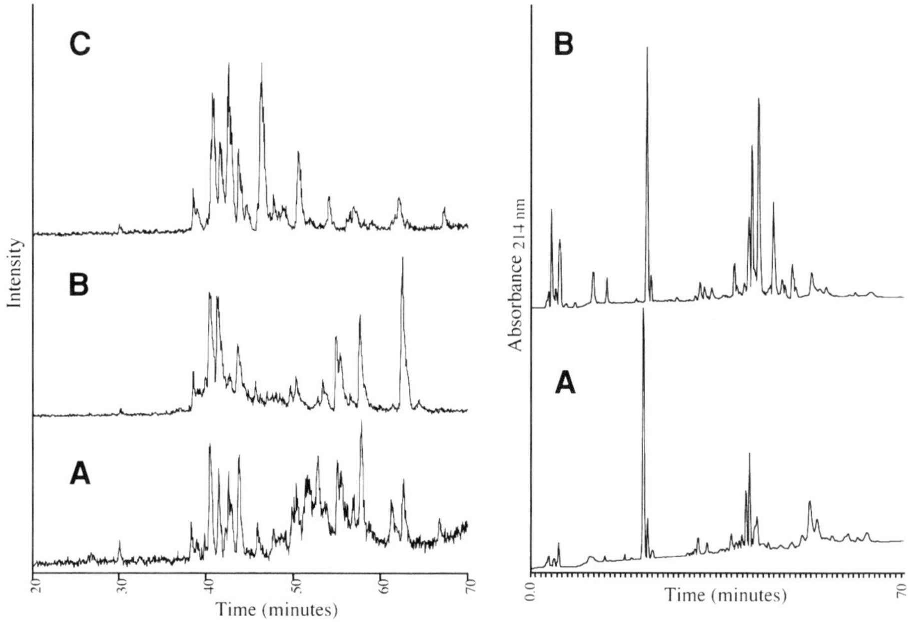

Fig. 3. A comparison of the RP-HPLC chromatograms from the juvenile, female, and mature male venom of *Atrax sp.:* Illawarra. **(A)** RP-HPLC/ESI-MS *(left panel)* and RP-HPLC/UV *(right panel)* chromatogram data obtained for the female *Atrax sp.:* Illawarra venom sample. **(B)** RP-HPLC/ESI-MS and RP-HPLC/UV chromatogram data obtained for the juvenile male *Atrax sp.:* Illawarra venom sample. **(C)** RP-HPLC/ESI-MS chromatogram data for the venom sample obtained from the *Atrax sp.:* Illawarra postmaturing moult mature male specimen.

4. Notes

1. An important aspect in venom analysis is to minimize the loss of sample. This is even more important in cases where the study is performed on individual specimens, as a single specimen may only supply one venom collection before it dies, and some specimens may be very difficult to obtain (e.g., a particular species, or gender of a particular species or geographical variant). In these cases, as much information as possible has to be extracted from the solitary sample. Remember, once it's gone, it's gone for good!
2. MALDI-MS allows high-sensitivity analysis of large molecular weight compounds, including biomolecules such as proteins. MALDI involves ionization and transfer of a sample from a condensed phase to the gas phase by vaporization of a non-volatile solid matrix with laser radiation *(18)*. This type of ionization is termed as "soft" ionization and results in molecule ionization without molecule fragmentation. The primary advantage of "soft" ionization is the ability to observe

biomolecules in their native state. The principal disadvantage of MALDI-MS, for rapid venom profile analysis, is the requirement of matrix addition to the sample eliminating the possibility of on-line infusion RP-HPLC coupled mass spectrometry (RP-HPLC/MS). Venom profile analysis by MALDI-MS does not permit accurate mass/retention time correlation determination because of the necessity to acquire the RP-HPLC fractions first, followed by lyophilization and addition of the solid matrix of the fractions prior to mass spectrometry analysis. Furthermore, the reproducibility of MALDI-MS experiments on an individual sample suffers from variations in ionization of components. This variation results from inconsistent component crystallization with the matrix between crystals. Therefore, the mass spectrum obtained varies with respect to ion presence from crystal to crystal, and consequently the point of focus of the ionizing laser.

3. ESI-MS is also a "soft" ionization technique and involves the production of gaseous ionized molecules from a liquid solution ***(18)***. This is achieved by creating a fine spray of sample laced solvent in the presence of a strong electric field resulting in the formation of highly charged droplets. The charged droplets are electrostatically attracted to the inlet of the mass spectrometer where dry gas evaporates the solvent from the surface of the drops causing ions to leave the particle as the droplet size decreases. ESI results in the formation of multiply charged ions and, because the mass spectrometer measures the mass-to-charge ratio *(m/z)*, allows the observation of relatively large molecules using a comparatively small mass range analyzer. Deconvolution of the multiply charged ion series can then be performed by a computer package and providing an accurate mass through averaging ***(18)***. The addition of a turbospray interface supplies heat and a greater dry gas flow to the solvent droplets and assists in the evaporation of the solvent. This increases the sensitivity of the mass spectrometer (up to 100×)*, particularly at higher solvent flow rates, and allows sufficient data collection from sample amounts down to a fraction of a single venom milking sample (<10 μg of crude venom). Because ESI-MS involves sample delivery in a solvent and a very "soft" ionization, on-line RP-HPLC coupled mass spectrometry of biomolecules can be performed ***(18)***. This ability to couple the RP-HPLC on-line with the mass spectrometer is the biggest advantage of RP-HPLC/ESI-MS over RP-HPLC/MALDI-TOF analysis of venom samples. RP-HPLC/ESI-MS allows direct flow of the eluting components to the mass spectrometer, in contrast to RP-HPLC/MALDI-TOF, which requires separate sample collection, lyophilization, and matrix preparation prior to MS. The ability to incorporate on-line RP-HPLC into the analysis of complex mixtures, such as venom samples, greatly improves the sensitivity of detection by removing problems associated with the dynamic range of the detector. This problem results from ion saturation arising in analysis of complex samples possessing components of different concentrations. The maximum detection intensity resides with the most intense ion in the spectrum. A saturated

*Applied Biosystems homepage (http://www.appliedbiosystems.com/ab-mds_sciex/products/ionsources/turboionspray.html).

ion will, therefore, result in the suppression of weaker concentration or less ionizable components, suppressing the signal into the baseline noise. Separation of the components prior to mass spectral analysis (as in RP-HPLC/ESI-MS) reduces the effects of ion suppression as each component is analyzed individually. Despite the advantages of RP-HPLC/ESI-MS over RP-HPLC/UV, the UV data is still required, as the intensity of the mass spectrometer peaks depend on the ability of the component to ionize in the mass spectrometer and not the relative concentration of the component in the venom.

4. The venom collection container must be suitable for the creature in question (e.g., snakes require much larger venom collection containers than Funnel-web spiders do).
5. It is important when choosing a column for RP-HPLC/ESI-MS that the sample will not overload or underload the RP-HPLC column because this results in a loss of resolution. Furthermore, it is important to ensure that the amount of component eluting from the RP-HPLC column will not saturate the ESI-MS signal. This would lead to decreased sensitivity of low abundance molecules and decreased resolution in the RP-HPLC/ESI-MS chromatogram. Because detection by MS is far more sensitive than UV, the column choice and resultant flow rates are critical in the experimental setup. The experimental setup should always cater for the minimum amount of sample likely to be used, as this will always be the lower limit for detection (the detection of components eluting off RP-HPLC columns does not rely solely on the initial concentration of the sample added; the actual quantitative amount of sample and the solvent flow rate are important) and it is far easier to dilute concentrated samples than increase the concentration when only small amounts of sample are available. The experimental setup used should be kept as constant as possible to aid in the reproducibility of results and to also aid in the direct comparison of different samples.
6. Live or freshly dissected specimens are preferred because some venomous creatures can expel the venom from their venom glands upon freezing, and some specimens can be difficult to dissect frozen. Furthermore, dissection or freezing involves killing the specimen and limits the amount of venom that can be obtained from that specimen to a single collection. Venom extraction from dissected venom glands is also fraught with contamination problems (such as contaminating cellular byproducts and contents), providing a venom profile of the expelled venom that is not completely accurate. It has also been shown that the venom composition of Cone snail venom ***(19)*** and Funnel-web spider venom ***(20)*** differs between the venom gland and the expelled venom. “Milking” venom (collection of venom expelled by the creature from the venom duct) is the most preferred method, but can also have its problems with contamination. Very few venomous creatures are willing to expel venom in a manner such that it can be easily collected (snakes and the Australian Funnel-web spiders are the exception). Tricks have been developed to coax creatures such as Stonefish and Cone snails to be milked, however, the majority remain difficult. Electrical stimulation of the venom glands has become a popular method of “milking” venom and works by providing a pulsed electrical current to the area housing the venom apparatus ***(16,21)***. This current

causes rapid pulsed contraction of the muscles controlling venom output of the venom apparatus. In creatures where the venom apparatus is close to the mouth region (such as spiders) care must be taken to avoid contamination of the venom sample with regurgitated digestive tract material.

7. The use of polyethylene/polypropylene materials avoids problems associated with sample loss owing to the affinity of some venom molecules for glass. Although it is possible to avoid this problem by using silanized glassware, polyethylene/polypropylene is preferred because its hydrophobic nature results in more complete sample transfer.
8. Lyophilization of the sample is preferably performed in something like a Savant speedvac to pellet crude venom. Standard driers without the added centrifugation step provided by the speedvac tend to result in lyophilized peptide up the sides of the tube, making reconstitution in small volumes difficult. Venom to be analyzed at a later date should be lyophilized and stored at –20°C, however, most highly disulfide bonded peptides are stable at room temperature for several days and even longer (lyophilized or in 30% organic solvent).
9. When reconstituting the sample to the desired analysis concentration, use a 30% solution of the organic solvent used in RP-HPLC. This concentration generally caters for all the hydrophilic and hydrophobic molecules, falling in the midrange of the extremes of typical component elution.
10. Where a group of molecules outside the range of interest are likely to interfere with those in the analysis range, the venom should be run through a size exclusion column (e.g., Microcon). However, this has the potential to alter the venom composition slightly through components binding to the column and should be avoided unless absolutely necessary.
11. The mass spectrometer signal must be very clean, with no contaminating flowthrough from the RP-HPLC columns, so the column must be washed thoroughly before use. Prior to commencing the RP-HPLC/ESI-MS analysis, it is a very good idea to attach the equilibrating RP-HPLC column to the mass spectrometer and equilibrate for a short period to ensure the column and mass spectrometer are clean. The mass spectrometer should also be calibrated prior to analysis. The narrow bore tubing used between the LC column, UV detector, and mass spectrometer should be suitable for the flow rate to be used (dependent on the column bore and packing) and kept as short as possible to ensure good correlation between retention times in the UV and MS data.
12. Contaminants dominating the signal intensity at either end of the mass range can be eliminated by reducing the mass scan range to just exclude the contaminating signal.
13. For RP-HPLC systems that routinely run at less than 200 µL/min with a small bore column (id <2.1 mm), a 5-µL injection of a 1 mg/mL solution gives very good data, therefore, acceptable data can be safely obtained from loading 3–5 µL of sample.
14. HPLC offers a couple of alternatives regarding the data that can be obtained. The UV detector can be removed if only MS data are required, however, it is typically

desirable to incorporate the UV data. It is also possible to introduce a flow splitter in-line following the UV detector. This allows larger amounts of sample to be loaded and run at higher flow rates on larger bore columns. By splitting the flow so only a small portion goes to the mass spectrometer, the remainder can be directed for fraction collection providing the scientist with the UV data, mass spectrometer data, and fractionation in one step. The time frame for collection of all this data fits into the normal time frame taken to run the RP-HPLC/UV, vastly improving productivity. Furthermore, the homogeneity of the isolated fractions is known from the mass spectrometer data and can be used in further experiments such as reduction/alkylation leading to amino acid sequencing.

15. If a peak is evident in the RP-HPLC/UV chromatogram and not the RP-HPLC/ESI-MS data, try scanning a wider mass range, or ionization in negative ion mode to obtain a mass.

References

1. Sutherland, S. K. and Tibballs, J. (2001) *Australian Animal Toxins: The Creatures, their Toxins and Care of the Poisoned Patient.* Second ed. Oxford University Press, South Melbourne, Australia.
2. Sheumack, D. D., Carroll, P. R., Hampson, F., et al. (1983) The isolation and N-terminal sequence of the lethal neurotoxin from the venom of the male Sydney Funnel-web spider. *Atrax robustus. Toxicon.* **Suppl. 3,** 397–400.
3. Collins, S. P., Comis, A., Tyler, M. I., Marshall, M., and Howden, M. E. H. (1995) Evidence for a high molecular weight pre-robustoxin molecule in the venom of the male Sydney funnel-web spider *(Atrax robustus). Comp. Biochem. Physiol.* **110C,** 89–93.
4. Cain, T. C., Lubman, D. M., and Weber, W. J. J. (1994) Differentiation of bacteria using protein profiles from matrix-assisted laser desorption/ionization time-of-flight mass spectrometry. *Rapid Commun. Mass Spectrom.* **8,** 1026–1030.
5. Chong, B. E., Wall, D. B., Lubman, D. M., and Flynn, S. J. (1997) Rapid profiling of *E. coli* proteins up to 500 kDa from whole cell lysates using matrix-assisted laser desorption/ionization time-of- flight mass spectrometry. *Rapid Commun. Mass Spectrom.* **11,** 1900–1908.
6. Krishnamurthy, T. and Ross, P. L. (1996) Rapid identification of bacteria by direct matrix-assisted laser desorption/ionization mass spectrometric analysis of whole cells. *Rapid Commun. Mass Spectrom.* **10,** 1992–1996.
7. Holland, R. D., Wilkes, J. G., Rafii, F., et al. (1996) Rapid identification of intact whole bacteria based on spectral patterns using matrix-assisted laser desorption/ionization with time-of-flight mass spectrometry. *Rapid Commun. Mass Spectrom.* **10,** 1227–1232.
8. Liang, X., Zheng, K., Qian, M. G., and Lubman, D. M. (1996) Determination of bacterial protein profiles by matrix-assisted laser desorption/ionization mass spectrometry with high-performance liquid chromatography. *Rapid Commun. Mass Spectrom.* **10,** 1219–1226.

9. Roda, A., Gioacchini, A. M., Seraglia, R., et al. (1997) A comparison of the analytical performance of sodium dodecyl sulfate-polycrylamide gel electrophoresis, electrospray and matrix-assisted laser desorption/ionization mass spectrometry in the study of the protein extract from *Bothrops jararaca* snake venom. *Rapid Commun. Mass Spectrom.* **11,** 1297–1302.
10. Jones, A., Bingham, J. P., Gehrmann, J., et al. (1996) Isolation and characterization of conopeptides by high-performance liquid chromatography combined with mass spectrometry and tandem mass spectrometry. *Rapid Commun. Mass Spectrom.* **10,** 138–143.
11. Bingham, J.-P., Jones, A., Lewis, R. J., Andrews, P. R., and Alewood, P. F. (1996) *Conus* venom peptides (conopeptides): Inter-species, intra-species and within individual variation revealed by ionspray mass spectrometry, in *Biochemical Aspects of Marine Pharmacology* (Lazarovici, P., Spiro, M., Zlotkin, E., eds.), Alaken Inc., www.alaken.com., Ft. Collins, CO, pp. 13–27.
12. Krishnamurthy, T., Prabhakaran, M., and Long, S. R. (1996) Mass spectrometric investigations on *Conus* peptides. *Toxicon.* **34,** 1345–1359.
13. Romi-Lebrun, R., Martin-Eauclaire, M. F., Escoubas, P., et al. (1997) Characterization of four toxins from *Buthus martensi* scorpion venom, which act on apamin-sensitive Ca2+-activated K+ channels. *Eur. J. Biochem.* **245,** 457–464.
14. Escoubas, P., Celerier, M. L., and Nakajima, T. (1997) High-performance liquid chromatography matrix-assisted laser desorption/ionization time-of-flight mass spectrometry peptide fingerprinting of tarantula venoms in the genus *Brachypelma*—chemotaxonomic and biochemical applications. *Rapid Commun. Mass Spectrom.* **11,** 1891–1899.
15. Escoubas, P., Whiteley, B. J., Kristensen, C. P., Celerier, M. L., Corzo, G., and Nakajima, T. (1998) Multidimensional peptide fingerprinting by high performance liquid chromatography, capillary zone electrophoresis and matrix-assisted laser desorption/ionization time-of-flight mass spectrometry for the identification of tarantula venom samples. *Rapid Commun. Mass Spectrom.* **12,** 1075–1084.
16. Escoubas, P., Chamot-Rooke, J., Stocklin, R., et al. (1999) A comparison of matrix-assisted laser desorption/ionization time-of-flight and liquid chromatography electrospray ionization mass spectrometry methods for the analysis of crude tarantula venoms in the *Pterinochilus* group. *Rapid Commun. Mass Spectrom.* **13,** 1861–1868.
17. Sheumack, D. D., Baldo, B. A., Carroll, P. R., Hampson, F., Howden, M. E. H., and Skorulis, A. (1984) A comparitive study of properties and toxic constituents of funnel-web spider *(Atrax)* venoms. *Comp. Biochem. Physiol.* **78C,** 55–68.
18. Siuzdak, G. (1996) *Mass Spectrometry for Biotechnology.* Academic, San Diego, CA.
19. Bingham, J. P. (1998) *Novel Toxins from* Conus*: From Taxonomy to Toxins.* University of Queensland, St. Lucia, Australia, p. 290.
20. Wilson, D. T. R. (2001) *The Identification and Characterisation of Australian Funnel-web Spider Venom.* University of Queensland, St. Lucia, Australia, p. 282.

21. Atkinson, R. K. and Wright, L. G. (1990) A study of the necrotic actions of the venom of the Wolf Spider, *Lycosa Godeffroyi*, on mouse skin. *Comp. Biochem. Physiol.* **95C,** 319–325.

22

HPLC and Mass Spectrometry of Intrinsic Membrane Proteins

Julian P. Whitelegge

1. Introduction

High-performance liquid chromatograpghy (HPLC) separations of membrane proteins can be conveniently divided into two categories. First, there are many methods available for isolation of functional membrane proteins. Typically, the proteins are maintained in configurations as close to their native state as possible through the use of mild detergents that provide solubility without denaturation, allowing convenient ion-exchange or size-exclusion chromatography, for example. Proteins or complexes isolated in this way are subsequently used for functional analysis or crystallization, and so on. These isolation techniques have been well reviewed and readers are referred to literature specific to the protein of interest. The second category of separations are those used to separate membrane proteins from detergents and salts for the purpose of protein chemistry; although tempting to call these methods "denaturing" there is substantial evidence that this is not always the case. The focus of this chapter is to review the latter category of HPLC techniques with specific reference to those methods that provide conditions compatible with mass spectrometric analysis, especially on-line electrospray ionization.

1.1. Biological Mass Spectrometry

The discovery of mild ionization techniques for mass spectrometry (MS) of biological macromolecules has revolutionized HPLC, because it is now convenient to monitor elution profiles not only by UV absorbance, but by MS as well. Matrix-assisted laser desorption ionization–time-of-flight (MALDI-TOF; *1*) uses samples cocrystallized with a small UV absorbing matrix molecule to

From: *Methods in Molecular Biology, vol. 251, HPLC of Peptides and Proteins: Methods and Protocols*
Edited by: M.-I. Aguilar © Humana Press Inc., Totowa, NJ

transfer molecules from a dried solid phase to the gas phase, and is thus not directly compatible with HPLC though spotting small aliquots (0.5 μL) of sequential fractions for MALDI-TOF remains a useful option. However, it was the application of electrospray-ionization (ESI; *2*) that introduced the option of coupling HPLC directly to MS and the first liquid chromatography–mass spectrometry (LC–MS) experiments provided an exciting breakthrough in technology *(3)*. The HPLC eluent is directed straight to the ESI source for continuous ionization while the mass spectrometer measures mass to charge ratio *(m/z)* of ions detected during the course of a scan through the *m/z* range; a process that typically takes from 1 to 10 s. The read-out of mass spectra with time is thus complementary to the UV profile providing the opportunity to rapidly distinguish between proteins and other UV absorbing molecules, for example. As a consequence of the very low tolerance of MS to salts, LC–MS has been most widely applied to reverse-phase (RP)-HPLC with volatile solvents (e.g., acetonitrile) and ion-pairing agents (e.g., TFA) though success is also achievable with size-exclusion HPLC provided salts are excluded. Variants of these methods have been successfully applied to a wide range of intrinsic membrane proteins with structures containing up to fifteen transmembrane helices *(4–7)*.

1.2. MS of Membrane Proteins

Both ESI and MALDI-TOF have been applied to membrane proteins and peptides derived from them *(4–13)*. The focus of this discussion will be ESI in the context of on-line LC–MS because it provides superior mass accuracy and resolution compared to MALDI-TOF, especially for intact proteins. The ability to use ESI-MS for analysis of intrinsic membrane proteins was first demonstrated in 1993 *(9,10)* using flow-injection methods. LC–MS was applied in 1997 *(14)* though the gradient elution with undiluted formic acid can be regarded as somewhat extreme and mass accuracies in the 0.01–0.05% range were lower than obtained for water-soluble proteins. The adoption of a reverse-phase system involving elevated concentrations of aqueous formic acid (60%) and gradient elution with isopropanol allowed routine achievement of 0.01% accuracy *(4)*. A size-exclusion LC–MS method for intrinsic membrane proteins was reported in 1999 *(5)* and has proved broadly applicable *(6,7)* again with routine achievement of 0.01% accuracy (+/- 9 Da at 90 kDa). **Figure 1** shows the results of a size-exclusion LC–MS experiment performed upon bacteri-

---

Fig. 1. *(see facing page)* ESI-MS of bacteriorhodopsin via online size-exclusion HPLC. Bacteriorhodopsin (Sigma, 50 μg) suspended in 1 m*M* CHAPS was acidified by addition of two volumes of formic acid/isopropanol (1/1, v/v) and loaded immediately onto an HPLC system with a size-exclusion column (Super SW 2000, 4.6 × 300 mm;

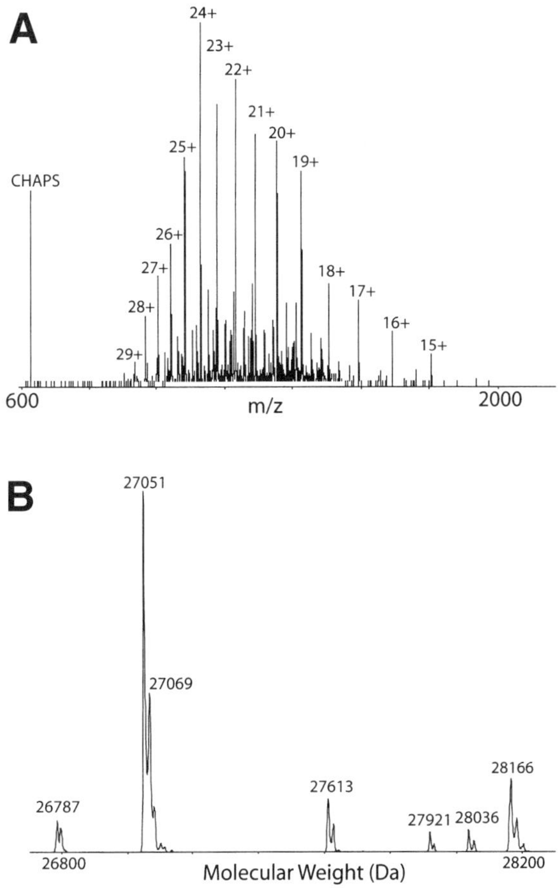

Fig. 1. *(continued)* Tosohaas) equilibrated in chloroform/methanol/1% aqueous formic acid (4/4/1, v/v; 250 µL/min.; 24°C). Eluent was passed through a UV detector and directly to an electrospray-ionization source (Ionspray™) of a triple quadrupole mass spectrometer (API III, Applied Biosystems). The instrument was operated with an orifice potential of 65 V and scanned from *m/z* 600–2300 with a step size of 0.3 and a dwell of 1 ms to give a scan time of 6 s. Scans collected during the elution of the first major protein peak were averaged **(A)** and processed by computer to generate a zero-charge spectrum **(B)**. Multiple charging during the electrospray process generates a cluster of ions that differ in *m/z* as a consequence of the number of charges (protons) associated with a particular protein molecule, in this case, from 15 to 29 with the most abundant at M + 24 H^+. The singly charged CHAPS molecules presumably remained associated with the protein during chromatography but were separated during ionization. Minor signals at the center of the mass spectrum result from minor isoforms of the protein that are more clearly visualized in the molecular weight spectrum (B). Note resolution of lesser subpopulations of all species that measure 16–18 Da larger and probably correspond to singly oxidized species (+ O, 16 Da) or possibly carry a single bound water (+ 18 Da). Running the column at the usual operating temperature of 40°C results in predominance of the apoprotein.

orhodopsin. The mass spectrum (*see* **Fig. 1A**) collected during elution of the protein shows a characteristic cluster of multiply-charged ions corresponding to the protein itself (15–29 H^+) and the singly charged ion derived from the detergent CHAPS. A computerized reconstruction of the zero-charge molecular weight spectrum (*see* **Fig. 1B**) shows predominance of the holoprotein (27051 Da; calculated 27050.06 Da), a small amount of apoprotein (26787 Da; calculated 26783.64 Da) and other minor species that probably correspond to incompletely processed forms of bacteriorhodopsin that have nevertheless bound the retinal cofactor.

1.3. Proteomics

Whereas most proteomics studies are using 2D gel electrophoresis as the primary separation technology, there are recognized disadvantages of applying these methods to some classes of intrinsic membrane proteins ***(15)***. Therefore, alternative technologies are sought with HPLC separations playing a potentially important role. **Figure 2** shows a modification of LC–MS for use in proteomics. By inserting a liquid-flow splitter between HPLC and mass spectrometer it is possible to collect fractions concomitant with MS (LC–MS^+) which can be used in downstream experiments for protein identification and characterization of posttranslational modifications. This approach was applied successfully to the cytochrome b_6f complex from higher plants, identifying a ninth subunit as ferredoxin/$NADP^+$ oxidoreductase (FNR; ***16***). Because typical membrane preparations are too complex to fully resolve by a single chromatographic separation, it is suggested that subfractions should be employed ***(17)***.

1.4. Solubility and Aggregation of Membrane Proteins

Following extraction from bilayers, membrane proteins have an extreme tendency to aggregate and precipitate unless stabilized in detergent micelles. For HPLC separations in the absence of detergent, it is then necessary to find solvent conditions compatible with both solubility and disaggregation of the analyte. During attempts to isolate hydrophobic peptides derived from proteolytic cleavage of the photoaffinity-labeled D1 herbicide receptor ***(18)*** it was observed that the labeled peptide of interest would only be *retained* in size-exclusion HPLC using 50% formic acid, 50% methanol as mobile phase. The observation that the peptide eluted in the void when 50% aqueous formic acid or 50% aqueous acetic acid or 50% acetic acid, 50% methanol were used as mobile phase emphasized that both solubility *and* disaggregation were necessary for chromatographic separation. Although formic acid is a very useful reagent for solubilization/disaggregation, it should be recognized that at high concentrations it will modify proteins with formyl groups that appear as multiple +28-Da adducts in the mass spectrum. Consequently, more recent proto-

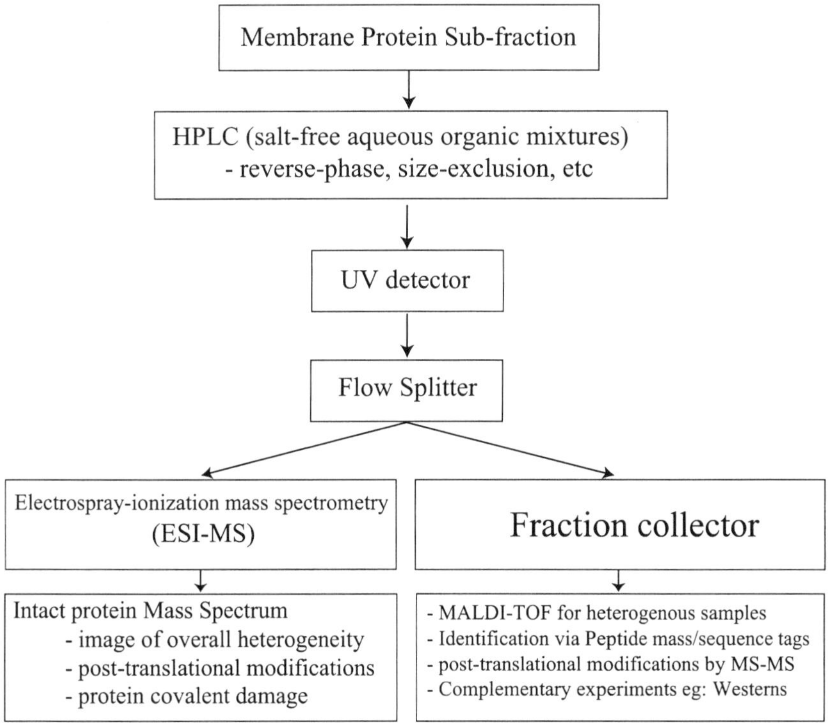

Fig. 2. LC with mass spectrometry and fraction collection (LC–MS$^+$). Use of a flow splitter after the HPLC detector allows concomitant mass spectrometry and fraction collection. Fractions can be analyzed after analysis of intact protein mass spectra to allow increased confidence of identification and characterization of posttranslational modifications.

cols, although using high concentrations of formic acid for initial solubilization, emphasize immediate transfer to more inert solvents.

2. Materials

2.1. Hardware

1. HPLC pumps/plumbing: The techniques described here use standard HPLC systems, with syringe pump instruments for flow rates less than 200 μL/min. High concentrations of formic acid (60–90%) are rinsed promptly from pumps/syringes but are probably no more corrosive than 1 *M* NaCl, for example. High-pressure lines use stainless steel tubing and ferrules but lower pressure lines use Peak(tm) tubing and Fingertight(tm) fittings (< 2000 psi). Fused-silica capillaries (PolyMicro) are used to interface LC to the ESI source using minimal length connections owing to the possibility of chromatography on the inner glass surface. All systems are modified to minimize liquid paths and dead volumes of con-

nections. Absorbance is measured at 214 m or 280 nm if solvent background is prohibitive at 214 nm (high [formic acid]; chloroform, and so on). Note that some flow cells are sensitive to excessive backpressure and that it is sometimes necessary to use splitters or wider bore tubing to avoid expensive accidents. HPLC systems typically need servicing after 2–3 yr of moderately heavy use. An oven is often used in order to run columns at elevated temperatures.

2. Syringe pumps: HPLC systems with syringe pumps are typically used for LC–MS at lower flow rates. Well-serviced machines provide good performance provided care is taken to ensure buffers/solvents are thoroughly degassed. Pistons and seals are replaced when necessary. Syringes are typically purged empty at the end of the day. Sometimes specific problems can be overcome by appropriate modification. For example, an ABI 120A being used for LC–MS work had a static mixer (a small C18 column) placed before the purge valve that resulted in an overpressure condition when purging high viscosity solvents such as 60% formic acid/isopropanol. The static mixer column was consequently situated after the purge valve such that it was eliminated from the purge cycle.
3. Columns: For RP chromatography, polystyrene-divinylbenzene copolymer is preferred over silica-based columns. This media can be regenerated more thoroughly and run at higher temperatures. PLRP/S (Polymer Labs) provides excellent chromatographic performance while PRP (Hamilton) provides an acceptable less-expensive alternative. For size-exclusion chromatography, the SW series of silica supports (Tosohaas [now Tosoh Biosciences]) provides the only suitable media. Several other companies have advertised equivalent performance but tests have shown that only the Tosohaas product performs adequately in chloroform/methanol/1% aqueous formic acid (4/4/1). Elevated concentrations of formic acid shorten column life significantly. Virgin columns provide ultimate chromatographic resolution.
4. Grounding and safety: Use of electrospray ionization involves high potential differences between source and mass spectrometer that can be conducted down solvent lines. Thus, the HPLC stack must be grounded to the mass spectrometer. Furthermore, source voltage must be removed prior to manipulation of solvent lines to protect the operator. Protective clothing and eyeware must be used when working with pressurized solvents and acids. An extractor at the ESI source is also useful to avoid accumulation of solvent/acid vapors.

2.2. Chemicals

The use of MS with HPLC makes the operator very aware of the quality of solvents and chemicals being used. Remarkable differences in chemical background can be observed with different suppliers. Thus, different batches or alternative supplies are tested carefully in the chromatographic systems of interest to ensure low background.

1. TFA is from Pierce (1 mL sealed ampules of Ionate™ grade).
2. Formic acid is from Fisher (90%, ACS grade) and is not stored for long periods (*see* **Note 1**).

3. Water is double distilled in Quartz or from a Millipore reverse osmosis system (>18 MΩ).
4. Organic solvents are from Fisher; methanol is HPLC grade, acetonitrile is Optima grade and chloroform/isopropanol are ACS grade. Solvents are mixed in glass (try LC–MS with a chloroform containing solvent that was mixed in plastic for your amusement) and degassed by sonication under vacuum. Solvent mixtures containing chloroform or isopropanol are made up daily whereas standard TFA buffers may be kept for at least 1 wk. Note that halogenated solvent waste is generated by nearly all the methods described and should be disposed of properly.

3. Methods

3.1. Sample Purification

There are numerous methods for isolation of membrane fractions, membrane protein complexes and monomeric membrane proteins, but these are not the focus of this chapter and are covered by many other authors.

3.2. Sample Preparation

Once a membrane protein fraction has been isolated in a functional state, it is typically accompanied either by native lipids or detergents. Sometimes it may be possible to load a sample in this state and use the chromatography to separate the protein from these contaminants. Often, however, small amphiphiles remain associated with and/or coelute with the protein and interfere with chromatography/mass spectrometry. Thus, it is usual to precipitate the protein with organic solvents in order to remove contaminating lipids/detergents, or at least deplete them, prior to chromatography.

3.2.1. Precipitation With Acetone

Acetone precipitation is achieved at 80% concentration by volume at –20°C. Either add acetone to 80% or add 80% acetone to a small sample volume. Stocks should be at –20°C. Precipitation usually occurs rapidly within 10 min at –20°C, but can take longer. Concentrated samples may work better if aliquoted out whereas minute samples may be lost. Some lipids will remain associated with precipitated protein. Detergents interfere with acetone precipitation dramatically and should be avoided. If detergents are present, protein should be precipitated at the interface of a chloroform/methanol/water phase separation. Precipitates are recovered by brief centrifugation (10,000*g*; 1 min).

*3.2.2. Precipitation With Chloroform/Methanol/Water Phase Separation **(19)***

The protocol, modified from Wessel and Flugge ***(19)***, for this precipitation is shown in **Fig. 3** and represents an important breakthrough in membrane protein chemistry. Precipitation at the interface is usually efficient even in the pres-

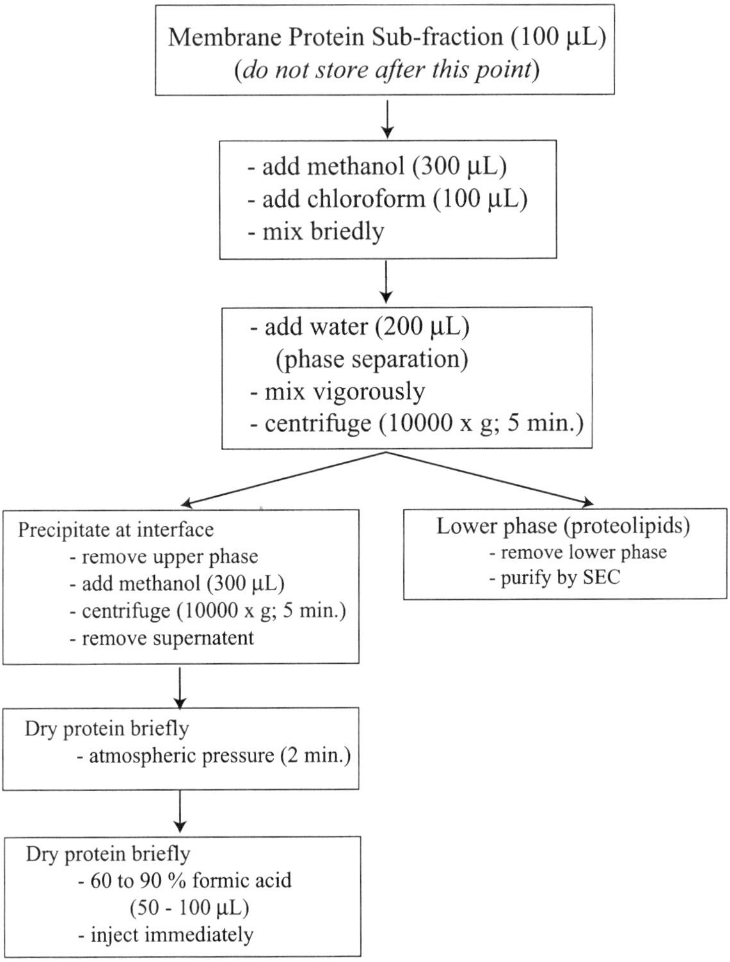

Fig. 3. Chloroform/methanol/water phase separation for precipitation of protein and removal of lipids and detergents. Protein can be precipitated from remarkably high concentrations of detergents. Procedure modified from Wessel and Flugge ***(19)***.

ence of 2% SDS/1% Triton X-100™ in the sample. Of course overall yields vary with sample type and amount of detergent. More lipid is removed than with acetone precipitation though some may remain. The most hydrophobic of proteins can partition into the lower phase; these are the proteolipids and can be recovered by size-exclusion chromatography (*see* **Subheading 3.3.**) or by precipitation with ether at low temperature.

3.2.3. No Precipitation

Precipitation can be disadvantageous and even lead to covalent modification of the sample. A good example is bacteriorhodopsin, a seven-transmembrane protein that carries a Schiff base linked retinal cofactor that is central to its function as a light-driven proton pump ***(20)***. Initial success in analysis of the protein by LC–MS was achieved after precipitation of purple membrane with acetone, solubilization in formic acid/water/isopropanol, and LC–MS in the 60% formic acid/isopropanol system (*see* **Subheading 3.2.**) though the majority of the protein was recovered as the apoprotein. When the sample, instead of acetone precipitation, was dispersed in detergent (1 m*M* CHAPS) and acidified with formic acid/isopropanol (three volumes of 1 : 1) prior to LC–MS, nearly complete recovery of the holoprotein resulted. It is believed that stabilization of the Schiff results because of retention of some secondary/tertiary structure ***(4)***, despite what seem rather harsh conditions.

3.2.4. Dissolution

Following precipitation with organic solvents, samples are typically dried for about 2 min at room temperature and atmospheric pressure. This is to allow evaporation of excess organic solvents but not water. Thorough drying at this stage results in pellets that are more resistant to redissolution. A small aliquot of solvent (50–100 μL) is added to the "wet" pellet for immediate redissolution and injection to HPLC. The solvent used most frequently is 60% formic acid, but 90% formic acid was necessary to dissolve the *Escherichia coli* lactose permease. Acetic acid can often be substituted for formic acid. TFA (50–100%) is also effective for solubilization though there is increased tendency for backbone hydrolysis in this strong acid. Solubility is sometimes influenced by additions of organic solvents, such as isopropanol, trifluoroethanol, or hexafluoroisopropanol at this stage, though it should be noted that such additions may lead to *decreased*, as well as increased solubility. Conditions for specific samples are best determined empirically. Furthermore, solubility is influenced by the precipitation protocol used, which influences degrees of denaturation and delipidation, as well as specifics such as the actual concentration of protein during precipitation. It is suggested that extreme attention to detail is applied to all aspects of the sample preparation protocol with any storage steps minimized especially after removal of lipids/detergents (*see also* **Note 3**).

In cases where samples were not precipitated, it is usual to acidify the sample prior to injection, with an equal volume of formic acid, for example. If the sample is not acidified it will, in specific cases, remain bound to the column until an aliquot of the appropriate acid is passed down the column. This behavior can be used to advantage for on-column delipidation or detergent removal, for example.

3.3. RP-HPLC

The compatibility of the solvent systems used in RP with LC–MS, combined with the often superior chromatographic resolution, makes this a frequently used option for HPLC of intrinsic membrane proteins, especially when a mixture is present. Many have described standard RP chromatographic systems and the first system to be described here is the author's version that includes a polymeric stationary phase and has given satisfactory results for many intrinsic membrane proteins.

3.3.1. TFA/Acetonitrile System

Although many intrinsic membrane proteins elute efficiently from a RP stationary phase with the TFA/acetonitrile system, there are some proteins that have very low elution efficiencies such that little or no protein is observed to elute at all. These are typically the most hydrophobic, such as bacteriorhodopsin. However, it should be appreciated that this behavior provides the opportunity to wash proteins of lipids and detergents on-column and to specifically enrich these molecules on-column, for subsequent elution with secondary gradients of more powerful eluents.

1. Buffer A is 0.1% TFA in water, buffer B is 0.1% TFA in acetonitrile. Buffers are degassed prior to use, typically by sonication and vacuum aspiration (with trap).
2. Column (PLRP/S, Polymer Lab.s; 300 Å, 5 μm; 2 × 150 mm; 40°C) is equilibrated in 5% B for 25 min (100 μL/min.)
3. Sample injection (100 μL loop) initiates gradient program. 5% B for 5 min followed by a linear gradient from 5–100% B over 60 min. System runs at 100% B for 5 mins and returns to 50% B (syringe pumps) or 5%B (regular pumps) over 5 min. Gradient may be adjusted for optimal separations of specific samples.
4. A preliminary run with no injection should show a flat baseline. A second preliminary run with injection of formic acid (100 μL) will then reveal potential ghosts from previously retained proteins as well as UV absorbing material contaminating the formic acid. Mass spectrometry (ESI-MS) can be used to distinguish ghosts from UV absorbing material of nonprotein origin. If ghosts are detected, the column must be regenerated/replaced.
5. Dual wavelength detection at 214 and 280 nm.
6. Column regeneration (*see* **Note 2**).

3.3.2. TFA/Acetonitrile/Isopropanol System

The addition of isopropanol to acetonitrile has been known to improve elution efficiency of many intrinsic membrane proteins in reverse-phase chromatography for many years ***(21)***, enabling the purification of rhodopsin, for example. Bacteriorhodopsin elutes quite efficiently allowing fully resolved spectra of many different isoforms of widely varied abundance.

1. As above, but buffer B is 0.05% TFA in acetonitrile/isopropanol (1 : 1 by volume). Buffer B is prepared freshly daily in glass by mixing equal volumes of 0.1% TFA in acetonitrile and isopropanol. Use glass stoppers on HPLC bottles rather than plastic lids that will leach plasticizers.
2. Gradients may be adjusted to focus upon the 50–100% B part of the chromatogram where the most hydrophobic of the intrinsic membrane proteins elute.
3. Column regeneration (*see* **Note 2**).

3.3.3. Formic Acid/Isopropanol System

Although addition of isopropanol to acetonitrile often improves elution efficiency, there will most usually be some residual intrinsic membrane protein remaining column bound. A solvent system with elevated formic acid concentration in the polar phase and isopropanol as the eluting solvent was originally described for separation of poliovirus proteins with quantitative recovery of material ***(22)***. Most intrinsic membrane proteins elute with close to 100% efficiency in the 60% formic acid/isopropanol system described here though exceptions are possible. The major subunit of *Torpedo* Vo ATPase (four transmembrane helices) has reduced elution efficiency and the lactose permease (twelve transmembrane helices) from *E.coli* was not recovered at all under these conditions. In both examples, elevation of formic acid concentration to 90% allowed full recovery of the proteins. The major disadvantage of the technique is the potential for protein formylation that results in multiple +28 Da adducts, very apparent in protein ESI mass spectra. Protein exposure to formic acid should be minimized or avoided. Acetic acid may provide a suitable alternative in some cases.

1. Buffer A is 60% formic acid, buffer B is isopropanol. Buffers are degassed prior to use, typically by sonication and vacuum aspiration (with trap).
2. Column (PLRP/S, Polymer Labs, 300 Å, 5 μm; 2 × 150 mm; 40°C) is equilibrated in 5% B for 10 min (100 μL/min.)
3. Sample injection (100 μL loop) initiates gradient program. 5% B for 5 min followed by a linear gradient from 5–100% B over 40 min. System runs at 100% B for 5 min and returns to 50% B (syringe pumps) or 5% B (regular pumps) over 5 min. Extended runs should be avoided because of the potential for sample formylation.
4. Detection is at 280 nm.
5. Column regeneration (*see* **Note 2**).

3.4. Size-Exclusion Chromatography (HPLC-SEC)

Whereas the 47 kDa *E.coli* lactose permease could be eluted in RP by increasing the formic acid concentration to 90%, the resulting mass spectrum showed multiple formylation adducts resulting in overestimate of mass by 100–150 Da. Thus, an alternative chromatographic separation was sought that would permit separation of the purified permease from residual detergent/lipid and the multitude of small molecules contaminating 90% formic acid. John E. Walker had previously reported success with ESI of mitochondrial inner membrane proteins using a solvent system containing chloroform/methanol/1% aqueous formic acid (4/4/1, v/v; ***13***) though separations were performed off-line prior to ESI-MS. Based upon previous success in size-exclusion HPLC with Tosohaas (now Tosoh Biosciences) silica SW media using formic acid/methanol (1/1, v/v; ***18***), the Walker solvent system was substituted. The precipitated permease was dissolved in 90% formic acid and immediately injected onto a SEC LC–MS system using an SW2000 column equilibrated in the 4/4/1 solvent mixture. Instead of precipitating, the protein was quickly transferred to the running solvent. Small molecule contaminants were retained and the permease eluted before them, close to the void, allowing spectra to be recorded with excellent signal/noise ratio ***(5)***. The SEC-MS system described has proved robust allowing general success with a variety of membrane and other proteins ***(6,7)***. The main limitation of the technique is the lack of chromatographic resolution afforded by SEC such that samples must be quite highly purified. The thylakoid cytochrome b_6f complex provided a good example of the complexity of the mixture that could be resolved. Five larger subunits coeluted but could be recognized in the rather complex mass spectrum because small subunits and small molecule contaminants were more highly retained and, thus, eliminated from the spectrum (***16,23***). A typical elution profile is shown in **Fig. 4**. Protein elutes first followed by smaller UV absorbing molecules (*see* **Fig. 4A**). The total ion chromatogram is a read-out of total mass spectrometer signal with time and shows the separation of the protein (concomitant with A_{280}) from other small molecules that give strong signals during mass spectrometry and would spoil the protein mass spectrum unless removed.

1. Buffer is chloroform/methanol/1% formic acid in water (4:4:1; by volume) and degassed as above. The solvent is prepared freshly each day and mixed/stored in glass containers.
2. Column (Tosohaas [now Tosoh Biosciences], Super SW2000, 4.6 × 300 mm; 40°C) is equilibrated for several column volumes (250 µL/min.) until ESI-MS reveals complete drop in chemical background. Note that the columns ship in sulfate/phosphate containing buffers which must be thoroughly washed away with water prior to transfer of the column first to 80% methanol and then the final solvent.
3. Sample is introduced with a 100 µL loop initiating the run which is complete within 35 min unless large amounts of small molecules such as detergents have been introduced. Complete removal of excessive sodium dodecyl sulfate (SDS) or salt contamination can take days of washing.

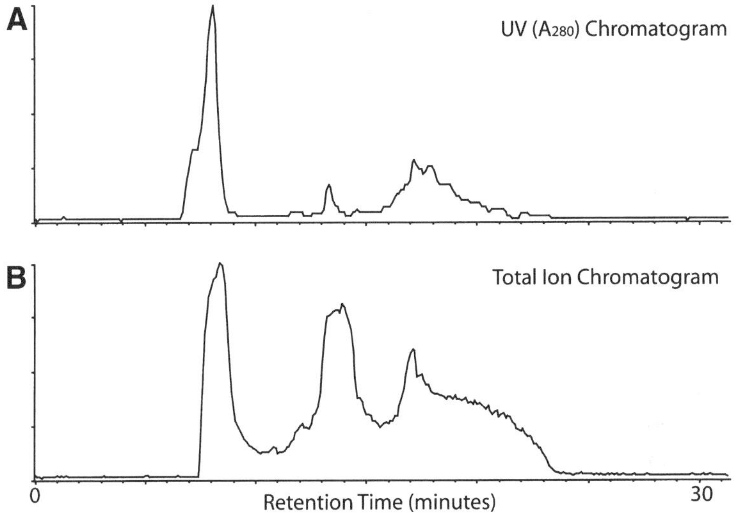

Fig. 4. Elution profile during size-exclusion liquid chromatography mass spectrometry (LC–MS). A protein sample (100 μg KcsA in 25 μL; a kind gift of Chris Miller, Brandeis University) was acidified (75 μL 90% formic acid) and subjected to size-exclusion chromatography mass spectrometry on Super SW2000 (Tosohaas; 4.6 × 300 mm) equilibrated in chloroform/methanol/1 aqueous formic acid (4/4/1; v/v) at a flow rate of 250 μL/minute and 40 °C. The absorbance profile at 280 nm **(A)** is compared to the total ion chromatogram **(B)**. The total ion chromatogram is generated by summing the total signal in one scan of the mass spectrometer (6 s) and plotting against time. Note the appearance of strong mass spectrometric response concomitant with the first UV peak as the protein elutes. The forward shoulder on the UV profile is probably a result of some protein running in an aggregated or oligomeric state and thus eluting in the void. The protein is clearly separated from other molecules that absorb UV light and exhibit strong response in mass spectrometry, with retention times greater than 10 min.

4. Columns may be connected in series for improved chromatographic resolution.
5. Detection is at 280 nm.
6. Column regeneration (*see* **Note 2**).

3.5. UV Detection With MS

The availability of mass spectrometers able to tolerate a continuous flow of liquid to the source enabled the advent of liquid chromatography mass spectrometry (LC-MS; ***3***). Not only has this development revolutionized biological MS but also HPLC. The eluent from the HPLC UV detector is directed to the electrospray ionization source of the mass spectrometer that thus becomes an on-line detector (*see* **Fig. 1**). The combination of UV and mass spectral data provides vastly improved detection because proteins are not only visualized by their general UV profile but also their unique mass spectral profile including covalent heterogeneity. Many protein modifications can change retention properties and thus mass spectral information becomes essential for interpreting data. An excellent illustration of this point is provided by a study that used changes in retention time to conclude that formic acid treatment leads to reduction of disulfide bonds within insulin and other proteins ***(24)***. When insulin was treated with formic acid there were changes in retention time to values similar to the individual α- and β-subunits. However, repetition of these experiments using LC–MS reveals formylation of the intact insulin molecule with corresponding changes of retention time (Tjon, Stevens, Faull, and Whitelegge, unpublished data). Reliance upon retention time alone had led to complete misinterpretation of the data. Though listed as a strong reducing agent by the Merck index, we have found no evidence of disulfide reduction by formic acid in several different proteins.

Many are tempted to collect fractions and perform mass spectrometry later. Although such an approach may be tolerated where necessary, many benefits of LC–MS are lost. Perhaps the most significant of these is the very low chemical noise that a well set up LC–MS system provides, though loss of material owing to adsorption, loss of solubility, and covalent modification can all contribute to dilution of benefits.

3.6. Sample Recovery and Tertiary Structure

If it is necessary to collect fractions for further experiments, the benefits of LC–MS can be retained through the use of a splitter as described (LC–MS$^+$; **Subheading 1.3.**). If yield must be maximized, then the whole sample can be collected at the expense of mass spectrometric detection. For structural measurements, samples can be dried onto Germanium chips for attenuated total reflectance Fourier transform infra red (ATR-FTIR) spectroscopy; both bacteriorhodopsin dried down from the 60% formic acid/isopropanol system (Whitelegge and Waring, unpublished), as well as the lactose permease dried down from the chloroform/methanol/1% aqueous formic acid system (Whitelegge, le Coutre, and Kaback, unpublished) showed FTIR spectra indicative of substantial α-helix, supporting the possibility of retention of structure in these solvents.

3.7. The Future

Although the techniques/protocols above provide a sound starting point for quite highly purified preparations, there is a great need for techniques that provide superior resolution in separations of all intrinsic membrane proteins and especially crude mixtures. There are a wealth of different surface chemistries and solvent mixtures available for HPLC and LC–MS providing a huge practical area for further exploration.

4. Notes

1. Formic Acid. Considerable attention should be paid to formic acid stocks used for protein chemistry. The reagent should be clear and colorless as supplied. Aging in the laboratory leads to a yellowish/brown appearance than can become intense after prolonged storage. Unfortunately, formic acid is typically shipped in plastic bottles that leach plasticizers readily. UV 280 absorbing contaminants as well as molecules that ionize readily during ESI accumulate steadily with storage, contributing to chemical background. Whereas chromatographic separations can be relied upon for separation of such contaminants from molecules of interest, it is preferable to replace stocks regularly and to test each new stock in case it sat in a warehouse for some months before delivery. Different sources of formic acid show varying potentials for protein formylation and the relationship between this activity and contaminants is not clear. Despite these drawbacks, formic acid is nevertheless most useful for membrane protein chemistry.
2. Column Regeneration. RP columns were originally regenerated by running gradients of formic acid/isopropanol (A: 90% formic acid, as supplied by Fisher, undiluted; B: isopropanol) from 5% B to 100% B with prolonged washing for 1 h at 100% B. Although effective for regenerating the columns, this protocol shortens the life of polymeric columns substantially. Now RP columns are equilibrated first in 80% methanol and then chloroform/methanol/1% aqueous formic acid (4/4/1). Aliquots of undiluted formic acid are injected in a repetitive fashion until LC–MS shows that such injections give no further release of protein. For regeneration of the silica Tosohaas SEC columns, the same latter protocol is used. If protein contamination is particularly tenacious with substantial elution of protein even after five formic acid injections, the column can be regenerated with 90% formic acid/[chloroform/methanol/water; 4/4/1] (8:1, v/v) for one column volume, though this wash is avoided accept where absolutely necessary.
3. Pressure-Assisted Solubilization. During analysis of samples of bovine rod outersegment membranes, highly enriched in the G-protein coupled receptor rhodopsin, it was common to observe a dramatic increase in system pressure upon injection of sample. Typically 2000–2500 psi could be reached before a sudden return of the system to standard operating pressure occurred. Although we have not investigated this phenomenon further, a reasonable explanation is that elevated pressure played a role in transferring the rhodopsin protein to the soluble/retained phase prior to successful elution in the gradient.

Acknowledgments

The techniques described here are evolved from huge amounts of groundwork by many others, especially Gobind Khorana and John Findlay. I specifically thank John R. Bowyer and Patrick Camilleri who understood and shared my interest in techniques compatible with MS and Chris Gerrish for his skillful help with HPLCs. My career in MS would not have been possible without the generous mentorship of Kym Faull and my colleagues Richard Stevens, Arv Fluharty, and Stephen Gómez. I have learned hugely from my peers and I acknowledge invaluable contributions from Sir John E.Walker, Ian Fearnley, H. Ron Kaback, Johannes le Coutre, and Bill Cramer among many others to whom I apologize for not mentioning. Financial support from DOE (DE-FG03-01ER15253) is gratefully acknowledged. NIH, NSF, the Pasarow Family, and the W. M. Keck Foundation are thanked for funds toward instrument purchases. Finally, I laud PE Sciex (now Applied Biosystems) for building a mass spectrometer at which I can spray anything. Correspondence: jpw@chem.ucla.edu.

References

1. Karas, M. and Hillenkamp, F. (1988) Laser desorption ionization of proteins with molecular masses exceeding 10,000 daltons. *Anal. Chem.* **60,** 2299–2301.
2. Fenn, J. B., Mann, M., Meng, C. K., Wong, S. F., and Whitehouse, C. M. (1989) Electrospray ionization for mass spectrometry of large biomolecules. *Science* **246,** 64–71.
3. Covey, T. R., Bonner, R. F., Shushan, B. I., and Henion, J. D. (1988) The determination of protein, oligonucleotide and peptide molecular weights by ion-spray mass spectrometry. *Rapid Commun. Mass Spectrom.* **2,** 249–256.
4. Whitelegge, J. P., Gundersen, C., and Faull, K.F . (1998) Electrospray-Ionization mass spectrometry of intact intrinsic membrane proteins. *Protein Sci.* **7,** 1423–1430.
5. Whitelegge, J. P., le Coutre, J., Lee, J. C., et al. (1999) Towards the bilayer proteome, electrospray-ionization mass spectrometry of large intact transmembrane proteins. *Proc. Natl. Acad. Sci. USA* **96,** 10,695–10,698.
6. le Coutre, J., Whitelegge, J. P., Gross, A., et al. (2000) Proteomics on full-length membrane proteins using mass spectrometry. *Biochemistry* **39,** 4237–4242.
7. Turk, E., Kim, O., le Coutre, J., et al. (2000) Molecular characterization of *Vibrio parahaemolyticus* vSGLT: a model for sodium-coupled sugar cotransporters. *J. Biol. Chem.* **275,** 25,711–25,716.
8. Schey, K. L., Papac, D. I., Knapp, D. R., and Crouch, R. K. (1993) Matrix-assisted laser desorption mass spectrometry of rhodopsin and bacteriorhodopsin. *Biophys. J.* **63,** 1240–1243.
9. le Maire, M., Deschamps, S., Moller, J. V., Le Caer, J. P., and Rossier, J. (1993) Electrospray ionization mass spectrometry on hydrophobic peptides electroeluted from sodium dodecyl sulfate-polyacrylamide gel electrophoresis application to the topology of the sarcoplasmic reticulum Ca2+ ATPase. *Anal. Biochem.* **214,** 50–57.
10. Schindler, P. A., van Dorsselaer, A., and Falick, A. M. (1993) Analysis of hydrophobic proteins and peptides by electrospray ionization mass spectrometry. *Anal. Biochem.* **213,** 256–263.
11. Rosinke, B., Strupat, K., Hillenkamp, F., et al. (1995) Matrix-assisted laser desorption/ionization mass spectrometry (MALDI-MS) of membrane proteins and non-covalent complexes. *J. Mass Spec.* **30,** 1462–1468.

12. `Ghaim, J. B., Tsatsos, P. H., Katsonouri, A., Mitchell, D. M., Salcedo-Hernandez, R., and Gennis, R. B. (1997) Matrix-assisted laser desorption ionization mass spectrometry of membrane proteins: demonstration of a simple method to determine subunit molecular weights of hydrophobic subunits. *Biochim. Biophys. Acta.* **1330,** 113–120.
13. Fearnley, I. M. and Walker, J. E. (1996) Analysis of hydrophobic proteins and peptides by electrospray ionization ms. *Biochem. Soc. Trans.* **24,** 912–917.
14. Schaller, J., Pellascio, B. C., and Schlunegger, U. P. (1997) Analysis of hydrophobic proteins and peptides by electrospray ionization mass spectrometry. *Rapid. Comm. Mass Spec.* **11,** 418–426.
15. Santoni, V., Molloy, M., and Rabilloud, T. (2000) Membrane proteins and proteomics: un amour impossible? *Electrophoresis* **21,** 1054–1070.
16. Zhang, H., Whitelegge, J. P., and Cramer, W. A. (2001) Ferredoxin: NADP+ oxidoreductase is a subunit of the chloroplast cytochrome b6f complex. *J. Biol. Chem.* **276,** 38159–38165.
17. Gómez, S. M. , Nishio, J. N., Faull, K. F., and Whitelegge, J. P. (2002) The chloroplast grana proteome defined by intact mass measurements from LC-MS. *Molec. Cell. Proteomics* **1,** 46–59.
18. Whitelegge, J. P., Jewess, P., Pickering, M. G., Gerrish, C., Camilleri, P., and Bowyer, J. R. (1992) Sequence analysis of photoaffinity-labelled peptides derived by proteolysis of photosystem-2 reaction centres from thylakoid membranes treated with [C-14]azidoatrazine. *Eur. J. Biochem.* **207,** 1077–1084.
19. Wessel, D. and Flugge, U. I. (1984) A method for the quantitative recovery of protein in dilute solution in the presence of detergents and lipids. *Anal. Biochem.* **138,** 141–143.
20. Khorana, H. G. (1993) Two light-transducing membrane proteins: bacteriorhodopsin and the mammalian rhodopsin. *Proc. Natl. Acad. Sci. USA* **90,** 1166–1171.
21. Tarr, G. E. and Crabb J. W. (1983) Reverse-phase high-performance liquid chromatography of hydrophobic proteins and fragments thereof. *Anal. Biochem.* **131,** 99–107.
22. Heukeshoven, J. and Dernick, R. (1982) Reversed-phase high-performance liquid chromatography of virus proteins and other large hydrophobic proteins in formic acid containing solvents. *J. Chromatog.* **252,** 241–254.
23. Whitelegge, J. P., Aguilera, R., Zhang, H., Taylor, R., and Cramer, W. A. (2002) Full subunit coverage liquid chromatography electrospray-ionization mass spectrometry (LCMS+) of an oligomeric membrane protein complex: cytochrome b_6f complex from spinach and the cyanobacterium, *M. laminosus. Mol. Cell Proteomics* **1,** 816–827.
24. Villa, S., De Fazio, G., and Canosi, U. (1989) Cyanogen bromide cleavage at methionine residues of polypeptides containing disulfide bonds. *Anal. Biochem.* **177,** 161–164.

23

IgG Purification

Maree S. Powell and Bruce D. Wines

1. Introduction

Immunization with a foreign antigen causes B cells of the immune system to produce antibodies of exquisite specificity toward the challenging antigen. This specific reactivity has made antibodies an essential tool for the detection and purification of protein in all fields of biological research. IgG is the most predominant class of serum antibody and is an integral part of many applications within the laboratory. The need to purify monoclonal or polyclonal antibodies is largely determined by the intended application of the antibody. Unpurified antibody is well suited to use in indirect flow cytometry assays, most enzyme-linked immunosorbent assays (ELISAs), for cytotoxicity assays or Western blot analyses. Purified antibody must be used, however, when accurate concentrations are required, chemical modifications such as labeling with fluorescent or radioactive probes are needed, when fragmentation of the antibody is required for binding or crystallization analysis or when antibody is directly coupled to a matrix for immunoaffinity chromatography.

The key determinants when choosing an antibody purification procedure will not only be influenced by what the intended use of the antibody is, but the available laboratory resources. Outlined are two protocols frequently used in the laboratory which will result in antibody of high purity; ion exchange chromatography (*see* **Subheading 3.1.**) and affinity chromatography using protein A or protein G (*see* **Subheading 3.2.**).

2. Materials

2.1. Ion Exchange Purification of IgG

This protocol purifies polyclonal IgG from rabbit serum using the non-denaturing protocol of ion-exchange chromatography (*see* **Note 1**).

From: *Methods in Molecular Biology, vol. 251, HPLC of Peptides and Proteins: Methods and Protocols*
Edited by: M.-I. Aguilar © Humana Press Inc., Totowa, NJ

2.1.1. Crude Fractionation by Precipitation

1. Saturated ammonium sulfate solution. This is prepared by warming 500 mL of double-distilled water, adding approx 450 g of ammonium sulfate and allowing to cool to room temperature. The presence of ammonium sulfate crystals when cooled to room temperature indicates the solution is saturated.

2.1.2. Dialysis

Buffer exchange into the chromatography buffer requires:

1. Dialysis tubing (e.g., Spectra Por, Spectrum Laboratories, TX). A molecular weight cutoff 6,000–8,000 Dalton, 8 cm width is ideal for various applications involving samples of several hundred mL.
2. A magnetic stirrer for mixing.

2.1.3. Column Equipment

1. DEAE-Sephacel (Amersham Biosciences, Australia).
2. A sintered glass funnel that fits a vacuum flask with side arm and access to a vacuum line is needed for washing and equilibrating the ion exchange media.
3. A glass column of approximate dimensions 30 cm long x 4 cm diameter is ideal for ion-exchange chromatography and are supplied by companies specializing in chromatography equipment, such as Amersham Pharmacia Biotech, Bio-Rad (CA) and Amicon (MA).
4. Tubing to plumb the chromatography system.
5. A fraction collector is the minimum requirement for this protocol. However, the inclusion of a peristaltic pump and an ultraviolet (UV) monitor makes for a more convenient and better controlled system.

2.1.4. Buffers

1. 500 m*M* Ethylenediaminetetra-acetic acid (EDTA) pH 8.0.
2. Five liters of 20 m*M* sodium phosphate pH 7.0.

2.2. Affinity Purification of IgG

This protocol describes the purification of IgG using protein A or protein G Sepharose affinity chromatography. Protein A and protein G are both bacterial proteins (from *Staphylococcus aureus* and Group G *Streptococcus*, respectively), which bind to the Fc portion of polyclonal, and monoclonal antibodies *(3–6)*. Immobilized onto Sepharose, they form extremely useful absorbents for the purification of monoclonal and polyclonal IgG classes, offering high binding capacity and ease of use (*see* **Table 1**). The protocol requires that the antibody is applied to the column at pH 7.0–8.0 and eluted from the column using a lower pH. The antibody is neutralized and then dialyzed against phosphate-buffered saline (PBS).

Table 1
Binding Capacities of IgG Proteins

Antibody	Protein A	Protein G
Human IgG	S	S
Human IgG_1	S	S
Human IgG_2	S	S
Human IgG_3	W	S
Human IgG_4	S	S
Mouse IgG	S	S
Mouse IgG_1	W	M
Mouse IgG_{2a}	S	S
Mouse IgG_{2b}	S	S
Mouse IgG_3	S	S
Rat IgG_1	W	W
Rat IgG_{2a}	NB	S
Rat IgG_{2b}	NB	W
Rat IgG_{2c}	S	S

W, weak binding; S, strong binding; M, medium binding; NB, no binding.

2.2.1. Buffers

1. Saturated ammonium sulfate: 450 g ammonium sulfate into 500 mL warm double-distilled water (ddH_20).
2. 500 m*M* EDTA pH 8.0.
3. 5 m*M* EDTA pH 8.0.
4. 20 m*M* sodium phosphate pH 7.0.
5. PBS: 3.5 m*M* NaH_2PO_4, 16 m*M* Na_2HPO_4, 150 m*M* NaCl.
6. 0.1 M Glycine or 0.1 M citrate at pH appropriate for the subclass of antibody.
7. 2 *M* Tris-HCl pH 8.0.
8. 3 *M* Potassium thiocyanate.

2.2.2. Equipment

The equipment required for the purification of IgG can vary from the basic column run under manual conditions to columns run on an automated system that offer regulated flow rates and more reproducible separations.

1. The minimal requirements for the purification of IgG are the use of a column and fraction collector.
2. A peristaltic pump to load the ascites or tissue-culture supernatant is also extremely useful.

3. Similarly, the collection of the eluted product with the use of an UV detector and chart recorder attached to the column is convenient but not always necessary.

2.2.3. Column

There are several varieties of column matrices available for the purification of IgG. Protein A and protein G Sepharose are supplied as freeze-dried powders, swollen gels, or as prepacked columns. The choice of column matrix is important as the binding capacity, flow rates, and resolution of eluted product can vary between products. For example; fast flow matrices offer the convenience of high binding efficiency (typically 20–40 mg for human IgG and 3–10 mg for mouse IgG per mL of packed gel) and flow rates (approx 10 mL/min) but the resolution of eluted protein peaks can be diminished. This is in contrast to Sepharose CL-4B, which binds equivalent amounts of IgG but has a slower flow rate (approx 2 mL/min) and offers improved resolution of the eluted protein peaks. It is recommended that suppliers such as Bio-Rad, Pharmacia, or Pierce be consulted for specific product information when preparing a column for use. **Subheading 3.1.4.** provides some general guidelines for packing columns.

2.2.4 Sample

The high affinity and specificity of protein A or G makes this method suitable for the purification of human and mouse immunoglobulin from ascites or cell culture supernatant. Refer to **Table 1** to determine which subclasses of IgG binds most efficiently to protein A or G. For example, protein A binds poorly to mouse IgG_1 and is purified more efficiently using protein G. IgG binds to protein A and G at near physiological pH and ionic strength. PBS is recommended as the binding buffer.

3. Methods

3.1. Ion-Exchange Purification of IgG

This method is adapted from that of Fanger and Smyth *(1)* and Mage *(2)*. IgG is a relatively robust protein and can be routinely purified at room temperature. It is not suitable for the purification of rat or mouse IgGs. These can be purified using other ion-exchange techniques but are more conveniently purified using protein A or G affinity chromatography (*see* **Subheading 3.2.**).

3.1.1. Precipitation

1. Transfer the rabbit serum to a measuring cylinder, add saturated ammonium sulfate in the ratio 2:3 (v/v), seal the cylinder with parafilm, and mix by inversion. (This step is equivalent to adding solid ammonium sulfate to 40% saturation, but minimizes the mixing required; *see* **Note 2**).

2. Transfer the sample to centrifuge buckets, balance, and collect the ammonium sulfate precipitated proteins by centrifugation at 10,000*g* for 20 min.
3. Discard the supernatant and note the color of the pellet. Usually, the pellet will be pink in color, rather than white. This indicates the presence of hemoglobin in the precipitate and that the precipitation step needs to be repeated to remove impurities.
4. Dissolve the pelleted proteins in 20 m*M* sodium phosphate pH 7.0 by adding aliquots of buffer to the centrifuge tube, which is then capped and gently inverted or swirled. The original volume of the serum sample should not be exceeded and vigorous mixing to dissolve the pellet is to be avoided as the resulting frothing of a protein solution is indicative of denaturation of the protein.
5. Transfer the redissolved proteins to the measuring cylinder add saturated ammonium sulfate to the sample in the ratio 2:3 (v/v) and mix gently.
6. Again, transfer the sample to centrifuge buckets, centrifuge to pellet the precipitated proteins, and discard the supernatant. If the pellet appears white, proceed to dialysis of the sample, but if it is still pink, the precipitation step will need to be repeated again.

3.1.2. Dialysis

1. Dissolve the pellet in buffer approx one-third the volume of the original serum sample. This will give a more concentrated IgG preparation and reduce the number of fractions that need to be collected.
2. Cut a piece of dialysis tubing to accommodate approximately twice the volume of the sample to avoid bursting the tubing during dialysis.
3. Boil a solution of 5 m*M* EDTA and place the tubing in this for 2 min to remove glycerin and metals. Tubing prepared in this way can be stored at 4°C in 20% ethanol for future use.
4. Close one end of the dialysis tubing by tying a knot or using a dialysis clip.
5. Transfer the sample to the tubing using a pipet or funnel. Air can be expelled from above the liquid in the tubing by pressing the sides of the tubing together. Secure the top of the tubing with a knot or clip.
6. Perform the dialysis in a beaker containing about 20 times the sample volume (1 L is ideal) with stirring at 4°C. The dialysis buffer can be discarded every 6 h and renewed at least three times over a 24-h period. A single dialysis against a larger buffer volume is usually not sufficient (*see* **Note 3**).

3.1.3. Preparation of DEAE Sephacel

1. It is convenient for rapid equilibration of the DEAE-Sephacel to use a sintered glass funnel (*see* **Note 4**). Transfer the DEAE-Sephacel slurry to a sintered glass funnel on a buchner flask with side arm attached to a vacuum line.
2. Drain the storage buffer from the media by briefly applying the vacuum.
3. Resuspend the media in the chromatography buffer and after 1 min measure the pH of the suspension. DEAE-Sephacel is a soft ion-exchange media and thus has considerable buffering capacity as the amino groups are titrated as the pH is

adjusted, and so several washes of buffer will be required before the media is equilibrated to pH 7.0 ready for chromatography.

4. Continue cycles of draining the slurry using the vacuum and resuspending in buffer until the media is adjusted to the required pH. This can be accelerated if the media is far from the desired pH by equilibrating with a stronger buffer solution (e.g., higher concentration of phosphate) or by directly adjusting the pH with 0.5 *M* HCl or NaOH, remembering to wear safety glasses and take care.
5. Fully equilibrate the media in 20 m*M* sodium phosphate pH 7.0.

3.1.4. Packing of the Column

1. Ion-exchange columns for low-pressure chromatography are relatively easy to pack with the aim being to create a column with an even bed of media without cracks or channels. Warm the media to room temperature to avoid the formation of bubbles in the column bed.
2. Set up the empty glass column vertically on a retort stand. Attach tubing to the outlet.
3. Swirl or gently suspended the slurry of equilibrated media with a stirring rod (never with a magnetic stirrer, which may damage the media), and pour into the column.
4. Drain the buffer from the outlet tubing by gravity or by using the peristaltic pump at a flow rate of at least 2 mL/min.
5. Add fresh media to top up the column as the media settles and the buffer drains. This needs more frequent attention as the column is progressively filled to the required level. If the column has an adjustable plunger, it is not important what this level is exactly, but if the top fitting is simply a fixed end the column will need to be filled to the top to avoid a dead space in which the loaded sample would be diluted.
6. Stop the buffer flow by using a clip/value or by raising the outlet tubing.
7. Place a 1-L reservoir of chromatography buffer on a shelf above the column, and insert the inlet tubing into the buffer and hold it in place by taping, or threading into a Pasteur pipet. Start buffer siphoning using a pipet or syringe.
8. Join this tubing to the column top end fitting (or plunger) and as buffer drips from the fitting attach it to the column. There should be only a small gap between the top end of the column and the media bed.
9. Allow buffer to flow from the outlet tubing and make a final check of the pH of the eluate to ensure the column is fully equilibrated (*see* **Note 5**).

3.1.5. Chromatography

The chromatography system used depends on equipment availability (*see* **Note 5**). The minimum requirement is a fraction collector. The flow rate through the column can be set by attaching to a peristaltic pump or by adjusting the height of the buffer reservoir above the end of the outlet tubing to control the siphon. In the latter case, collect the eluate over 10 min into a small measuring cylinder to calculate flow rate. A flow rate of about 1 mL/min is required. Check plumbing and program the fraction collector for 5-mL fractions.

1. Clarify sample by centrifugation at 10,000*g* for 15 min.
2. Transfer the supernatant into a measuring cylinder. Place the inlet tubing into the bottom of the measuring cylinder to load the sample. Set a stopwatch timer to alert 15 min before the sample should be completely loaded. The final loading of the sample needs to be closely monitored. If the sample completely empties, air will be introduced into the column, cracking the column bed and ruining the separation. Stop the buffer flow as the last of the sample loads into the inlet tubing by stopping the pump or, if using a siphon, by raising the end of the outlet tubing or using a clip/value.
3. Transfer the inlet tubing from the empty sample reservoir to the reservoir containing the chromatography buffer. Recommence the buffer flow and start the fraction collector.
4. The IgG does not bind to the ion exchange column but most other serum proteins from the ammonium sulfate precipitate will bind or be retarded. Thus, the breakthrough peak contains the purified IgG. Once this first peak has been obtained, the chromatography is completed. This is easily assessed if the system includes a UV monitor and chart recorder, if not, the peak fractions will need to be tested manually by determining the absorbance at 280 nm using a spectrophotometer. Following the elution of IgG, proteins retarded on the column are eluted, first serum albumin, and depending on the separation these impurities will contaminate the "tail" of the IgG breakthrough peak.
5. The fractions across this peak should be pooled as purified IgG, remembering the later "tail" of the peak will be increasingly contaminated with other serum proteins. This can be monitored by sodium dodecyl sulfate-polyacrylamide gel electrophorese (SDS-PAGE).
6. A solution of IgG with an absorbency of 1.0 contains 0.71 mg/mL of antibody. For most applications this provides IgG of sufficient purity.

3.1.6. Clean Up

1. Dissassemble the column and transfer the DEAE-Sephacel to the sintered glass funnel on the vacuum flask.
2. Wearing safety glasses (*see* **Note 6**), suspend the media in several volumes of 0.5 *M* HCl, wait 5 min and briefly apply the vacuum to drain the wash solution.
3. Suspend the media in several volumes of water and drain.
4. Then resuspend in 0.5 *M* NaOH, wait 5 min and again briefly apply the vacuum to drain. Repeat this NaOH wash twice more.
5. Rinse the media several times in water and finally resuspend the slurry in 20% ethanol for storage.

3.2. Affinity Purification of IgG

1. For ascites, remove lipid by passing ascites through a funnel lined with filter paper. Clarify the sample by centrifuging the filtered ascites for 30 min in a SS34 rotor at 13,000 rpm (20,000*g*) at 4°C. Decant and save the supernatant.
2. For monoclonal antibody produced in tissue culture, centrifuge the media containing antibody at 13,000 rpm (20,000*g*) at 4°C for 30 min, decant, and save the supernatant.
3. Dialyze the ascites or supernatant in previously prepared dialysis tubing (boiled for 2 min in 5 m*M* EDTA pH 8.0). The dialysis should be performed against 20 times the sample volume in PBS at 4°C. The dialysis buffer should be discarded every 6 h, with the dialysis buffer renewed at least three times. Alternatively, if the ascites or supernatant is a small volume (typically less than 10 mL) dilute the sample 1:10 with PBS. If the sample is turbid filter the sample through a 0.45 μ*M* filter to avoid clogging the column.
4. This method can be performed at room temperature or 4°C.
5. Equilibrate column in 5 column volumes of PBS.
6. Apply the sample to the column using a peristaltic pump or by gravity flow and allow the sample to enter the column. Reserve the eluate from the loading of the sample to ensure that the all the antibody has bound to the column and has not been overloaded.
7. Wash the column with approx 10 column volumes of PBS to remove unbound contaminants and to ensure that baseline has been reached.
8. Elute the antibody from the column using 5 column volumes of 0.1 *M* glycine or 0.1 *M* citrate at a suitable pH (human IgG subclasses pH 3.0; mouse IgG_1 pH 6.0; IgG_{2a} pH 4.5–5.0; IgG_{2b} 2.5–3.5; IgG_3 pH 2.5). Collect 2.5 mL fractions.
9. Neutralize protein-containing fractions by adding approx 50–100 μL 2 *M* Tris-HCl pH 8.0 to the collection tube. The addition of 2 *M* Tris-HCl can be done prior to the elution of the IgG or if the system includes a UV monitor and chart recorder the 2 *M* Tris-HCl pH 8.0 can be added as the IgG is being eluted. It is important to note that prolonged exposure of the IgG to acidic pH can damage the protein. Thus, it is important to neutralise the protein promptly after elution from the column.
10. Identify fractions corresponding to the eluted IgG using the UV monitor trace or by reading the absorbance at 280 nm (A_{280}) of the individual fractions using a spectrophotometer.
11. Assess the purity of the eluted IgG by SDS-PAGE and pool the protein containing fractions. IgG run under non-reducing conditions on a SDS-PAGE gel will appear as a 150-kDa band. Alternatively, under reducing conditions the heavy and light chains of IgG appear as 50- and 25-kDa bands (*see* **Note 7**).
12. Place the eluant into dialysis tubing and dialyse against 20 times the sample volume in PBS for 24 h with 3–4 changes of buffer.
13. Determine the concentration of the purified IgG. A solution of IgG with an A_{280} of 1.0 contains 0.71 mg/mL of antibody. For most applications this provides IgG of sufficient purity for all future applications. If, however, the concentration of

the pooled IgG is not sufficient a concentration device can be used to concentrate the protein. It is recommended that suppliers such as Millipore be consulted when determining what concentration membrane and device is best suited to the IgG preparation.

14. Regenerate the column using 5 column volumes of 3 *M* potassium thiocyanate.
15. Reequilibrate the column with 10 column volumes of PBS. When not in use, the column should be stored in PBS containing an antibacterial agent, e.g., 0.02% sodium azide.

3.3. Concluding Comments

Ion exchange or affinity chromatographies are the classic methods for the purification of IgG. As purified IgG has such a broad spectrum of applications within the laboratory, there are many other techniques and products available to purify this protein. Therefore, it is highly recommended that the product range of suppliers such as Bio-rad, Pharamcia, or Pierce be regularly consulted for new products that offer alternatives to these methods that maybe better suited to the individual needs of the researcher. For example, MEP Hypercel (Bio-rad), a hydrophobic interaction chromatography media has recently been introduced for the purification of antibodies.

4. Notes

1. Polyclonal IgG can be prepared from the serum of other species using this method. For human IgG, substitute the chromatography buffer for 20 m*M* sodium phosphate pH 8.0 *(2)*.
2. As an alternative to using ammonium sulfate, use the addition of solid ammonium sulfate to 40% w/v or anhydrous sodium sulfate at 25°C to 18% w/v with stirring. For large samples, this is more convenient than increasing the volume by addition of saturated ammonium sulfate solution. It is important not to cool the sodium sulfate solution, avoiding the use of a chilled centrifuge, as the sodium sulfate itself will precipitate. The crude IgG fraction prepared by precipitation may be sufficient for some applications.
3. The time taken to exchange a buffer for the sample will depend on the sample size and dimensions of the tubing. A sample in tubing with a large diameter will equilibrate more slowly than a sample in thin tubing. If in doubt, it is better to make another change and leave the dialysis overnight. Too much salt or the wrong pH will ruin an ion exchange separation.
4. This method is structured assuming the DEAE-Sephacel has been regenerated at the end of its last use and only requires equilibration in the chromatography buffer. If it has not, or you are unsure of the history of the media, proceed first to **Subheading 3.1.6.**, clean up, to regenerate the media before equilibration. A column that is already packed can have the media equilibrated by passing chromatography buffer through the column. These steps would then take longer than equilibrating batch wise using the sintered glass funnel.

5. The more expensive automated chromatography workstations will have in line conductivity and pH monitors. Although these are luxury items for this simple method, they do provide important information on the chromatography conditions that makes troubleshooting easy. For example, column not equilibrated, sample insufficiently dialyzed, buffer incorrectly made up, and so on.
6. Regenerating a used packed column with acid or base washes is not recommended because of the risk of spray from a leaky fitting. A packed DEAE Sephacel column that has not been used extensively can be regenerated with several column volumes of 2 *M* NaCl.
7. There may also be trace contamination of serum albumin (66 kDa) in the eluted protein. This can be further purified by ion-exchange or gel-filtration chromatography depending on how pure the IgG needs to be. Typically, the small amounts detected do not interfere with chemical modifications or fragmentation of the antibody but this may have significant effects on crystallisation of the protein.

References

1. Fanger, M.W. and Smyth, D.G. (1972) The oligosaccharide units of rabbit immunoglobulin G. Asymmetric attachment of C2-oligosaccharide. *Biochem J.* **127,** 767–74.
2. Mage, M.G. (1980) Preparation of Fab fragments from IgGs of different animal species. *Methods Enzymol.* **70,** 142–50.
3. Goding, J.W. (1978) Use of staphylococcal protein A as an immunological reagent. *J. Immunol. Methods* **20,** 241–253.
4. Lindmark, R., Thoren-Tolling, K., and Sjoquist, J. (1983) Binding of immunoglobulin to protein A and immunoglobulin levels in mammalian sera. *J. Immunol. Methods* **62,** 1–13.
5. Akerstrom, B. and Bjorck, L. (1986) A physicochemical study of protein G, a molecule with unique immunoglobulin G-binding properties. *J. Biol. Chem.* **261,** 10,240–10,247.
6. Akerstrom, B., Brodin, T., Reis, K., and Bjorck, L. (1986) Protein G: a powerful tool for binding and detection of monoclonal and polyclonal antibodies. *J. Immunol.* **135,** 2589–2592.

24

DNA-Binding Proteins

LC–MS to Identify Key Domains in RNA Polymerase–Promoter Interactions

Malcolm Buckle

1. Introduction

Many cellular processes are ultimately regulated by protein: DNA interactions at the level of transcription. It is increasingly evident that accessing the information stocked in the genetic code is in itself an intrinsic part of the regulation of gene expression. The basal transcription machinery consists of a promoter sequence that is recognized by its cognate RNA polymerase and begins to processively copy the template strand of the DNA into RNA *(1,2)*. Evolution has modeled this process so as to enable modulation at one or more of a number of control points. In most cases, this involves the participation of accessory proteins that bind to specific DNA sequences and affect either the initial recruitment of RNA polymerases or ensuing isomerisations and rearrangements that lead to a functional transcription complex.

In prokaryotes, RNA polymerase is a relatively large multisubunit enzyme consisting of three subunits $[(\alpha)_2\beta'\beta]$ in the core enzyme) and four subunits $[(\alpha)_2\beta'\beta, \sigma]$ in the holoenzyme. The core enzyme (378,784 Da) consists of the two β (150,625 Da) and β' (155,145 Da) subunits and the dimeric form of the α (36,507 Da) subunit involved in interactions with activators and specific regions on the DNA. The holoenzyme (449,044 Da), the form that actively recognizes promoter sequences responsible for housekeeping genes contains the σ^{70} subunit (70,260 Da). The mechanism by which a σ^{70} containing RNA polymerase operates in order to locate its DNA counterpart, locally separate the two DNA strands, and precisely position the catalytic center at the start site for tran-

From: *Methods in Molecular Biology, vol. 251, HPLC of Peptides and Proteins: Methods and Protocols*
Edited by: M.-I. Aguilar © Humana Press Inc., Totowa, NJ

scription, is not understood at the molecular level. Without a σ subunit, *Escherichia coli* (*E. coli*) RNA polymerase is unable to recognize a promoter sequence ***(1–3)***. In agreement with the direct contact model originally proposed by Losick and Pero ***(4)***, several genetic studies have indicated that σ^{70} interacts in a bipartite manner, via its conserved regions 2.4 and 4.2, with two hexameric consensus sequences (separated by 17 [±1 bp]) located around positions –10 and –35 relative to the transcription start site. More recent biochemical evidence has corroborated the interaction with the –10 region ***(5,6)*** and direct crosslinking data have revealed the presence of intimate contacts made by σ^{70} with the DNA in the –35 region ***(7)***. Whereas the nature of contacts on the nucleic acid is becoming increasingly clear, direct evidence for those subunits, domains, or amino acids of the protein that are involved in the interactions is still lacking despite several biochemical studies ***(8,9)***. The elucidation of the three-dimensional structure of RNA polymerases at increasing degrees of resolution ***(10–13)*** gives useful insights into how the enzyme may interact with the DNA. However, because the assembly of a functional transcription apparatus is a kinetic, dynamic process no static analysis (for example, X-ray crystallography) will ever completely describe the situation prevalent at the outset of transcription. Time-resolved analysis ***(7,14)*** using nonperturbing crosslinking techniques gives access to key intermediates in a process and thus allows characterisation of rat limiting steps in the overall process. In general, however, such techniques generate small (fmole) quantities of material that then require isolation and characterization. This, therefore, requires unambiguous identification and sequencing of the resulting peptide domains.

In general, in a multisubunit protein-DNA complex, three major questions need to be addressed?

1. Which protein or subunit is in contact with the DNA?
2. Which peptide domains are involved in the interaction?
3. Which amino acids are involved in the contacts?

The judicious use of specific crosslinking techniques provides a relatively easy procedure to address the first of these questions. Provided that one has confidence in the crosslinking procedure then a comparative sodium dodecyl sulfate polyacrylamide gel electrophoresis (SDS-PAGE) analysis of the products of the reaction should allow identification of the domains or subunits involved. A particularly elegant demonstration of this is given in ***(7)*** where after crosslinking (using a high energy ultraviolet [UV] laser) the DNA was digested to leave a single radioactively labeled nucleotide covalently bound to the protein. This base did not affect the subsequent migration of the attached subunit which could be identified by autoradiography.

The answers to the latter two questions require a more precise and resolutive approach that realistically is currently accessible through isolating the resulting crosslinked peptide then either sequencing the peptide (Edman degradation) or using an accurate and sensitive mass spectrometry technique (liquid chromatography–mass spectroscopy [LC–MS], trandem mass spectroscopy). In either case, present available state-of-the-art analytical methods require pmole quantities of relatively pure material.

Because of the relative ease of manipulation of nucleic acids, and the rapidity with which molecular biology has advanced, much of the work on these key nucleoprotein complexes has centered on the nucleic acid component. Consequently, in general, it is relatively straightforward to identify globally protein binding sites on DNA by genetic and biochemical procedures. It is equally feasible to determine by footprinting techniques those bases that are in contact in a given complex, and even, because of the advent of time-resolved footprinting ***(15)***, to follow changes at the nucleic acid level during the formation of a complex. On the other hand, direct knowledge of those amino acids, peptide domains, subunits, or even protein components of a multiprotein complex has proven difficult to obtain. The reasons for this are in part owing to the complexity of the proteins involved, the lack of the equivalent of end labeling and direct sequencing techniques available for DNA and RNA, the difficulties of detecting and analysing subpicomolar quantities of material and the inadequacy of separation techniques to resolve complex mixtures of proteins, peptides and nucleoprotein complexes. Although separation methodologies have changed relatively little, large improvements have been made in the chemistry and, thus, the resolving power of many of the supports available for separation, especially in reversed phase chromatography. The architecture of separation columns has also improved accommodating a larger range of flow rates, loading capacities, pressures, and so on. Chromatographic systems have similarly evolved to be more sensitive, more modular, and thus, more flexible and adaptable to on-line attachment to a variety of emerging separation and detection techniques. In my laboratory, we have used a state-of-the-art high-performance liquid chromatography (HPLC) Ettan LC™ (Amersham Biosciences) interfaced with a surface plasmon resonance (Biacore) apparatus or an electrospray mass spectrometer. In this chapter, I will illustrate the use of the HPLC to analyze peptides derived from RNA polymerase and describe the strategy to identify domains of the holoenzyme that are involved in the recognition of promoter sequences.

2. Materials

2.1. Preparation of RNA Polymerase

1. Stock Solutions:
   a. 1 *M* Tris-HCl, pH 8.0 (1 L).
   b. 1 *M* Imidazole (200 mL).

c. 1 *M* $MgCl_2$ (100 mL).
d. 0.5 *M* $NiSO_4$ (200 mL).
e. 10 *M* Isopropyl-D-thiogalactopyranoside (IPTG) kept at –20°C.

2. Luria Broth (LB): 1%(w/v) tryptone, 0.5% (w/v) yeast extract, 1% (w/v) NaCl, pH 7.0.
3. Buffer A (disruption buffer): (100 mL) 50 m*M* Tris-HCl, pH 8.0, 5% glycerol, 300 m*M* NaCl, 1 m*M* Di-isopropylfluorophosphate or PEFA block, 0.05% deoxycholate.
4. Buffer B (binding buffer): (500 mL) 20 m*M* Tris-HCl, pH 8.0, 10 m*M* $MgCl_2$, 300 m*M* NaCl.
5. Buffer C: (200 mL) 200 m*M* $NiSO_4$.
6. Buffer D (elution buffer): (500 mL) 20 m*M* Tris-HCl, pH 8.0, 10 mM $MgCl_2$, 300 m*M* NaCl, 0.5 *M* imidazole.
7. Buffer E (Size exclusion buffer): (1 L) 20 m*M* Tris-HCl, pH 8.0, 200 m*M* NaCl, 5% (v/v) glycerol, 1 mM ethylene diamine tetraacetic acid (EDTA), 100 m*M* dithiothreitol (DTT).
8. Buffer F (anion-exchange gradient buffer): (1 L) 20 m*M* Tris-HCl, pH 8.0, 1 *M* NaCl, 5% (v/v) glycerol, 1 m*M* EDTA, 100 m*M* DTT.
9. Buffer G (storage buffer): (1 L) 20 m*M* Tris-HCl, pH 8.0, 150 m*M* NaCl, 50% (v/v) glycerol, 1 m*M* EDTA, 100 m*M* DTT.

2.2. SDS-PAGE

1. Solubilizing buffer (X2): 20 m*M* Tris-HCl, pH 8.0, 2 m*M* EDTA, 5% SDS, 10% 2-mercaptoethanol, 0.02% bromophenol blue.
2. SDS gels: A range of precast polyacrylamide gels (13 mm stacking gel zone and 32 mm separating gel zone with an overall thickness of 0.45 mm) containing 0.112 *M* acetate, 0.112 *M* Tris-HCl, pH 6.4 are available (Amersham Biosciences); in general, a wide range (20–150 kDa) proteins may be separated using a gradient gel (10–15% or 8–25% acrylamide). PhastGel SDS buffer strips of 3% agarose containing 0.2 *M* tricine, 0.2 *M* Tris-HCl, pH 8.1, and 0.55% SDS supply the running buffer.
3. Coumassie brilliant blue stock solution (0.2%): dissolve one tablet of PhastGel Blue R (Amersham Biosciences) in 80 mL of distilled H_2O, add 120 mL methanol, and filter across a 0.45-μm filter. Take one part (100 mL) stock solution and dilute with one part (100 mL) 20% acetic acid in H_2O.
4. Destaining solution: 30% methanol, 10% acetic acid, 60% H_2O.
5. Preserving solution: 10% glycerol, 10% acetic acid.

2.3. In Situ *Tryptic Digestion*

1. 1 *M* Ammonium bicarbonate.
2. Trypsin 5 mg/mL in 1 m*M* HCl frozen and stored at –20°C.
3. 2% Tween-20.
4. 10% Trifluoroacetic acid (TFA).

2.4. HPLC Separation of Peptides Produced From Tryptic Digest

1. Equilibration buffer: 3% acetonitrile, 0.05% TFA.
2. Elution buffer: 70% acetonitrile, 0.05% TFA.

3. Methods

3.1. Preparation of RNA Polymerase

3.1.1. Preparation of pRL663 (C-(His) 6–β′) Cells for RNA Polymerase Purification

To prepare 2 L of culture:

1. Take a Petri dish containing isolated cultures of pRL663 ***(16)*** and remove a single culture spot using a sterile toothpick. Resuspend the bacteria in a small (5 mL) volume of LB and then inoculate a 2-L Erlenmeyer flask containing 100 µg/mL of ampicillin. Agitate at 37°C until the optical density at 560 nm is around 0.2–0.4 (approx 1 h after inoculation). Remove a small aliquot (50 µL) for subsequent SDS-PAGE analysis.
2. Add IPTG to a final concentration of 1 m*M*. (Stock solution = 10 *M* thus add 50 µl/500 mL of culture.) Agitate vigorously and monitor the optical density at 550 nm to 0.75. Recover the wet cells by centrifugation (16,000*g*, 15 min @ 4°C).
3. Rapid freeze the cells in liquid nitrogen and store at –20°C.

3.1.2. Isolation and Purification of RNA Polymerase

3.1.2.1. Cell Disruption

1. Resuspend the frozen cells (10 g) in Buffer A (20 mL) and disrupt the cells by means of a French press.
2. Place the resulting viscous mixture in a beaker and gently stir at room temperature. Add a cocktail of DNase I (2 µg/mL) and Benzonase (2 µg/mL) (Boehringer-Mannheim) and stir the mixture for 60 min (*see* **Note 1**). Remove cellular debris by centrifugation at 16,000*g* for 60 min at 4°C.

3.1.2.2. Ni^{2+} Affinity Fractionation

1. Filter the supernatant from the previous step across a 0.45-µm filter and apply to a Hitrap Ni^{2+} chelating column (Amersham Biosciences) in the following manner using an AKTA™prime (Amersham Biosciences). Arrange the input tubes as follows: Port 1: Buffer B; Port 2: H_2O; Port 3: Buffer C; Port B: Buffer D.
2. Place a HiTrap I mL Ni^{2+} chelating column (Amersham Biosciences) in the AKTA™prime apparatus (*see* **Note 2**). Use the program shown in **Table 1**. Set the flow rate to 1 mL/min, collection is 1 mL per tube and the OD_{280} maximum is 1.0. Apply a gradient of 0 to 500 m*M* imidazole over 20 mL. S in **Table 1** refers to the applied sample volume.

Table 1
Program for Binding and Elution of Proteins to a Chelating Column Using the AKTA™prime Systems

Volume	Conc. %B	Flow	Fraction	Buffer valve	Inject valve	Comment
0	100	40	0	Pos 2	WASTE	Priming
25	100	40	0	Pos 2	WASTE	End priming
25 + 0.1	0	40	0	Pos 2	WASTE	Priming Port 2
60	0	1	0	Pos 2	LOAD	H_2O column wash
65	0	40	0	Pos 3	WASTE	Priming buffer C
100	0	1	0	Pos 3	LOAD	Ni^{2+} application to column
101	0	40	0	Pos 2	WASTE	Priming H_2O
116	0	1	0	Pos 2	LOAD	H_2O column wash
121	0	40	0	Pos 1	WASTE	Binding buffer
156	0	1	0	Pos 1	LOAD	Equilibration
166						Auto zero
166	0	1	0	Pos 1	INJECT	Sample application
166 + S	0	1	0	Pos 1	LOAD	Binding buffer Wash
176 + S	0	1	1	Pos 1	LOAD	Elution
196 + S	100	1	1	Pos 1	LOAD	Elution wash out
213 + S	100	1	0	Pos 1	LOAD	End Wash
213.1 + S	0	40	0	Pos 2	WASTE	Priming H_2O
228 + S	0	1	0	Pos 2	LOAD	Re-equilibration
233 + S						End Method

3. Use SDS-PAGE to determine those fractions that contain the RNA polymerase holoenzyme and precipitate pooled fractions by the addition of ammonium sulphate (35 g/mL) and stir at 4°C overnight.

3.1.2.3. Gel Permeation Chromatography on Sephacryl 300

1. Centrifuge the precipitated protein at 14,000 rpm for 30 min at 4°C.
2. Resuspend the pellet in buffer E (<4 mL final volume)
3. Centrifuge at 10,000 rpm for 10 min at 4°C.
4. Apply the supernatant (<4 mL) at a flow rate of 0. 5 mL/min to a Sephacryl S300 column (Amersham Biosciences) preequilibrated in buffer E.
5. Collect fractions (3 mL/tube).
6. Pool fractions containing RNA polymerase holoenzyme.

Table 2
Program for Purification of Proteins on a Mono_Q_HR_5/5 Column Using an AKTA Purifier

Volume (mL)	Instruction	Value
0.00	Flow	1 mL/min
0.00	Wavelength	280 nm, 254 nm
0.00	SetCondScale0%	15.7 mS/cm
0.00	SetCondScale100%	72.800 mS/cm
0.50	AutoZeroUV	
1.00	Injection Valve	Inject
8.00	Injection Valve	Load
10.00	Gradient	100.00% B, 40.000 mL
11.00	Fraction Collector	0.5 mL
52.00	Fractionation Stop	
62.00	End	

3.1.2.4. Anionic Chromatography on Mono Q

1. Directly inject the pooled fractions from the S 300 column containing the RNA polymerase holoenzyme to a Mono_Q_HR_5/5 column preequilibrated in buffer E on an AKTA Purifier system (Amersham Biosciences). The elution buffer is buffer F. Apply the method outlined in **Table 2**.
2. Pool fractions containing the holoenzyme and dialyse overnight at 4°C against 1 L of storage buffer (buffer G).

Typical profiles from the Sephacryl 300 and MonoQ columns are shown in **Fig. 1**.

3.2. SDS-PAGE of Isolated RNA Polymerase and In Situ *Tryptic Digestion of Isolated Subunits*

3.2.1. SDS-PAGE Separation and Band Excision

SDS-PAGE is carried out using the PhastGel electrophoresis system (Amersham Biosciences) on 10–15% acrylamide gels.

1. Suspend samples (2 µL) in 2 µL of solubilizing buffer.
2. Apply samples using the direct contact combs and run according to a preset method for 10–15% acrylamide gels. A typical gel is shown in **Fig. 2**.
3. After staining in coomassie carefully remove the gel containing the band of interest is from the back plastic support of the gel and transfer to a sterile 1.5-mL Eppendorf tube.

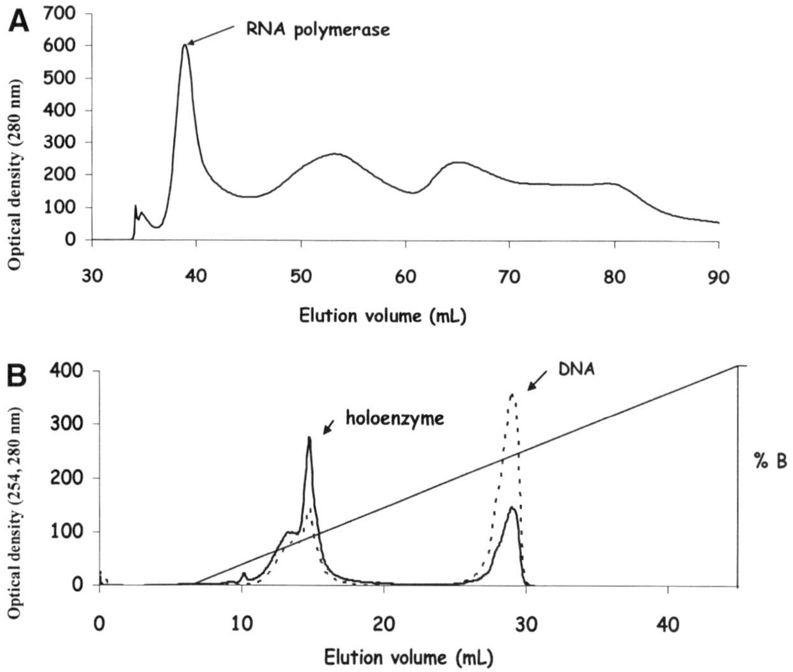

Fig. 1. Gel permeation and anionic exchange chromatography of fractions containing RNA polymerase. **(A)** Sample (4 mL) is applied to a sephacryl S300 column (preequilibrated in Tris-HCl 20 m*M* pH 8.0, NaCl 200 m*M*, glycerol 5% (v/v), ethylenediamine tetraacetic acid (EDTA) 1 m*M*, dithiothreitol (DTT) 100 m*M*) on an AKTA Purifier system at a flow rate of 0. 5 mL/min. Fractions (1 mL) containing the RNA polymerase are shown by the arrow. **(B)** The combined fractions from the S300 chromatography shown in (A) are applied to a Mono_Q_HR_5/5 column preequilibrated in Tris-HCl 20 m*M* pH 8.0, NaCl 200 m*M*, glycerol 5% (v/v), EDTA 1 m*M*, DTT 100 m*M* on an AKTA Purifier system. Elution is effected by applying a linear gradient of the same buffer to 1 *M* NaCl. Fractions containing the RNA polymerase holoenzyme are pooled. Note that DNA fragments elute at the end of the gradient.

3.2.2. In-Gel Digestion of Isolated Subunits

There are a number of techniques available for this depending upon the goal of the experiment ***(17–20)***. The technique described is one that we are developing for the analysis of RNA polymerase subunits (*see* **Note 3**).

1. Add a volume of 0.2 *M* ammonium bicarbonate 0.2% Tween-20; sufficient to cover the gel slices in an Eppendorf (centrifuge to put slice at bottom of the tube).

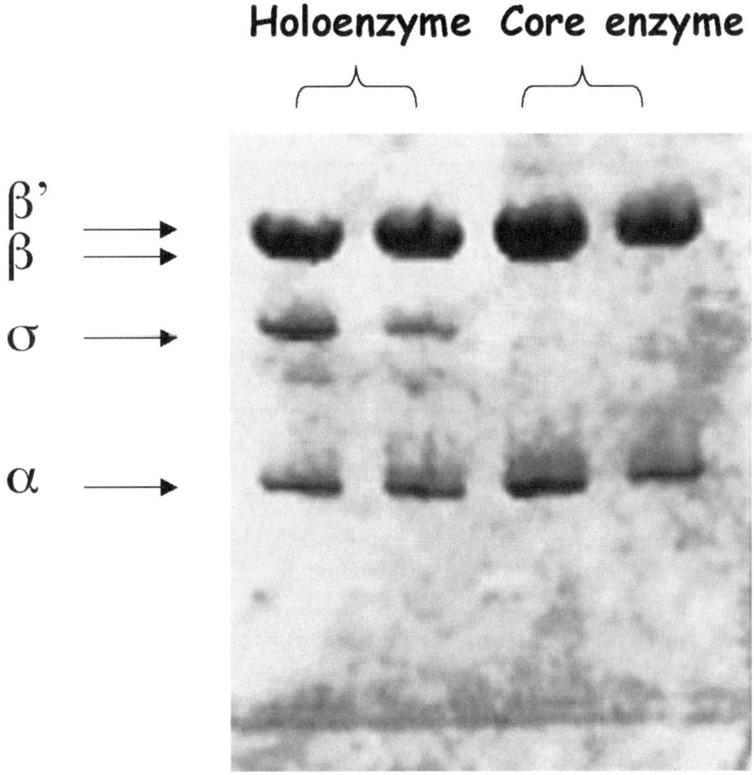

Fig. 2. SDS-PAGE analysis of RNA polymerase. RNA polymerase isolated as described is applied to a 10–15% acrylamide gel that allows separation of the four subunits [$(\alpha)_2\beta'$ β, σ). Approx 0.5 µg of total protein is present in each lane. The gel is visualized by staining with Coomassie blue.

2. Incubate for 1 h at 30°C with agitation.
3. Remove the supernatant and repeat wash.
4. Dry the gel in a speed vac.
5. Dilute the trypsin solution in 0.2 *M* ammonium bicarbonate to give a final concentration of trypsin of 0.5 µg/mL.
6. Add 100 µL per gel slice.
7. Incubate overnight at 30°C with shaking.
8. Add TFA to final 0.1%
9. Remove and keep supernatant.
10. Extract the gel slices twice with 500 µL of acetonitrile (60%)/ 0.1% TFA for 1 h at 30°C with shaking.
11. Remove supernatant, dry in speed vac.

Table 3
Programs for the Separation of Peptides on a Reverse Phase Column on an Ettan™ LC

Time (min)	Instruction	Value
0.00	Flow	0.100 mL/min
0.00	Wavelength	215 nm
0.00	SetCondScale0%	0.48 mS/cm
0.00	SetCondScale100%	2.58 mS/cm
0.00	AutoZeroUV	
0.20	Injection Valve	Inject
5.00	Injection Valve	Load
15.00	Gradient	100.00% B, 180.000 min
15.00	Fraction Collector	0.50 mL
200.00	Fractionation Stop	
220.00	End	

3.3. Reverse Phase Separation of Digested Peptides

The dried supernatant from the digest is resuspended in 100 μL of equilibration buffer and applied to a μRPC_C2/C18_SC_2.1/10 column (Amersham Biosciences) on an Ettan™ LC (Amersham Biosciences). The following protocol **(Table 3)** is used where buffer A is the equilibrium buffer and buffer B the elution buffer.

Samples are collected in a FRAC 950 fraction collector containing 4, 8 × 12, well microtiter plates. A typical digest pattern for the sigma subunit of RNA polymerase is given in **Fig. 3**. For further fractionation and identification (*see* **Notes 3** and **4**) the eluted fractions are evaporated to dryness and individually resuspended in the equilibration buffer and applied to a 5-μL 200 Å, 0.3 mm × 150 mm C18 MAGIC column (Michrom Bioresources, Inc.) on an Ettan™ LC using a flow splitter after the pumps and before the columns. The protocol is shown in **Table 4**.

A typical separation of four peptides is shown in **Fig. 4** with the resulting mass analysis on the Ettan LC–MS spectrometer.

4. Notes

1. Use of nucleases in the isolation of DNA-binding proteins. This may, at first sight seem strange. However, in reality many of the proteins present in a cell, even when the protein is overexpressed may be associated with DNA, which is going to compete with many classical chromatographic procedures. Many techniques use competing charged macromolecules such as heparin or polyamines. The disadvantages

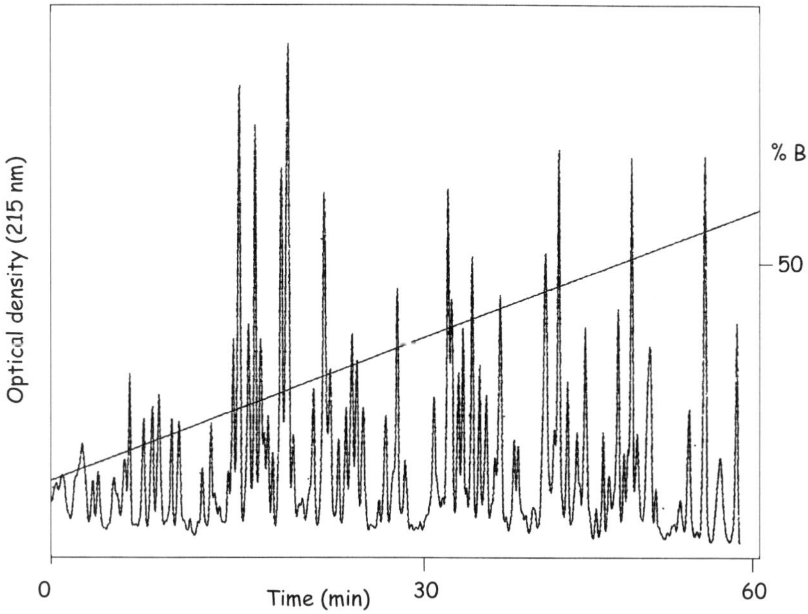

Fig. 3. Separation of tryptic digest of the sigma subunit of RNA polymerase by RP chromatography. Samples in 3% acetonitrile, 0.05% TFA are applied to a µRPC_C2/C18_SC_2.1/10 column (Amersham Biosciences) on an Ettan LC. A linear gradient to 70% acetonitrile, 0.05 % TFA allows separation of the around 72 tryptic products which can be collected or analyzed directly on MS.

I have found with these techniques are that binding to weaker cations, such as heparin, is not always an efficient way of displacing native DNA and conversely, while strong anions such as polymine efficiently compete DNA, subsequent removal of the polyamine introduces long and tedious chromatographic steps. The solution of using a cocktail of endonucleases we have found to be very efficient and reduces the DNA into short chains of <5 nucleotides. These small nucleotides and the endonucleses themselves are subsequently removed by the size exclusion and anionic exchange procedures, respectively.

2. Our laboratory is equipped with the AKTA range of chromatographic apparatus and it is thus natural that the protocols detailed here are for use with these. Of course, the techniques are generally transferable to most standard chromatographic platforms. A point to stress is that a key ingredient in all these separations is speed. The judicious arrangement of subsequent steps so as to transit rapidly between different separation procedures obviates at least in all the proteins that we are studying, the necessity to work at low temperatures, to carry out lengthy dialysis or ammonium sulphate precipitations.
3. In-gel digestion. In cases where a great deal of protein is available (i.e., strong colouration with Coomassie on a gel thus in the region from 0.5 to 1 mg/mL) one

Table 4
Program for Separation of Peptides on a 0.3-mm × 150-mm C18 MAGIC Column on an Ettan LC

Time (min)	Instruction	Value
0.00	Flow	0.200 mL/min
0.00	Wavelength	215 nm; 205 nm
0.00	SetCondScale0%	0.48 mS/cm
0.00	SetCondScale100%	2.58 mS/cm
0.00	AutoZeroUV	
1.0	Injection Valve	Inject
5.00	Injection Valve	Load
15.00	AutoZeroUV	
25.00	Gradient	70%B, 15 min
60.00	End	

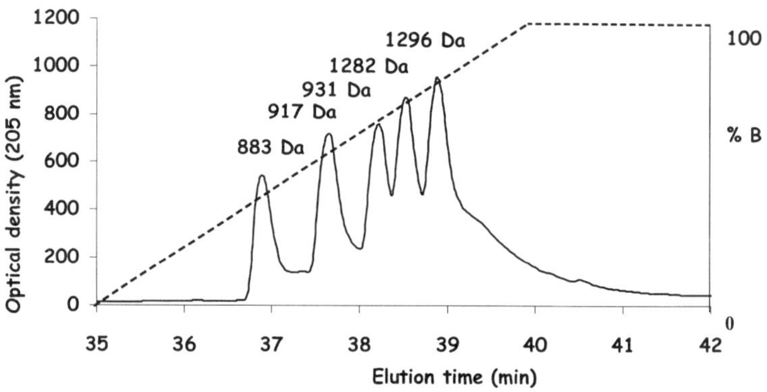

Fig. 4. Separation of peptides by RP chromatography. A mixture of peptides (2.0 µL of 10 ng/µL) is applied to a 5 µL 200 Å, 0.3 mm × 150 mm C18 MAGIC column (Michrom BioResources Inc.) on an Ettan LC. The gradient from 3% acetonitrile, 0.05% TFA to 70% acetonitrile, 0.05% TFA allows separation of the following peptides which in this instance are identified on line using an ESIToF mass spectrometer from Amersham Biosciences. 883 Da, RVYVHPI; 917 Da, RVYVHPF; 931 Da, RVYIHPF; 1282 Da, QRVYVHPFHL; 1296 Da, QRVYIHPFHL.

generally obtains efficient digestion in the sense that the detection procedure can visualize the expected peptide products. However, it is our experience that when small amounts (<pmole) of protein are present in a gel it becomes extremely difficult to identify unambiguously specific products against a background of partially digested protein. In this case, it is worth trying a number of different recipes chang-

ing the rehydration/solvation procedures, and the extraction techniques. Furthermore the presence of adducts such as a crosslinked DNA molecules on a protein alter the accessibility to proteases even under denaturing conditions, and greatly change the separation characteristics of the peptide on chromatographic supports. One hint on how to aid separation and identification is to decrease as much as possible the size of the crosslinked adduct by the use of endonucleases in the case of DNA (*see [7]*) or more simply a cocktail of RNAses in the case of RNA.

4. Separation of peptides. It is laudable but ambitious and often impossible to try to separate and identify all the peptides in a given digest of a specific protein especially one such as the sigma subunit of RNA polymerase which produces in excess of 70 peptides. It is far more practical, especially when the detection system is an electrospray mass spectrometer, to collect grouped peptides from a preliminary separation on RP-HPLC and then subject each of these to a more resolutive separation allowing less unambiguous identification of the separate peptides. This is especially important in the case of peptides carrying a cross-linked adduct. In this instance, a dedicated fraction collector that allows efficient recovery of samples and lends itself to automation is a great asset.

References

1. deHaseth, P. L. and Helmann, J. D. (1995) Open complex formation by Escherichia coli RNA polymerase: the mechanism of polymerase-induced strand separation of double helical DNA. [Review]. *Mol. Microbiol.* **16,** 817–824.
2. deHaseth, P. L., Zupancic, M. L., and Record, M. T., Jr. (1998) RNA polymerase-promoter interactions: the comings and goings of RNA polymerase. *J. Bacteriol.* **180,** 3019–3025.
3. Helmann, J. D. and Chamberlin, M. J. (1988) Structure and function of bacterial sigma factors. [Review] [162 refs]. *Ann. Rev. Biochem.* **57,** 839–872.
4. Losick, R. and Pero, J. (1981) Cascades of Sigma factors. *Cell* **25,** 582–584.
5. Roberts, C. W. and Roberts, J. W. (1996) Base-specific recognition of the non-template strand of promoter DNA by *E. coli* RNA polymerase. *Cell* **86,** 495–501.
6. Huang, X., Lopez, de Saro, Fj, and Helmann, J. D. (1997) Sigma factor mutations affecting the sequence-selective interaction of RNA polymerase with –10 region single-stranded DNA. *Nucl. Acids Res.* **25,** 2603–2609.
7. Buckle, M., Pemberton, I. K., Jacquet, M. A., and Buc, H. (1999) The kinetics of sigma subunit directed promoter recognition by E. coli RNA polymerase. *J. Molec. Biol.* **285,** 955–964.
8. Panaghie, G., Aiyar, S. E., Bobb, K. L., Hayward, R. S., and de Haseth, P. L. (2000) Aromatic amino acids in region 2.3 of Escherichia coli sigma 70 participate collectively in the formation of an RNA polymerase-promoter open complex. *J. Molec. Biol.* **299,** 1217–1230.
9. Panaghie, G., Aiyar, S. E., Bobb, K. L., Hayward, R. S., and de Haseth, P. L. (2000) Aromatic amino acids in region 2.3 of Escherichia coli sigma 70 participate collectively in the formation of an RNA polymerase-promoter open complex. *J. Molec. Biol.* **299,** 1217–1230.

10. Polyakov, A., Severinova, E., and Darst, S. A. (1995) Three-dimensional structure of E. coli core RNA polymerase: promoter binding and elongation conformations of the enzyme. *Cell* **83,** 365–373.
11. Landick, R. and Roberts, J. W. (1996) The shrewd grasp of RNA polymerase [comment]. [Review] [19 refs]. *Science* **273,** 202–203.
12. Malhotra, A., Severinova, E., and Darst, S. A. (1996) Crystal structure of a sigma 70 subunit fragment from E. coli RNA polymerase. *Cell* **87,** 127–136.
13. Zhang, G., Campbell, E. A., Minakhin, L., Richter, C., Severinov, K., and Darst, S. A. (1999) Crystal structure of Thermus aquaticus core RNA polymerase at 3.3 A resolution [see comments]. *Cell* **98,** 811–824.
14. Buckle, M., Fritsch, A., Roux, P., Geiselmann, J., and Buc, H. (1991) Kinetic studies on promoter-RNA polymerase complexes. *Methods Enzymol.* **208,** 236–258.
15. Sclavi, B., Woodson, S., Sullivan, M., Chance, M. R., and Brenowitz, M. (1997) Time-resolved synchrotron X-ray "footprinting", a new approach to the study of nucleic acid structure and function: application to protein-DNA interactions and RNA folding. *J. Molec. Biol.* **266,** 144–159.
16. Kashlev, M., Martin, E., Polyakov, A., Severinov, K., Nikiforov, V., and Goldfarb, A. (1993) Histidine-tagged RNA polymerase: dissection of the transcription cycle using immobilized enzyme. *Gene* **130,** 9–14.
17. Hellman, U., Wernstedt, C., Gonez, J., and Heldin, C. H. (1995) Improvement of an "In-Gel" digestion procedure for the micropreparation of internal protein fragments for amino acid sequencing. *Anal. Biochem.* **224,** 451–455.
18. Mirgorodskaia, O. A., Kazanina, G. A., Mirgorodskaia, E. P., et al. (1997) [Proteolysis of human proinsulin catalysed by native, modified, and immobilized trypsin]. *Bioorg. Khim.* **23,** 91–97.
19. Shevchenko, A., Chernushevich, I., Ens, W., et al. (1997) Rapid 'de novo' peptide sequencing by a combination of nanoelectrospray, isotopic labelling and a quadrupole/time-of-flight mass spectrometer. *Rapid Commun. Mass Spectrom.* **11,** 1015–1024.
20. Shevchenko, A. and Shevchenko, A. (2001) Evaluation of the efficiency of in-gel digestion of proteins by peptide isotopic labelling and MALDI mass spectrometry. *Anal. Biochem.* **296,** 279–283.

25

Sensitive Enzymatic Analysis of Histidine Decarboxylase Using HPLC

Jordi Gómez-Ramirez, Isaac Blanco, and Jordi Ortiz

1. Introduction

High-performance liquid chromatography (HPLC) can be used in enzymatic analysis in order to achieve a good purification of the reaction product. In this chapter, we will see how to use HPLC to measure histamine formation, a process catalyzed by the enzyme histidine decarboxylase (EC 4.1.1.22). Often the tissue under analysis is a poor source of this enzyme, expressed only by a few cell types such as histaminergic neurons, mast cells, and gastrointestinal enterochromaffin-like cells. For this reason, a method of high sensitivity is needed to determine histamine synthesis in tissue slices or histidine decarboxylase activity in low homogenate volumes. However, it may be difficult to eliminate the precursor histidine from histamine purifications because of their similar chemical properties. The method we describe here is based on an HPLC purification of histamine *(1)* that eliminates histidine much more selectively than previous methods based on cation-exchange gravity columns *(2–5)*. A radiolabeled substrate is used to obtain to the maximum sensitivity and specificity of the assay.

Because decarboxylation of histidine spares the imidazole ring, a suitable radiolabelled precursor would be commercial [2,5-^{3}H]L-histidine. Unfortunately, imidazole rings show a partial tautomerism that favors tritium exchange. This causes a decomposition rate of this substrate of about 1–3% per month or even higher according to the manufacturers. This problem can be overcome if substrate specific activity is routinely monitored after storage. We show in **Subheading 3.1.** how to purify periodically by HPLC one aliquot of original commercial [^{3}H]-histidine and assess its specific activity. This is also

From: *Methods in Molecular Biology, vol. 251, HPLC of Peptides and Proteins: Methods and Protocols*
Edited by: M.-I. Aguilar © Humana Press Inc., Totowa, NJ

important because the sensitivity of histamine purification can be limited if the blanks contain excessive disintegration per minute (dpm) from tritium contamination.

Approximately 0.01–0.5% of the substrate histidine is converted into histamine *(1,4)*. Such a low degree of conversion may cause difficulties in the purification of the product histamine in the presence of a large excess of substrate. To facilitate purification of histamine, we eliminate most of the histidine after incubation by binding to an anion-exchange resin added to deproteinized samples (*see* **Subheading 3.3.**). As the carboxylate of the histidine binds to the resin, unfortunately there is also a low degree of nonspecific binding of histamine to the resin. We overcome this problem by estimating histamine recovery in every sample through the addition of nonradiolabeled histamine as an internal standard. The use of unlabeled histamine as an internal standard ensures a very high reproducibility and also facilitates the HPLC fraction collection.

For HPLC purification of histamine, we have chosen an inexpensive reversed phase C_{18} column and a mobile phase with an ion-pair reagent, but it should also be possible to use ion-exchange HPLC. Our mobile phase has a high content of salt and methanol to facilitate quick elution of histidine in samples. Histamine retention in the column is caused by the high ion-pair concentration in the mobile phase which is run isocratically at low pH. The low dpm obtained in blank samples (*see* Table 2) makes further development of histamine purifications using gradient elution unnecessary. The HPLC is automated by a sample injector and a fraction collector. Collection is started by ultraviolet (UV) detection of the histamine internal standard. HPLC separation is straightforward provided that routine checks of tritium in the system and adequate clean-ups are performed, especially if the same HPLC system has been used previously for [^{3}H]-histidine purification. Finally, dpm are counted in the histamine fraction by scintillation.

Our method can also be applied with minor modifications to study histamine synthesis and release in synaptosomes and slices. Because the tissue slices maintain relative synaptic integrity, histidine decarboxylase activity in homogenates may not correlate with histamine synthesis quantified in slices *(6)*. Slices must take up ^{3}H-histidine into histaminergic terminals before being decarboxylated. Also, newly formed histamine in slices could be subject to regulatory processes *(7)*.

2. Materials

2.1. [^{3}H]-Histidine Purification

1. Ring-labeled [2,5-^{3}H]L-histidine stocks (1 mCi, 50 Ci/mmol) obtained from Amersham and stored at 4°C. Stocks can be kept for several years if they are purified 1–3 mo before use.

2. Nonradiolabeled histidine (Sigma).
3. Mobile phase for HPLC gradient: prepare fresh solutions A and B in Milli-Q water and vacuum filter them through nylon membranes of 0.2 µm pore size. Solution A: 25 m*M* NaH_2PO_4 pH 3, 0.1 m*M* octanesulfonic acid (Sigma), and 2% methanol (HPLC grade). Solution B: 0.1 m*M* octanesulfonic acid and 2% methanol.
4. HPLC system: Kontron 325 pump, Rheodyne injector model 7125 and Kontron 432 UV detector.
5. Reverse-phase C_{18} column, 25 × 0.46 cm (Tracer Extrasil ODS-2, of 5 µm particle size; Teknokroma, Spain), equipped with a 2 × 20 mm guard column (Upchurch; Teknokroma).
6. HPLC cleaning solutions: 0.1 *M* nitric acid and methanol/Milli-Q water (70/30%) (*see* **Note 1**.)

2.2. Enzymatic Assays and Slice Preparations

1. 10 m*M* Potassium phosphate buffer, pH 7.4 (stored as 1 *M* stock at 4°C for several months).
2. 0.2 m*M* Pyridoxal phosphate (stored aliquoted at –20°C).
3. 10% (v/w) Trichloroacetic acid (stored as 50% stock solution at 4°C for months).
4. 10 m*M* Histamine (make 1 mL aliquots and freeze).
5. Modified Krebs–Ringer bicarbonate medium (KRM) made of 120 m*M* NaCl, 0.8 m*M* KCl, 0.67 m*M* $MgSO_4$, 1.2 m*M* KH_2PO_4, 2.6 m*M* $CaCl_2$, 27.5 m*M* $NaHCO_3$, 10 m*M* glucose, dissolved in Milli-Q water (*see* Note **2** for precautions on preparation). Bubble O_2/CO_2 (95:5) through solution before use and periodically during use.
6. Depolarizing medium: 2 NaCl m*M*, 119 m*M* KCl, 0.67 m*M* $MgSO_4$, 1.2 m*M* KH_2PO_4, 2.6 m*M* $CaCl_2$, 27.5 m*M* $NaHCO_3$ and 10 m*M* glucose, dissolved in Milli-Q water (*see* **Note 2**). Bubble O_2/CO_2 (95:5) through solution before and during use. (Depolarizing medium is a KRM rich in KCl and poor in NaCl to maintain isomolarity.)
7. McIlwain tissue chopper (The Mickle Laboratory Engineering Co., Surrey, England).
8. Sonic-dismembrator DYNATECH model 300 (ARTEK Systems Co., Formingdale, NY).

2.3. Elimination of Excess Substrate

1. Amberlite IRA-900 strong anion-exchange resin (Supelco) was purchased from Teknokroma (Spain) and stored at room temperature. It should be prepared as described in **Subheading 3.3.1.** before use.
2. Microspin filter Ultrafree-MC tubes with low-binding durapore membrane 0.45 µm pore size, obtained from Millipore (Spain).
3. Eppendorf tube mixer 5432 (Netheler + Hinz GmbH, West Germany).

2.4. HPLC Purification of Histamine

1. Mobile phase: Make an aqueous solution of 0.3 *M* NaH_2PO_4 and 10 m*M* octanesulfonic acid in Milli-Q water, adjust it to pH 3.0 and store it (for months) at room

temperature. On the day of use, mix up 79% of aqueous solution with 21% methanol and filter it through nylon filter membranes 0.2 µm pore size. Methanol may evaporate if mixed with aqueous solutions longer than a week before use, which would change the retention time of histamine.
2. HPLC cleaning solution: methanol/Milli-Q water (70/30%). Add 10 m*M* octanesulfonic acid to the water to facilitate posterior equilibration of column with the mobile phase.
3. 0.1 *M* Nitric acid for the HPLC autosampler (*see* **Note 5**).
4. OptiPhase "HiSafe"-3 liquid scintillation cocktail (Wallac, EG&G Company, Turku, Finland).
5. HPLC MERCK-HITACHI: L-6200A Pump, L-4000 UV Detector, L-5200 Fraction Collector, and L-7200 Autosampler (Merck KgaA, Darmstadt, Germany).
6. Hercule 2000 Chromatography Interface with Borwin Chromatography Software (JMBS Developments, France).
7. Reversed-phase C_{18} column 25 × 0.46 cm (Tracer Extrasil ODS-2 of 5 µm particle size; Teknokroma) equipped with a 2 × 20 mm guard column (Upchurch, Teknokroma, Spain).

3. Methods

3.1. Purification of [2,5-^{3}H]L-Histidine

1. In order to avoid excessive radiolysis and check its specific activity, aliquots of the commercial [^{3}H]-histidine standard should be purified 1–3 mo before use.
2. The HPLC system is set to perform a linear gradient from 1 to 6 m*M* of sodium phosphate buffer in 12.5 min (solution A from 4% to 24%) at a flow rate of 1 mL/ min. A low percent of methanol (2%) and octanesulfonic acid (0.1 m*M*) are constant throughout the gradient because they are present in both A and B solutions. Under these conditions, histidine elutes at 9–10 min (*see* **Fig. 1A**).
3. Set UV wavelength at 225 nm and sensitivity at maximum.
4. Stabilize the column in initial conditions for 15 min before every injection.
5. Perform several injections of 0.2–20 nmol of nonradiolabeled histidine as external standard. Then inject about 50 µCi (2 nmol) of [^{3}H]-histidine. The whole histidine fraction should be collected manually into a polypropylene tube while watching histidine peak signal in the UV detector display.
6. Count 5 µL of the collected fraction by scintillation counting and calculate total dpm recovered.
7. To determine the amount of [^{3}H]-histidine, calibrate the [^{3}H]-histidine peak area against a linear regression of nonradiolabeled histidine areas.
8. To calculate specific activity, divide total dpm in histidine fraction by the amount of histidine detected. Specific activity should be similar to that reported by the vendor, of about 50 Ci/mmol. For cleaning of the HPLC, *see* **Note 1**.

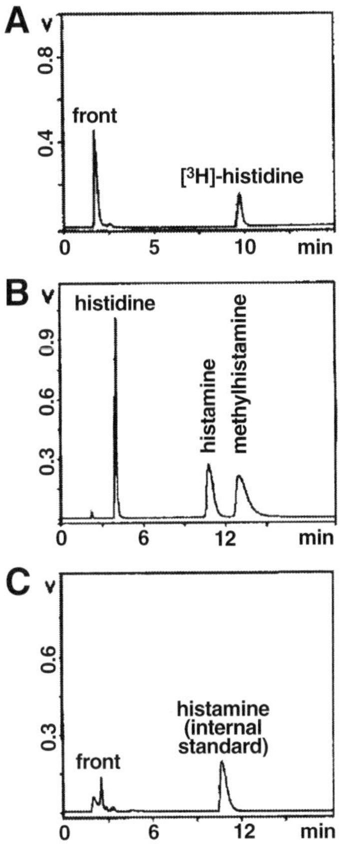

Fig. 1. Chromatograms of **(A)** [^{3}H]-histidine purification, **(B)** standard mixture, and **(C)** sample.

3.2. Incubation of Tissue

3.2.1. Histidine Decarboxylase Enzymatic Assay in Homogenates

1. Obtain a fresh brain from a rat, dissect the region of interest, and homogenize it manually using a glass-Teflon homogenizer in 10 vol ice-cold 10 m*M* potassium phosphate buffer (pH 7.4).
2. Determine protein content of homogenates by the method of Lowry, using bovine serum albumin as a standard. Keep samples at 0–4°C.
3. Prepare deproteinizing mixture containing one volume of 10% trichloroacetic acid with 2 vol 10 m*M* histamine to be used as internal standard. Total volume should be at least 15 µL × total number of samples.

4. Working on ice, mix aliquots of approx 0.2 mg protein (20–40 μL of homogenate) with 10 μM pyridoxal 5′-phosphate (5 μL 0.2 m*M* stock) and homogenization phosphate buffer (pH 7.4) to make a final incubation mix of 100 μL.
5. Add to blank samples 15 μL of previously prepared deproteinizing mixture.
6. Add prepurified [^{3}H]-histidine to all samples (to make 0.25 μ*M*, or 1.25 μCi per sample).
7. Incubate at 37°C for 1 h in a shaking water bath. Do not incubate blank samples, just keep them on ice.
8. Stop incubation by placing all samples on ice and immediately add 15 μL of deproteinizing mixture. Each sample should contain 100 nmol histamine and 0.5% trichloroacetic acid.
9. Vortex thoroughly and centrifuge at 12,000*g* at 4°C for 10 min.
10. Recover supernatants and treat them as described in **Subheading 3.3.**

3.2.2. Histamine Synthesis in Slices of Rat Brain Cortex

1. Obtain a fresh brain from a rat and place it into ice-cold modified Krebs Ringer bicarbonate medium (KRM) for 5 min.
2. Working in a cold environment (4°C), remove meninxs, dissect both cortical lobes and remove white matter.
3. Place cortices over the tissue chopper platform and chop at 0.3 mm.
4. Turn platform 90° and chop again to obtain miniprisms.
5. Place miniprisms in a glass bottle containing 20 mL of ice-cold KRM.
6. Wash miniprisms three times with ice-cold KRM to remove debris of damaged cells and released proteases.
7. Finally, settle miniprisms at the bottom of the bottle and remove excess KRM.
8. Prepare a deproteinization mixture containing 2.5 vol 10% trichloroacetic acid mixed with 1 vol of 10 m*M* histamine to be used as internal standard. Final volume of mixture should be at least 35 μL × number of samples.
9. Distribute 100 μL aliquots of the slice suspension (2–3 mg protein, but *see* **Note 3**) into polypropylene 2 mL tubes.
10. Bubble O_2/CO_2 (95 : 5) through solutions, cap and preincubate for 25 min at 37°C in a shaking water bath.
11. Add to blank samples 35 μL of deproteinization mixture and place them on ice.
12. Then add prepurified [^{3}H]-histidine to all samples (6.25 μCi, to make a final concentration of 0.5 μ*M*) and vortex gently.
13. Five minutes later, make final volumes equal to 250 μL by adding either KRM or depolarizing medium.
14. Maintain incubation/depolarization for 30 min unless otherwise desired. The slices should be treated with O_2/CO_2 periodically during the procedure.
15. Stop incubations by placing all samples on ice and immediately add 35 μL of deproteinization mixture. Each sample should contain finally 100 nmol histamine and 1% trichloroacetic acid.
16. Sonicate samples for 10–20 s at 4°C.

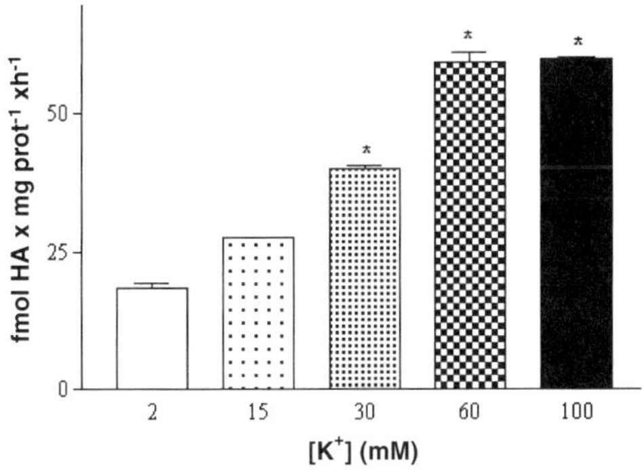

Fig. 2. K^+ - induced depolarization stimulates histamine synthesis in rat brain miniprisms. * $p < 0.01$ as compared to $K^+ = 2$ m*M*. Each column represents the mean ± SEM of 4 to 12 experiments.

17. Then remove a 60-µL aliquot of each sample and determine protein content by the Lowry method, using bovine serum albumin as a standard (*see* **Note 4**).
18. Centrifuge samples at 12,000*g* for 10 min at 4°C.
19. Recover supernatants and treat them as described in **Subheading 3.3.**

The stimulation of histamine synthesis by depolarization is shown in **Fig. 2**.

3.3. Elimination of Excess Substrate

3.3.1. Preparation of Amberlite Anion-Exchange Resin

The strong anion exchange Amberlite resin (IRA 900, mesh 16–50) bears a quaternary amine that efficiently binds carboxylic groups such as that of histidine. However, the functional group of the resin is originally supplied equilibrated with chloride as a counterion, which cannot be efficiently displaced by carboxylate ions. In contrast, carboxylates displace hydroxyl groups, so the resin is regenerated to replace chloride by OH^- as the counterion as follows:

1. Pass 20 vol of 1 *M* NaOH through a large resin column (50 mL) and rinse subsequently with 20 vol Milli-Q water.
2. Recover the regenerated resin from the column and store it in Milli-Q water at room temperature.
3. Calibrate the efficiency of histidine removal by the resin by mixing 100 µL of resin with nonradiolabeled histidine and histamine (*see* **Table 1**). The resulting resin binding capacity (estimated about 0.5–1 meq/g) remains stable for six months of storage simply in Milli-Q water at room temperature.
4. Immediately prior to use, gently rinse the resin 10× in fresh Milli-Q water.

Table 1
Histidine Removal by the Amberlite Anion-Exchange Resin[a]

Resin volume (μL)	Histidine area (% recovery)	Histamine area (% recovery)
0	1920 (100%)	1727 (100%)
20	1459 (76%)	1737 (100%)
40	1559 (81%)	1712 (99%)
60	581 (30%)	1324 (76%)
80	392 (20%)	1278 (74%)
100	305 (16%)	1256 (72%)
120	255 (13%)	1246 (72%)
140	227 (11%)	1200 (70%)

[a]100 nmol of Histidine and histamine were mixed with the indicated resin volume, vortexed for 10 min and injected into the HPLC. The results are means of 2–3 injections per group. SEM did not exceed 1.5% of the mean.

3.3.2. Elimination of [^{3}H]-Histidine

1. Pipet 100 μL of resin bed into the top half of Ultrafree microspin tubes (use a cut yellow pipet tip as for miniprisms; *see* **Note 3**). This is equivalent to 15 mg dry-weight resin.
2. Add the deproteinized supernatants obtained in **Subheading 3.2.** and vortex for 10 min in the multitube shaker at room temperature. During this step, the resin binds most of the [^{3}H]-histidine through its carboxylate group, thereby clearing most excess precursor from the samples. The anion-exchange resin also binds the trichloroacetate ion, raising the pH of the sample from 2.0 to 8.0.
3. Centrifuge the tubes at low speed (4000*g*) for 5 min.
4. Recover the filtrate in the bottom half of the tube, to be injected into the HPLC system.
5. The efficiency of [^{3}H]-histidine removal by the resin can be controlled for each sample by counting a 10-μL aliquot prior to injecting the remainder into the HPLC. It should be noted that resin dilution by addition of deproteinized supernatant affects efficiency of removal. Smaller supernatant volumes (the case of histidine decarboxylase assays) work better than bigger volumes (as is the case for brain miniprisms).

3.4. Histamine Purification by HPLC

1. To purify [^{3}H]-histamine formed, set up HPLC with the mobile phase for histamine at 1 mL/min and let stabilize column for 30 min. Avoid using a column previously used for [^{3}H]-histidine purification.
2. Set up detector wavelength at 225 nm (imidazole ring absorbance).

Table 2
Typical dpm Obtained in Histamine Fractions Under Different Conditions[a]

	Nonlabeled histamine	Incubation blank	Sample	
Rat brain homogenate	12 ± 0.6 (6)	25 ± 0.9 (7)	633 ± 4 (7)	
			K+ = 2 m*M*	**K+ = 60 m*M***
Cortical miniprisms	123 ± 3 (13)	359 ± 2 (49)	1347 ± 10 (30)	2710 ± 21 (30)

[a]The results are means ± SEM (N). More dpm are used in cortical miniprism experiments than in homogenate incubations, which explains the higher blank values. Nonlabeled histamine represents dpm carried out from a HPLC injection to the next.

3. Inject 100 nmol of each nonradiolabeled histidine and histamine.
4. Measure histamine peak area, as it will be used later to calculate recovery. Histamine retention time should be 10–11 min (*see* **Fig. 1B**). If it is not, adjust composition of mobile phase: Higher methanol content will decrease retention, whereas higher octanesulfonic acid concentration will increase it.
5. Check dpm in the postcolumn eluate before starting sample injection (typically less than 30 dpm/mL). In preliminary assays, it is advisable to monitor dpm baseline by recovering fractions at different times after injection of samples containing [^{3}H]-histamine. For routine determinations, it is sufficient to control for residual dpm in the system by injecting one nonradiolabeled standard of histamine every 10 samples.
6. Typical chromatograms are shown in **Fig. 1C**. Automated fraction collection is started upon detection of the histamine internal standard (8 min after injection, program will activate collection if slope exceeds 10 mV/min), and is completed automatically when the baseline is reached.
7. At the end of the assay, mix all fractions with Optiphase scintillation cocktail and count.
8. For each sample, compare internal standard histamine peak area with external standard histamine to obtain recovery. Use recovery to estimate dpm of histamine synthesized and subtract dpm in blank samples. Use specific activity of substrate to find out the amount of histamine synthesized. Use protein content and incubation time to quantify histidine decarboxylase activity in homogenates or histamine synthesis in slices.

The sensitivity of the assay will depend directly on the dpm in obtained for the blank samples. **Table 2** shows typical blanks obtained in homogenates and slices of rat brain cortex. Given that the specific activity of substrate is about 80 dpm/fmol, these blanks permit us to quantify as low as 2–5 fmol of histamine formed.

4. Notes

1. To clean up the HPLC after purification of [^{3}H]-histidine, remove the column and flow a few milliliters of 0.1 *M* nitric acid (pH = 1.0) through the HPLC system. This step will remove residual tritium in injection loops and the spectrophotometer. Next, wash with filtered Milli-Q water, and then with methanol (100%), to remove nitric acid. Check the system for residual dpm by scintillation counting. Next, reinstall the column and clean it with methanol/Milli-Q water (70/30%) at a flow rate of 0.8 mL/min until residual tritium is under 1000 dpm/mL. Avoid exposure of the column to nitric acid which would result in loss of the C_{18} material. Do not use the same column for [^{3}H]-histidine and [^{3}H]-histamine purifications.
2. It is possible store a solution of NaCl, KCl, $MgSO_4$,$7H_2O$, and KH_2PO_4 at 4°C for up to 2 mo. When required, add the rest of the solid reagents of the modified KRM. Do not add $CaCl_2$ and $NaHCO_3$ at the same time because Ca_2CO_3 will be produced and precipitate in the solution. The modified KRM has an approx pH 7.4 after gassing with O_2/CO_2.
3. Miniprisms in suspension can be pipeted into a standard yellow pipet tip cut at 1 cm from the tip. Although excessive variability should not be expected, it is still advisable to control for pipeting accuracy by determining the protein content in each tube after incubation. This will also be useful in order to express final results of histamine synthesis as per mg of protein.
4. The Lowry procedure requires an alkaline medium. When samples are acidic, add NaOH as required for the Lowry procedure and check alkaline pH with a paper pH indicator before adding other Lowry reagents.
5. To clean up between consecutive sample injections, the system washes the injection loop, needle, and syringe five times with 0.1 *M* nitric acid at pH 1.0. These washing steps do not introduce nitric acid into the column. Residual dpm carried from sample to sample can be controlled by placing a nonradiolabeled histamine standard to be injected in between (*see* **Table 2**). After the sample injection it is recommended to replace nitric acid solution in the injector by Milli-Q water in order to avoid corrosion and rusting of metals in the automatic injector. Finally, clean the column with methanol/water with 10 m*M* octanesulfonic acid (70/30%) at a flow rate of 0.8 mL/min for 2–3 h. This should remove residual [^{3}H]-dpm from the HPLC system, but this can be checked by counting dpm in the eluate (residual dpm are typically 20–30 dpm).

References

1. Ortiz, J., Gómez, J., Torrent, A., Aldavert, M., and Blanco, I. (2000) Quantitative radioisotopic determination of histidine decarboxylase using high-performance liquid chromatography. *Anal. Biochem.* **280,** 111–117.
2. Schwartz, J.-C., Lampart, C., and Rose, C. (1970) Properties and regional distribution of histidine decarboxylase in rat brain *J. Neurochem.* **17,** 1527–1534.
3. Palacios, J. M., Mengod, G., Picatoste, F., Grau, M., and Blanco, I. (1976) Properties of rat brain histidine decarboxylase. *J. Neurochem.* **27,** 1455–1460.

4. Baudry, M., Martres, M.-P., and Schwartz, J.-C. (1973) The subcellular localization of histidine decarboxylase in various regions of rat brain. *J. Neurochem.* **21,** 1301–1309.
5. Keeling, D. J., Smith, I. R., and Tipton, K. F. (1984) A coupled assay for histidine decarboxylase: in vivo turnover of this enzyme in mouse brain. *Naunyn-Schmiedeberg's Arch. Pharmacol.* **326,** 215–221.
6. Arrang, J.-M., Garbarg, M., and Schwartz, J.-C. (1987) Autoinhibition of histamine synthesis mediated by presynaptic H3 receptors. *Neuroscience* **23,** 149–157.
7. Gomez-Ramirez, J., Ortiz, J., and Blanco, I. (2002) Presynaptic H_3 autoreceptors modulate histamine synthesis through cAMP pathway. *Mol. Pharmacol.* **61,** 239–245.

26

Automated vs Manual Profiling of Peptide Libraries by Mass Spectrometry

Jean-Louis Aubagnac, Christine Enjalbal, Jean Martinez, Pierre Sanchez, and Gilles Subra

1. Introduction

Combinatorial chemistry *(1–7)* has evolved from the synthesis of very large mixtures to the preparation of collections of isolated compounds by parallel syntheses. Peptides constitute ideal materials because their preparation is completely automated using solid-phase methodology allowing high throughput syntheses.

Library profiling requires both compound identification and quantification. In addition, the analysis must be as fast as possible to keep pace with the synthesis and screening. Such requirements impart considerable constraints on the analytical method. Specificity, sensitivity, and rapidity are prerequisite for efficient control of large collections of molecules.

Among the various spectroscopic methods reported in the literature on that topic *(8–10)*, simultaneous ultraviolet (UV) and mass spectrometric detections were found the most convenient. Indeed, the two techniques are complementary. UV detection is sensitive, but not specific. Furthermore, liquid chromatography (LC)/UV experiment allows compound quantification if a suitable wavelength has been chosen. On the contrary, mass spectrometry (MS) exhibits both specificity and sensitivity, but quantification is not directly achievable because of possible discrimination during compound ionization. The recourse to hyphenated LC–MS technique allows gathering all the aforementioned data (UV chromatogram and mass spectrum) in a single experiment, thus reducing the analysis time.

From: *Methods in Molecular Biology, vol. 251, HPLC of Peptides and Proteins: Methods and Protocols*
Edited by: M.-I. Aguilar © Humana Press Inc., Totowa, NJ

Data management is of crucial importance. Specific software has been designed to allow automatic data acquisition and interpretation. Samples can be run continuously through the use of autosamplers able to handle vials or plates. Optimization of such systems can be effected by diminishing injection-to-injection cycle times and increasing the sensitivity. LC columns dedicated to combinatorial library analysis are commercially available. Their reduced dimensions (50×2.1 mm, 3.5 μm) enable the flow rate to be decreased from 1 to 0.25 mL/min while maintaining good chromatographic separation. Reduced analysis time and better sensitivity in electrospray MS are thus achieved.

Combinatorial peptide libraries are routinely analyzed by such fast LC–MS techniques, the molecule of interest and the eventual side-products being characterized but also quantified. Above a defined purity threshold, the synthesis is considered successful whereas all experiments which failed to contain enough compound are discarded. Such pass/fail criterion is usually visualized by green and red spots in automated result reporting using commercially available softwares. The selected samples can then be screened as such or further purified prior to bioassays.

We found that such automated data interpretation must be considered with caution. Provided that all ions related to the compound including adducts and multiply charged species are monitored, a negative response should indicate an unsuccessful synthesis. Manual inspection of such presumably failed syntheses showed that the desired peptides were present but in mixtures with various byproducts. The systematic recourse to double coupling avoids contamination by deleted chains. However, the final deprotection step, which is conducted rapidly to avoid oxidation of methionine residues, is not sufficient to cleave simultaneously all side-chain protecting groups. Whereas the presence of truncated peptides irreversibly affects the product purity, partially deprotected chains can be easily converted into the desired sequences by carrying out a second deprotection reaction. This chapter outlines a method for analysis of a 96-member library by LC–MS techniques.

2. Materials

2.1. Chemicals

1. High-pressure liquid chromatography (HPLC) grade trifluoroacetic acid (TFA) is available from Merck (Schuchardt, Germany).
2. HPLC grade acetonitrile is available from Panreac Chimica (Barcelona, Spain) or from Riedel-de-Haën (Seelze, Germany).
3. Water was glass-distilled in-house and further purified using a Milli-Q water purification system (Millipore, Bedford, MA).

4. Fmoc-protected amino acids and polystyrene Rink amide resin are available from Advanced Chemtech (Louisville, KY).
5. HBTU is available from SENN Chemicals (Dielsdorf, Switzerland).

2.2. Solid-Phase Peptide Synthesis

1. Solid-phase synthesis is performed on an ACT 496 Ω multiple organic synthesizer in Fmoc strategy using PS Rink amide resin (0.8 mmol/g).
2. Coupling solution (120 mmol/L Fmoc-amino acid, 120 mmol/L HBTU, 240 mmol/L DIEA) is prepared in dimethyl formamide (DMF).
3. Two successive coupling steps of 2 h are used at each stage of the synthesis to avoid deletion products.
4. Fmoc removal is performed in 20 min with a 80/20 DMF/piperidine solution (v/v).
5. A TFA/anisole/thioanisole/ethanedithiol/water (90/2.5/2.5/2.5/2.5, v/v) solution is used for the final resin cleavage and side-chain deprotection step.
6. After cleavage, peptides are precipitated several times in diethyl ether and lyophilized.

2.3. LC/ESI-MS Instrumentation

1. Identification of the peptides is performed by high performance liquid chromatography coupled to an electrospray mass spectrometer (LC/ESI-MS).
2. The mass spectrometer (Platform II from Waters-Micromass, Milford, MA) fitted with a quadrupole is calibrated in the positive ion mode using a mixture of *NaI* and *CsI*.
3. Voltages are set at +3.5 kV for the capillary and +0.5 kV for the skimmer lens.
4. The source is heated at 120°C. Nitrogen is used as nebulizing and drying gas at 15 and 250 L/h, respectively.
5. Mass spectrometric data are acquired in the positive mode according to one of the following methods (*see* **Note 2**):

   Method A: scan mode from *m/z* 200 to *m/z* 1200 in 3 s.

   Method B: scan mode from *m/z* 400 to *m/z* 1900 in 4 s.
6. Three to five scans are summed to produce the final spectrum at the selected retention time.
7. Voltage of the sampling cone is adjusted to 30 V.
8. Data are processed by Micromass MassLynx system.
9. An Alliance 2690 (Waters, Milford, MA) LC system equipped with an autosampler is used to deliver the mobile phase, continuously degassed, at a flow rate of 250 µL/min.
10. The eluent is split after the LC column to reach only 35 µL/min of mobile phase infused into the mass spectrometer source and 215 µL/min into the UV cell.
11. UV detection is set up at 214 nm.
12. The Symmetry Shield RP18 column (50 × 2.1 mm, 3.5 µm) is available from Waters-Micromass.

13. The best chromatographic separation is achieved at a flow rate of 250 µL/min using a gradient from 0 to 50% acetonitrile in water containing 0.1% TFA in 10 min followed by a second rapid gradient from 50 to 80% acetonitrile in 0.5 min.
14. Isocratic elution is carried out for 1.5 min at 80% acetonitrile to flush lipophilic compounds.
15. The initial conditions are then reached in 0.5 min and the column is equilibrated for 4.5 min before injection of the next sample.
16. The autosampler requires 1 min to select the vial and load the injection loop. So the overall run cycle per sample is 18 min (*see* **Note 3**).
17. All samples are dissolved in acetonitrile/water (50/50, v/v) at a concentration of 30 mmol/L.
18. Each solution is placed in one autosampler vial fitted with a 200-µL restrictor.
19. In all experiments, 1 µL of sample is injected at 25°C.

3. Methods

1. Place the vials in the autosampler.
2. Equilibrate the LC column under initial conditions.
3. Load the elution conditions file (LC file).
4. Choose the MS scan method for each sample according to the molecular weight of the peptide: Method A for molecules possessing a molecular weight inferior to 1000 atomic mass unit (amu) and Method B for molecules possessing a molecular weight superior to 1000 amu.
5. Fill the sample list (file name, LC file that is identical for all samples, MS file that varies according to the chosen scan mode).
6. Start the run overnight (electrospray probe into the source, gas supply on, whole sample list selected and automatic shutdown activated) because analysis of the whole library roughly lasts 29 h.
7. Effect manual inspection of all samples in the morning (quantification from UV integration and identification from regenerated spectra).

In this example, a 96-member library designed to contain fully deprotected peptides of molecular weight ranging from 700 to 3000 was analyzed by LC–MS. Automated data interpretation was effected by targeting the protonated and sodiated molecular ions as well as the doubly and triply charged molecule. Approximately one-quarter of the samples were correct, i.e., the main UV signal that was integrated corresponded to more than 80% (*see* **Fig. 1B**) was the expected peptide. The mass spectrum regenerated at 8.87 min (*see* **Fig. 1C**) exhibited two intense signal at *m/z* 970 and 647 attributed to the doubly and triply protonated peptide, respectively. The protonated molecule was scarcely detected and always with a very low abundance. The other samples exhibited multiple signals in UV and TIC chromatograms as displayed in

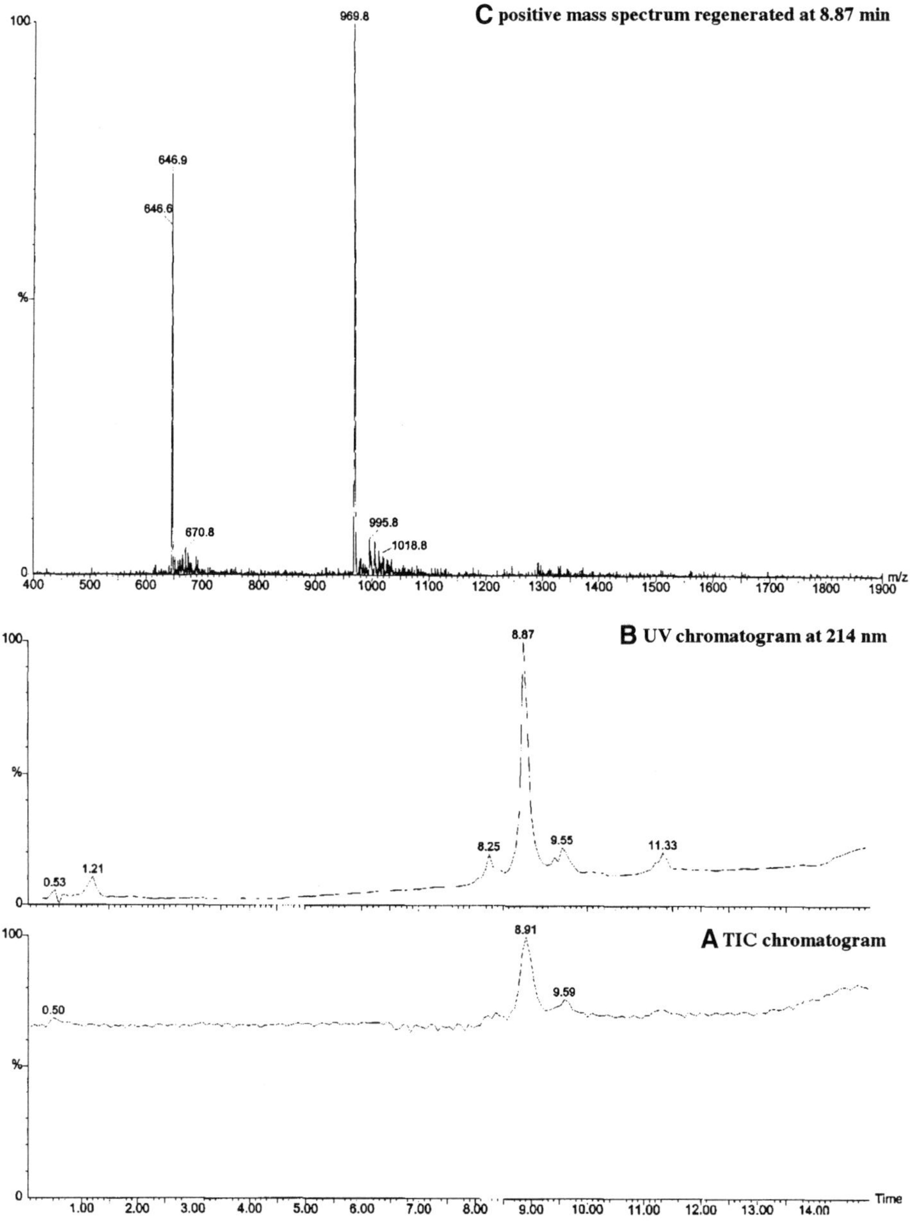

Fig. 1. **(A)** TIC chromatogram in the positive mode. **(B)** UV chromatogram at 214 nm. **(C)** Positive mass spectrum regenerated at 8.87 min showing the expected peptide: $(M+2H)^{2+} = 969.8$ and $(M+3H)^{3+} = 646.9$.

Fig. 2. The main peak at 8.98 min corresponded to the expected peptide [*see* **Fig. 3A**: $(M+H)^+ = 1281$ and $(M+2H)^{2+} = 641$] but three major byproducts were detected at 9.66, 11.85, and 12.33 min. The first eluted byproduct at 9.66 min [*see* **Fig. 3B**: $(M+2H)^{2+} = 669$] possessed one t-Bu protected side chain because the mass increment on the doubly charged species equaled 28 Th. Accordingly, the compound at 11.85 min corresponded to a sequence containing a side-chain-protected arginine residue [*see* **Fig. 3C**: $(M+2H)^{2+} = 774$, mass increment of 133 Th on the doubly charged species]. The more lipophilic molecule eluted at 12.33 min contained a t-Bu and a Pmc protecting group [*see* **Fig. 3D**: $(M+2H)^{2+} = 802$]. Most of the unsuccessful syntheses were a result of partial removal of the Pmc protection of the guanidinium moiety, especially in the case of arginine rich sequences (*see* **Note 1**). In some extreme cases, the desired fully deprotected peptide was hardly detected and more lipophilic protected structures were eluted at very high concentrations of acetonitrile. In several samples, partially deprotected sequences were detected at different retention times implying that regioisomers were present. Thus, the whole library was submitted again to a deprotection step and analyzed by LC–MS a second time.

Of the samples that contained partially deprotected chains after the first synthesis, only 25% of them were pure enough to undergo the screening process. Resubmitting the collection of samples to a second deprotection procedure raised this figure to 99%. So, the manual inspection of the data saved time by preventing the time consumed by resynthesis of the library.

4. Notes

1. Since the Pmc protection of the arginine side chain was not completely removed after two TFA treatments in some peptides, the Pbf moiety was preferred to mask the guanidinium functionality due to its faster deprotection rate *(11)*.
2. Two MS scan modes were required due to the heterogeneity of the studied library in terms of peptide length. The low-molecular-weight sequences (<1000 amu) exhibited mostly singly protonated molecular ion, whereas longer chains (>1000 amu) produced to multiple ions at different charge state. A wider mass range was thus necessary for the former samples to detect all ions. The upper limit of 1900 amu was determined according to preliminary studies which showed that peptides possess sufficient basic sites to promote multiple protonations thus decreasing the recorded *m/z* values.
3. More efficient library profiling was achieved by using a shorter LC column. The XTerra RP18 column (30 × 2.1 mm, 3.5 μm) was purchased from Waters and showed comparable chromatographic separation as the Symmetry Shield stationary phase. The flow rate was kept at 250 μL/min but the elution was performed faster using a gradient from 0 to 50% acetonitrile in water containing 0.1% TFA in 6.5 min (instead of 10 min) followed by a second rapid gradient from 50 to 80%

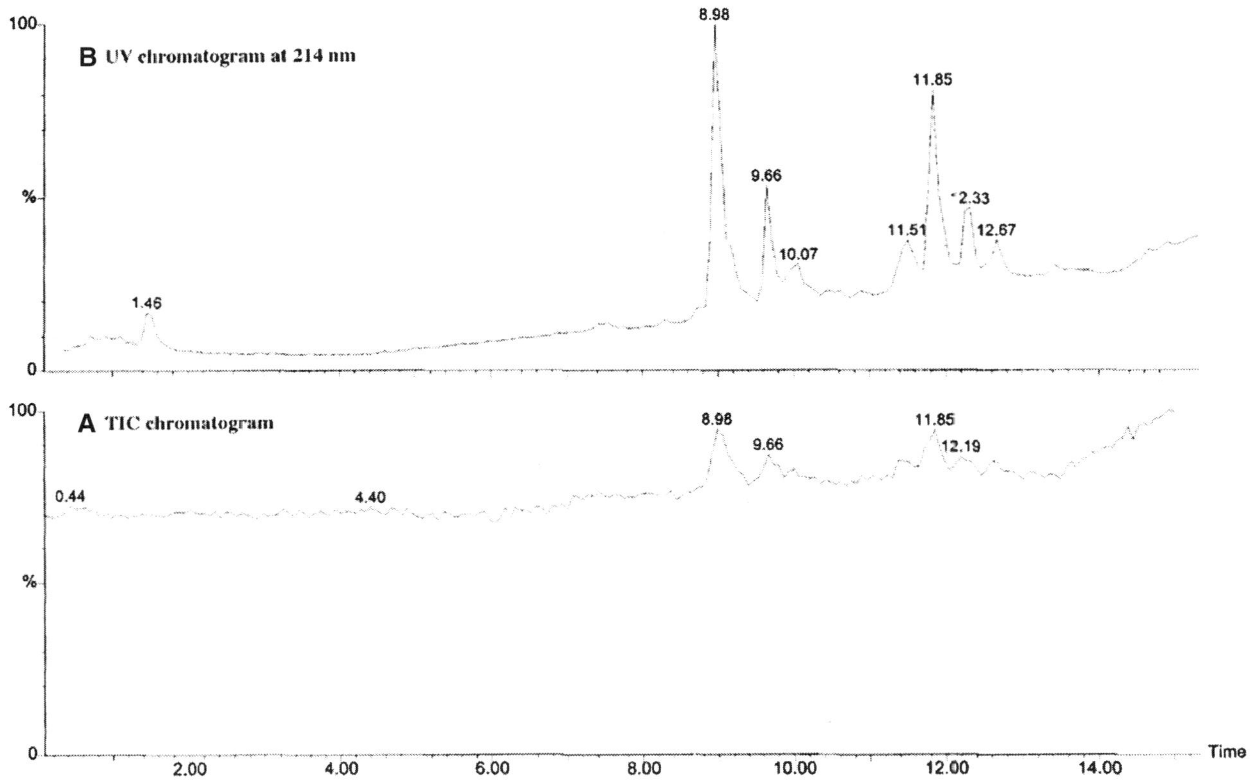

Fig. 2. (**A**) TIC chromatogram in the positive mode. (**B**) UV chromatogram at 214 nm.

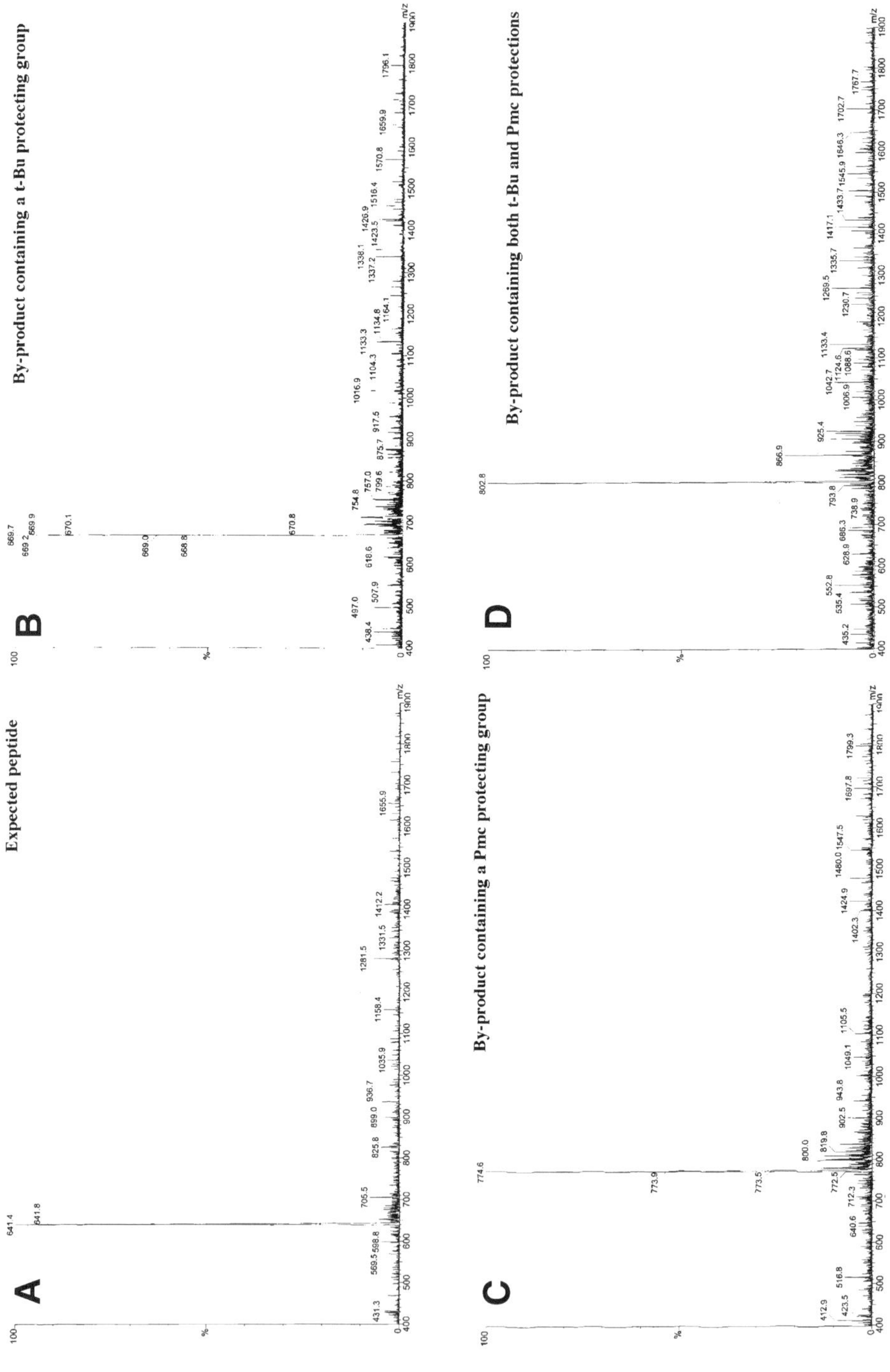
A
Expected peptide
B
By-product containing a t-Bu protecting group
C
By-product containing a Pmc protecting group
D
By-product containing both t-Bu and Pmc protections

acetonitrile in 0.5 min. Isocratic elution was carried out for 1.0 min at 80% acetonitrile to flush lipophilic compounds. The initial conditions were then reached in 0.5 min and the column was equilibrated for 4.0 min before injection of the next sample (1 min delay to load the injection loop). The injection-to-injection cycle time was thus reduced from 18 to 13.5 min. The same 96-member's library was analyzed in less than 22 h.

References

1. Czarnik, A. W. and Dewitt, S. H. (1997) *A Practical Guide to Combinatorial Chemistry*. American Chemical Society, Washington, DC.
2. Wilson, S. R. and Czarnik, A. W. (1997) *Combinatorial Chemistry—Synthesis and Application*. Wiley, New York.
3. Bunin, B. A. (1998) *The Combinatorial Index*. Academic, London.
4. Terrett, N. K. (1998) *Combinatorial Chemistry*. Oxford University Press, Oxford, UK.
5. Gordon, E. M. and Kervin, J. F. (1998) *Combinatorial Chemistry and Molecular Diversity in Drug Discovery*. Wiley, New York.
6. Obrecht, D. and Villalgordo, J. M. (1998) *Solid-Supported Combinatorial and Parallel Synthesis of Small Molecular Weight Compound Libraries*. Tetrahedron Organic Chemistry Series, Vol. 17, Pergamon, Elsevier, Oxford, UK.
7. Jung, G. (1999) *Combinatorial Chemistry—Synthesis, Analysis, Screening*. Wiley-VCH, Weiheim, RFA.
8. Shapiro, M. J. and Wareing, J. R. (1998) NMR methods in combinatorial chemistry. *Curr. Opin. Chem. Biol.* **2,** 372–375.
9. Fitch, W. L. (1999) Analytical methods for quality control of combinatorial libraries. *Mol. Diversity* **4,** 39–45.
10. Enjalbal, C., Martinez, J., and Aubagnac, J-L. (2000) Mass spectrometry in combinatorial chemistry. *Mass Spectrom. Rev.* **19,** 139–161.
11. Bray, A. M., Jhingran, A. G., Valerio, R. M., and Maeji, N. J. (1994) Simultaneous multiple synthesis of peptide amides by the multipin method—Application of vapor-phase ammonolysis. *J. Org. Chem.* **59,** 2197–2203.

Fig. 3. *(see opposite page)* Positive mass spectra regenerated from the TIC chromatogram of Fig. 2 at different retention times: **(A)** 8.98 min: expected peptide, $(M+2H)^{2+}$ = 641.4. **(B)** 9.66 min: byproduct containing a t-Bu protecting group, $(M+2H)^{2+}$ = 669.7. **(C)** 11.85 min: byproduct containing a Pmc protecting group, $(M+2H)^{2+}$ = 774.6. **(D)** 12.33 min: byproduct containing both t-Bu and Pmc protecting groups, $(M+2H)^{2+}$ = 802.8.

27

Proteome Analysis

Matthew J. Powell and Aaron T. Timperman

1. Introduction

1.1. General Information

A major goal of proteomics is qualitative and quantitative analysis of all the proteins expressed in a cell, tissue, or organism. Changes in protein expression owing to a stimulus or condition are measured in a systematic manner, and are used to elucidate mechanisms of cell function and signaling. A strength of proteomics is that a "shot-gun" approach requiring no prior knowledge of the system is often used and does not assume a model prior to data collection. Therefore, proteomics provides the ability to deal with the complexity of biological systems with minimal experimental bias. The complexity of biological systems arises from the numerous parallel signaling pathways that interact with each other. The ability to monitor many proteins simultaneously yields a global view of protein expression and posttranslation modifications, which is much more informative than monitoring a few proteins. Methods that follow a few proteins and assume a model are more likely to miss interactions and yield biased results.

For the understanding of cell function, proteomics is essential, as such mechanisms cannot be deduced directly from genomic data. First, it is difficult to predict what portion of the genome actually encodes for protein and corresponds to an open reading frame *(1)*. Second, it has been noted recently that protein and mRNA levels do not correlate, so increases in mRNA levels cannot be used to predict protein levels *(2,3)*. Third, proteins are often changed between active and inactive forms by changes in posttranslational modifications, such as phosphorylation, glycosylation, and acylation. Fourth, protein–substrate binding

From: *Methods in Molecular Biology, vol. 251, HPLC of Peptides and Proteins: Methods and Protocols*
Edited by: M.-I. Aguilar © Humana Press Inc., Totowa, NJ

events are integral to many cell signaling and function pathways. To date, there are no means to predict protein modifications and protein–substrate interactions directly form genomic data.

Currently, proteomics is both technology driven and technology limited. Recently, it has become apparent that it is difficult to observe low-abundance proteins using the most common and robust methods for proteomic analysis ***(2,4–9)***. This inability to detect low-abundance proteins is a serious limitation, as the proteins that are involved directly in cell signaling and function pathways are typically low-abundance proteins. The proteins that are observable through current methods of proteomic analysis are typically structural or housekeeping proteins and their expression levels provide little or no information about the mechanisms of cell signaling and function.

Because of these initial instrumental limitations of proteomics, many researchers have moved away from such "shot-gun" approaches to proteomics and toward more experimentally tractable and focused experiments. Examples of such focused experiments include characterization of protein–protein interactions and characterization of posttranslational modifications. Both of these examples require the use of selective-enrichment or affinity purification to increase the relative abundance of the protein or proteins of interest by many orders of magnitude. With use of selective-enrichment schemes, low-abundance proteins can be observed.

The two most commonly used methods for protein identification and proteome analysis are peptide mass fingerprinting by matrix-assisted laser desorption ionization mass spectrometry (MALDI-MS) ***(10–13)*** and peptide sequencing by electrospray ionization tandem mass spectrometry (ESI-MS/MS) ***(14)***. As always, each method has its advantages and disadvantages, and the relative importance of these characteristics varies with the application. The main advantage of the peptide mass fingerprinting by MALDI-MS is that it is much faster than capillary high-performance liquid chromatography (HPLC) ESI-MS/MS, and sample through-put can be 100-fold greater. However, MALDI peptide fingerprinting does not provide amino acid sequence level information which gives rise to a number of limitations. Without direct amino acid sequence information, MALDI peptide fingerprinting: (1) cannot be used for *de novo* sequencing, and therefore the protein of interest must be in a database, (2) it is not used for analysis of posttranslation modifications, and (3) does not work well with protein mixtures. Recently, MALDI sources have been combined with quadrupole time-of-flight mass analyzers (Q-TOF), and have been used to acquire the same MS/MS spectra that the capillary HPLC ESI-MS/MS method acquires, at the speed of the MALDI peptide fingerprinting experiments ***(15,16)***. However, the MALDI-QTOFs are expensive and the method development is not mature. Peptide sequencing has been achieved with MALDI-Q-TOFs, but some differences in fragmentation have been noted because MALDI produces peptides with lower charge states than ESI.

ESI is a most effective means of producing ions from the liquid phase ***(17)***, and consequently has been used to couple liquid phase techniques, such as nanospray and capillary HPLC to the mass spectrometer. Peptide sequencing by nanospray ESI-MS/MS refers to the introduction of unseparated peptide mixtures into the MS at nanoliter/min flow rates ***(18)***. Since the introduction of capillary HPLC-ESI-MS/MS ***(19–23)***; it has become a well-established and commonly used method for proteome analysis, ***(5,24)*** because the capillary HPLC separation provides substantial benefits. First, the greatest advantage of capillary HPLC is that the peptides can be extracted from a dilute solution and sprayed into the MS at much higher concentrations. Because ESI-MS is a concentration sensitive detector, the concentration of capillary HPLC provides a concomitant improvement in the signal/noise. Second, the narrow concentrated bands from the various peptides are injected into the MS at different times, allowing the MS to focus on collection of the MS/MS spectra of each of these peptides individually. Third, hydrophilic contaminants are not extracted onto the column and are washed away before the data collection is initiated.

An overview of the methods used for proteomic analysis is shown schematically in **Fig. 1**. Proteomic analysis typically begins with a cell lysate, which is a complex mixture of thousands of proteins whose individual concentrations span six orders of magnitude. After lysis and a crude purification of the cells, the protein mixture is separated by 2DE, and the protein spots are visualized on the gel by staining with Coomassie blue, silver, or fluorescent dyes (*see* **Note 1**). The protein spots are excised from the gel manually with a razor or automatically with a robotic spot cutter, and each spot is placed in a separate tube. The proteins are digested in-gel using enzymatic digestion and/or chemical cleavage, and are eluted from the gel by liquid extraction (*see* **Note 2**; ***25***). The samples are dried down and redissolved in a small volume of LCA buffer. A pressure vessel is used to individually load the protein digests onto a capillary HPLC column. The peptides from the digest are then separated by gradient RP-HPLC. The MS acquires data continually in the fullscan MS mode to identify ions of high enough intensity to collect CID spectra from. When an ion exceeds a preset threshold the MS automatically switches to the fullscan MS/MS mode for collection of the CID spectra. After collection of the CID spectra it returns to the fullscan MS mode. Protein identification is performed by correlating the CID spectra with theoretical spectra of known sequences from either protein or DNA databases. Additionally, *de novo* sequencing can be used to determine the peptide sequence directly from the spectra without the use of a database.

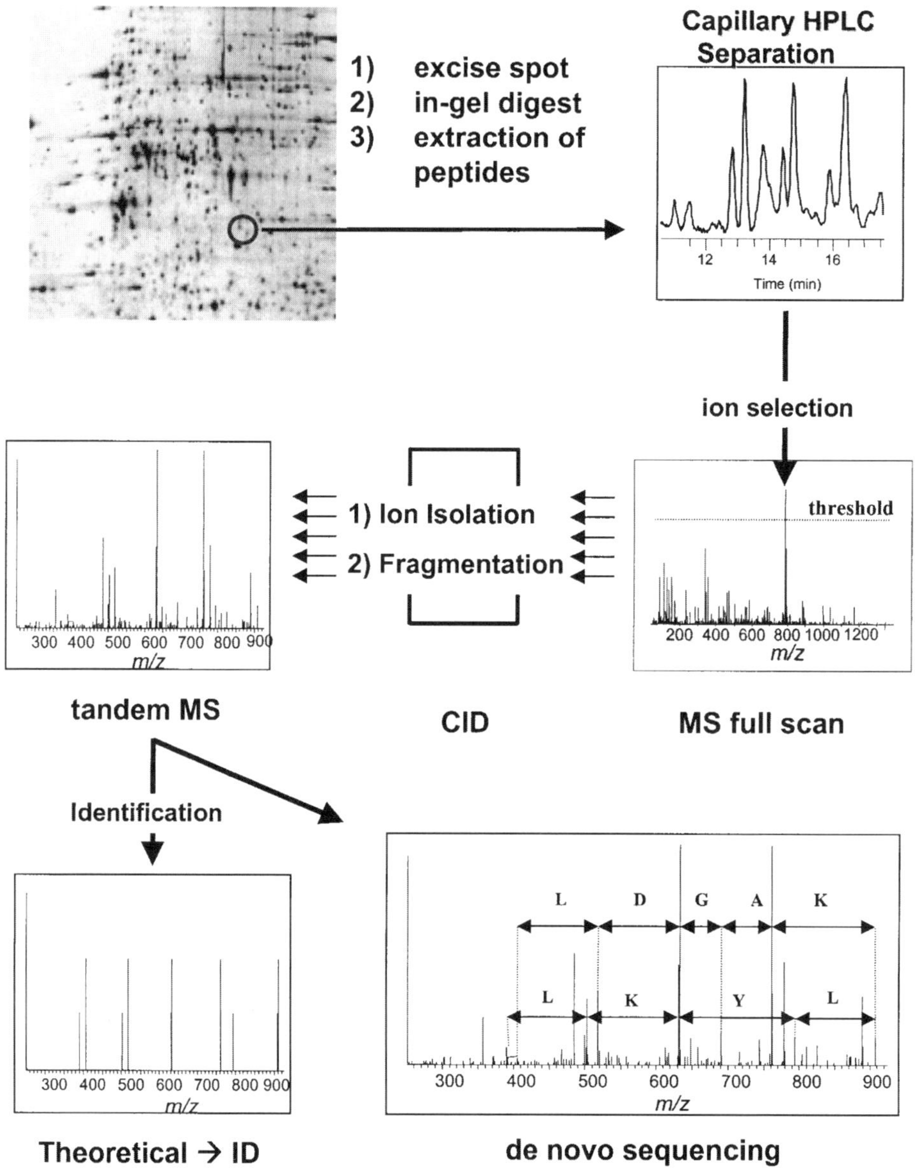

Fig. 1. Overview of the process of proteome analysis by capillary HPLC-MS/MS.

In this chapter, we will provide an overview of the entire method for proteome analysis by capillary HPLC-ESI-MS/MS, and then we will focus on performing the capillary HPLC separation in detail.

2. Materials

1. Deionized water, 18 $M\Omega$, used for ALL solution preps and sample preps (Fisher Scientific, Pittsburgh, PA).
2. HPLC—Agilent model 1100, with a binary pump, high pressure mixer, and vacuum degasser (Agilent Technologies, Wilmington, DE).
3. HPLC buffers (*see* **Note 3**):
   a. LCA buffer: 0.005% heptafluorobutyric acid (HFBA) (Sigma-Aldrich, St. Louis, MO; cat. no. 16,419-4). Make fresh weekly.
   b. 0.2% Acetic acid (cat no. 33, 882-6; Sigma-Aldrich, St. Louis, MO).
   c. 4% Acetonitrile (Optima grade; Fisher Scientific, Pittsburgh, PA).
   d. LCB buffer: 100% Acetonitrile (Optima grade; Fisher Scientific, Pittsburg, PA)
4. Standard fused silica capillary, 360-µm OD, 75-µm id (internal diameter) available from Polymicro (Phoenix, AZ).
5. All unions, sleeves, cross, and other accessories are available from Upchurch Scientific (Murietta, CA).
   a. Microcross with 0.006-inch through-holes plus fittings—P# P·777
   b. PEEK capillary sleeves, 380-µm ID—P# F·185x.
6. Tips, fritted tips, and columns are available from New Objectives (Woburn, MA).
7. Fisherbrand Disposable Micro Pipettes (Fisher Scientific, Pittsburgh, PA).
8. Butane minitorch—model B (Microflame, Minneapolis, MN).
9. Column packing.
   a. C_{18} beads; cat. no. 218MSB5 (Vydac, Hesparia, CA).
   b. Isopropanol (Sigma-Aldrich, St. Louis, MO).
10. High-pressure device (Mass Evolution, Spring, TX).
    a. For packing columns and loading samples.
    b. Requires a tank of nitrogen gas and appropriate gages and fittings.
    c. Alternatively, a Micro Autosampler (LC Packings, Amsterdam, Netherlands) can be used to load samples onto the columns.
11. EIMS with Tandem MS capability.
    a. LCQ-DECAXP or current model (ThermoFinnigan, San Jose, CA).
12. Microsprayer.
    a. Can be obtained from MS manufacturer or Mass Evolution (Spring, TX).

3. Methods

3.1. Capillary HPLC Separation

3.1.1. Fabrication of the Capillary Columns

To fabricate the capillary HPLC column, a frit is formed at one end of a 30-cm-long capillary to form an outlet frit *(26)*. After washing the frit, the column is slurry packed with a suspension of C_{18} beads (*see* **Note 4**) in isopropyl alcohol to a form a 7- to 10-cm long bed *(27)*. The slurry is injected into the capillary at 1000 psi using the high-pressure vessel (HPV). The extra length of the capillary allows for necessary connections to be made and to prevent the loss of beads, as no exit frit is used. Packed columns are prepared on the HPLC by a 10-min rinse with 80% LCB (organic) buffer, followed by a 10-min rinse with 0% LCB buffer. Freshly packed columns must be conditioned to optimize column separation efficiency. Columns are conditioned with 2–3 injections of 1 pmol of a solution digest of a standard protein, such as BSA, or 1 pmol of a peptide standard such as angiotensin. Presumably, conditioning improves column performance by irreversibly binding to sites of specific adsorption, and blocking these sites from future nonideal interactions. The performance of the column is monitored by running the MS in the full scan mode and checking the peak width. Peak full-width at half-maximum (FWHM) should be about 10 s for the parameters suggested herein.

3.1.2. Loading and Column Prep

Dried samples are resuspended by sonication in a low volume (5–20-µL) of LCA buffer and loaded onto the packed column at 1000-psi by use of the HPV. The sample is placed in the HPV in a 0.5-mL Eppendorf tube, and the top to the HPV carefully replaced and tightened securely. The column end is removed from the cross, slid through the orifice in the top of the HPV, secured in this position by tightening the nut with a wrench (tight enough to hold the column without destroying the ferrule), and the HPV is pressurized to load the sample (*see* **Note 5**). After loading the sample, the column is then reinserted into cross (*see* **Figs. 2** and **3** for setup information) and rinsed briefly with 100% LCA buffer (2 min).

3.1.3. Solvent Flow Rate

Capillary HPLC and microspray interface require low rates in the 200- to 400-nL/min range (*see* **Note 6**). Most HPLC pumps in use today do not have the capability to provide such low flow rates, requiring the use of a precolumn splitter. The splitter is assembled from a cross, and the extra junction is used to connect the ESI high-voltage electrode, as shown in Fig. 2. Back-pressure is determined by the splitter capillary id and length. Small adjustments in flow rate can be made by changing the length of the splitter, and large adjustments are made by changing the splitter capillary id. Initially, start with a splitter length of 1.5 m and an ID of 50-µm. The HPLC is set to the lowest flow rate that it can produce a reliable gradient at (0.1-mL/min for the Agilent 1100). The flow rate through the capillary column is measured directly by collecting the solution as it exits the spray tip with a graduated glass capillary for a set period of time. Flow rates for packed columns are adjusted to approx 350-nL/min for a 75-µm id column and 200-nL for a 50-µm column.

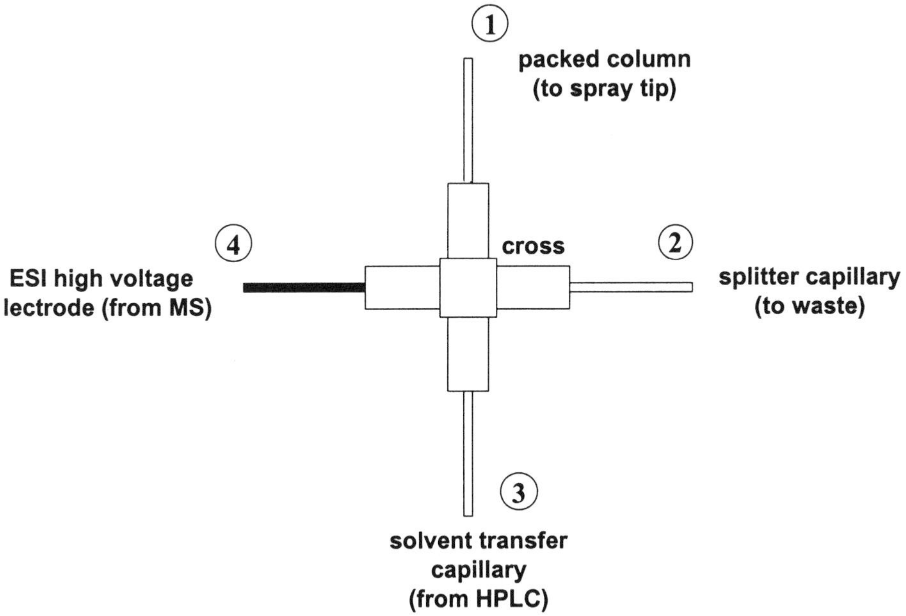

Fig. 2. Schematic diagram of the precolumn splitting cross. As illustrated in the diagram, the cross has four ports: 1, the packed column leading from the cross to the electrospray tip; 2, the splitter capillary which regulates the backpressure at the cross; 3, the solvent transfer capillary carrying the buffer solutions from the HPLC; and 4, the ESI high voltage electrode from the MS to induce electrospray.

3.1.4. Capillary HPLC and ESI Interface

After the column has been prepared, the inlet (nonfritted end) is inserted into the precolumn splitter cross. The outlet (fritted end) of the packed column is attached to a microspray tip by a zero dead volume (ZDV) union housed in a stable base on the microspray device as shown in **Fig. 3B**. Alternatively, columns with pulled tips on the end are commercially available. When using the integrated column and tip, the capillary is passed directly through the ZDV union, and the ZDV union is simply used as stable holding device. The home-made column with separate tip is less expensive, but the commercially available column with integrated tip is easier to use (*see* **Note 7**), when care is taken to minimize any dead volumes in the connecting union.

Last, the spray tip must properly be aligned with the heated capillary of the mass spectrometer for accurate spray direction, typically by means of an x-y-z translational stage.

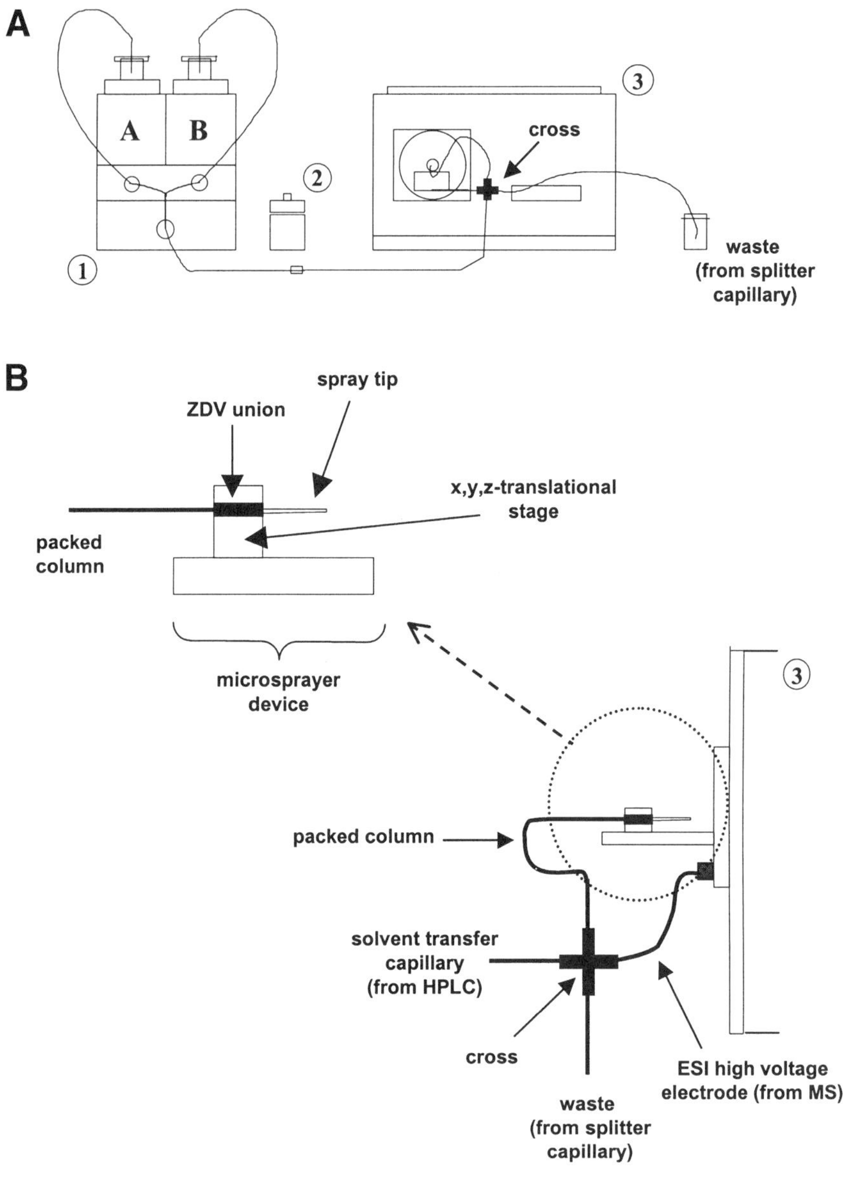

Fig. 3. Schematic diagram of the HPLC-MS/MS laboratory setup. Front view (**A**) and side view (**B**) of the capillary HPLC-MS/MS setup. As the gradient is being run, the capillary HPLC column is connected directly to the cross as shown. For the sample injection the inlet end of the capillary column is removed from the cross and inserted into the HPV (**2**). When the sample loading is complete the HPV is depressurized and the capillary column is returned to the cross so the HPLC-MS/MS analysis can be initiated. Shown in the sideview is a blow up of the microspray device. Notice the ZDV union joining the packed column and spray tip. Key: 1, HPLC. 2, HPV. 3, ESI mass spectrometer.

Table 1
Sample HPLC Gradient Conditions for a 50-min MS Scan Sequence

Time (min)	% LCB Buffer	% Acetonitrile	Event initiated
0	0.0	4	Wash sample
1	0.0	4	Steep gradient
3	11.5	15	Shallow gradient
43	42.7	45	Steep gradient
45.5	58.3	60	Steeper gradient
46	79.2	80	Clean column
51	80	80	End column cleaning
52	0	4	Re-equilibration
58	0	0	Stop

3.1.5. HPLC Gradient

The HPLC gradient must be programmed with a peptide-elution gradient. Typical gradients are 20- to 90-min in duration and include column cleaning and reequilibration steps (*see* **Note 8**). An example of an HPLC program method for a 50-min sample scan follow is shown in **Table 1**.

3.1.6. Elution and Analysis

After the sample has been loaded, the column readied in the cross, and the HPLC method selected, the MS scan is started. The HPLC method can be started automatically by the MS, or both instruments can be started separately. Precise retention times are not necessary, as all information is contained within the CID spectra. The gradient is run and data is collected in a data-dependent manner through the CID spectra.

Once the scans are complete, analysis is begun by using a database searching and correlation program, such as Sequest or Mascot, to match the acquired spectra with predicted spectra for known sequences in a database for protein identification. Most database searching routines will always return the peptide sequence from the database, which correlates best with the CID spectra. It is critical that care is taken in the data interpretation to ensure that the match is reliable. Correlation scores are assigned to the results as an indicator of the accuracy and uniqueness of the sequence. SEQUEST gives a cross-correlation scores based on the spectra's comparison to theoretical spectra, and a cross correlation score of 2.0 or greater is considered a good match with a protein in the database. Delta correlation scores (range 0–1.0) are based on a sequences "uniqueness," compared to other sequences within the database. A delta correlation score greater than 0.1 usually indicates the first ranked match is much better than the second ranked match for a large database. However, the delta correlation score is influenced by the size of the database searched and will have higher values for databases with fewer sequences, although the match is not any more reliable *(28)*. Sequences found by database searching should be carefully examined and manually sequenced to confirm the peptide's identity. The reliability of a match also increases greatly if more than one peptide from a protein is matched.

Additionally, the CID spectra can be manually *de novo* sequenced to reveal amino acid sequence information about the peptides and confirm the protein identification. New *de novo* sequencing software to decrease analysis time is currently in developmental stages.

4. Notes

1. When staining the gel from the electrophoretic separation of the protein mixture to be analyzed, it is important to use staining solutions with minimal concentrations of crosslinking agents, such as gluteraldehyde and formaldehyde. Silver staining recipes that use very low concentrations of formaldehyde and are compatible with MS analysis have been developed *(25)*.
2. Cleanliness before and during digestion is important because contaminant proteins will also be digested. During protein extraction from the gel pieces after enzymatic digestion, care must be taken to reduce contamination of the peptide collections. Particulates can clog the packed column or spray tip, and require additional device maintenance. To minimize clogging, centrifuge the pooled extractions for 5 min at 10,000*g*, and carefully decant the top approx 80% of solution, being careful not to disturb particulates at the bottom of the tube. The heavier contaminants are in this remaining solution and can be discarded.
3. Solvent purity is vitally important to reduce noise in the mass spectra, as less pure solvents will increase the background. Plastics and polymers should be avoided when handling solvents. Glass and stainless steel syringes should be used in place of micropipetters for measuring and handling the concentrated HFBA and acetic acid use in making the LCA buffer.
4. Reversed-phase separation on C_{18} beads is the method of choice for peptide mass spectrometry because these systems provide good separation of peptides and use buffers that are compatible with the mass spectrometer. Vydac low TFA beads are recommended because they require very little ion pairing agent (HFBA). Beads should be 3- to 5-µm, and pore sizes should range from 60 to 300-Å, with 300-Å being the most recommended selection for working with unknown peptides. In practice, column ID should be at least ten times larger than the bead diameter being used for the column (for a 50-µm column, beads should 5-µm or less in diameter) to minimize clogging and aggregation during packing of the columns.
5. Sample loading with a HPV must be carefully performed to avoid injury. Exercise caution with use of the HPV, as the high internal pressures of compressed gas make

its use inherently dangerous. Always secure the HPV top securely and wear goggles when using to load samples. When resolubilizing the sample in the LCA buffer, use small volumes (6- to 20-μL) to reduce the sample solution volume loaded, and consequently the time required for loading the sample. The capillary should be inserted fully into the HPV so that it touches the bottom of the microcentrifuge tube containing the sample. Caution must be taken not to attempt to load too much sample, as helium will be forced into the column packing material and cause chromatography problems. In practice, some of the sample (approx 3 μL) is lost during the injection; so more than 5- to 6-μL of a 10-μL sample should not be loaded. The volume of sample solution loaded is measured directly by collecting the solution exiting the capillary tip with a graduated glass capillary.

6. It is important to maintain a consistent flow rate through the capillary and column system. Microspray flow rates normally range between approx 200- and approx 350-nL/min for 50 and 75 μm columns, respectively. Smaller id columns yield lower detection limits, because electrospray MS is a concentration sensitive detector, but smaller id columns are more prone to clogging. Seventy-five micrometers columns usually provide the optimum balance between detection limits and ease of use for silver stained spots. The volume of sample solution loaded is measured directly by collecting the solution exiting the capillary tip with a graduated glass capillary.
7. Spray tips must be selected with id's in the range of 5 to 15 μm to insure low flow rate at the tip and a narrow electrospray. These tips are commercially available, or may be reproducibly made in the lab with a capillary laser puller. Alternatively, tips can be produced by cutting a small length of capillary, suspending it vertically with a weight (office binder clip) on the bottom end, and heating the middle with a butane minitorch. A the capillary is heated, it will be pulled and will separate into two tapered tips. The end of each tip should be trimmed so that the id is about 5–10 μm and wiped clean with methanol. Although these tips are suitable for routine work, it can be difficult to maintain a stable electropspray. Poor chromatography and unstable spray is often a result of a clogged spray needle.
8. Proper selection of the parameters for an HPLC gradient are essential for optimizing the peptide analysis. The slope of the gradient is the change in organic solvent (acetonitrile) concentration divided by the time in minutes. Within upper and lower limits, a steeper gradient slope gives better detection limits, whereas a shallower gradient slope gives better resolution *(29)*. For analysis of in-gel protein digests, the samples are usually very complex, requiring higher resolution and a shallower gradient. However, if the slope is too shallow it will degrade the limits of detection.

   Typical gradient times for the most active region for peptide desorption of 15–40% acetonitrile are between 15 and 90 min. Gradients for peptide elution usually consist of multiple linear segments with different slopes to optimize the time spent in different regions of the organic solvent concentrations. Regardless of the time spent in the 15–40% acetonitrile range, the times for the column washing and reequilibration should remain constant, as shown in **Table 1**.

References

1. Kuster, B., Mortensen, P., and Mann, M. (1999) in *Proc. 47th ASMS Conf. Mass Spectrom. and Allied Top.*, Dallas, TX, p. 1897.
2. Gygi, S. P., Rochon, Y., Franza, B. R., and Aebersold, R. (1999) Correlation between protein and mRNA abundance in yeast. *Mol. Cell. Biol.* **19,** p. 1720–1730.
3. Tew, K. D. (1996) Glutathione-associated enzymes in the human cell lines of the National Cancer Institute Drug Screening Program. *Mol. Pharmacol.* **50,** p. 149–159.
4. Boucherie, H., Sagliocco, F., Joubert, R., Maillet, I., Labarre, J., and Perrot, M. (1996) Two-dimensional gel protein database of *Saccharomyces cerevisiae. Electrophoresis* **17,** 1683–1699.
5. Ducret, A., Van Oostveen, I., Eng, J. K., Yates, J. R., and Aebersold, R. (1998) High throughput protein characterization by automated reverse-phase chromatography electrospray tandem mass spectrometry. *Protein Sci.* **7,** p. 706–719.
6. Garrels, J. I., Futcher, B., Kobayshi, R., et al. (1994) Protein Identifications for a Saccharomyces cerevisiae protein database. *Electrophoresis* **15,** 1466–1486.
7. Gygi, S. P., Corthals, G. L., Zhang, Y., Rochon, Y., and Aebersold, R. (2000) Evaluation of two-dimensional gel electrophoresis-based proteome analysis technology. in *Proc. Nat. Acad. Sci. USA* **97,** p. 9390–9395.
8. Shevchenko, A. O., Jensen, O. N., Podtelejnikov, A. V., et al. (1996) Linking genome and proteome by mass sepctrometry: large scale identification of yeast proteins from two-dimensional gels. *Proc. Nat. Acad. Sci. USA* **93,** p. 14,440–14,445.
9. Wilkins, M. R., Gasteiger, E., Sanchez, J.-C., Bairoch, A., and Hochstrasser, D. F. (1998) Two-dimensional gel electrophoresis for proteome projects. The effects of protein hydrophobicity and copy number. *Electrophoresis* **19,** 1501–1505.
10. Yates, J. R. 3rd, Speicher, S., Griffin, P R; Hunkapiller, T. (1993) Peptide mass maps: a highly informative approach to protein identification. *Anal. Biochem.* **214,** p. 397–408.
11. Henzel, W. J. B., Todd. M., Stults, J. T., Wong, S. C., Grimley, C., and Watanabe, C. (1993) Identifying proteins from two-dimensional gels by molecular mass searching of peptide fragments in protein sequence databases. *Proc. Nat. Acad. Sci. USA* **90,** p. 5011–5015.
12. James, P., Manfredo, Q., Carafoli, E., and Gonnet, G. (1993) Protein identification by mass profile fingerprinting. *Biochem. Biophys. Res. Commun.* **195,** p. 58–64.
13. Pappin, D. J. C., Hojrup, P., and Bleasby, A. J. (1993) Rapid identification of proteins by peptide-mass fingerprinting. *Curr. Biol.* **3,** p. 327–332.
14. Hunt, D. F. Y., Shabanowitz, J., Winston, S., and Hauer, C. R. (1986) Protein sequencing by tandem mass spectrometry. *Proc. Nat. Acad. Sci. USA* **83,** p. 6233–6237.
15. Loboda, A. V., Krutchinsky, A. N., Bromirski, M., Ens, W., and Standing, K. G. (2000) A tandem quadrapole/time-of-flight mass spectrometer with a matrix-assisted laser desorption/ionization source: design and performance. *Rap. Commun. Mass Spectrometry* **14,** p. 1047–1057.

16. Shevchenko, A., Loboda, A., Ens, W., and Standing, K. G. (2000) MALDI quadrupole time-of-flight mass spectrometry: A powerful tool for proteomic research. *Anal. Chem.* **72,** p. 2132–2141.
17. Fenn, J. B. M., Meng, C. K., Wong, S. F., and Whitehouse, C. M. (1989) Electrospray ionization for mass spectrometry of large biomolecules. *Science* **246,** p. 64–71.
18. Wilm, M. and Mann, M. (1996) Analytical properties of the nanoelectrospray ion source. *Anal. Chem.* **68,** p. 1–8.
19. Deterding, L. J. M., Arthur, M., Tomer, Kenneth, B., and Jorgenson, J. W. (1991) Nanoscale separations combined with tandem mass spectrometry. *J. Chromatogr.* **554,** p. 73–82.
20. Hail, M., Lewis, S., and Jardine, I. (1990) Detection and sequence analysis of tryptic peptides by microcolumn liquid chromatography-tandem mass spectrometry using an electrospray interface. *J. Microcolumn Sep.* **2,** p. 285–292.
21. Huang, E. C. and Henion, J. D. (1991) Packed-capillary liquid chromatography/ion-spray tandem mass spectrometry determination of biomolecules. *Anal. Chem.* **63,** p. 732–739.
22. Hunt, D. F., Alexander, J. E., McCormack, A. L., et al. (1990) Mass spectrometric methods for protein and peptide sequence analysis, in *Pap. Annu. Symp. Protein Soc., 4th*, Villafranca, J. J., ed., Academic, New York, p. 441–454.
23. Griffin, P. R., Coffman, J. A., Hood, L. E., and Yates, J. R. III (1991) Structural analysis of proteins by capillary HPLC electrospray tandem mass spec. *Int. J. Mass Spectrom. Ion Processes* **111,** p. 131–149.
24. Gygi, S. P., Han, D. K. M., Gingras, A. C., Sonenberg, N., and Aebersold, R. (1999) Protein analysis by mass spectrometry and sequence database searching: Tools for cancer research in the post-genomic era. *Electrophoresis* **20,** p. 310–319.
25. Shevchenko, A., Wilm, M., Vorm, O., and Mann, M. (1996) Mass spectrometric sequencing of proteins from silver-stained polyacrylamide gels. *Anal. Chem.* **68,** p. 850–858.
26. Cortes, H. J., Pfeiffer, C. D., Richter, B. E., and Stevens, T. S. (1987) Porous ceramic bed supports for fused silica packed capillary columns used in liquid chromatography. *J. HRC&CC* **10,** p. 446–448.
27. Kennedy, R. T. and Jorgenson, J. W. (1989) Preparation and evaluation of packed capillary liquid chromatography columns with inner diameters from 20 to 50 micrometers *Anal. Chem.* **61,** p. 1128–1135.
28. Yates, J. R. (1998) Database searching using mass spectrometry data. *Electrophoresis* **19,** p. 893–900.
29. Snyder, L. R., Stadalius, M. A., and Quarry, M. A. (1983) Gradient elution in reversed-phase HPLC separation of macromolecules. *Anal. Chem.* **55,** p. 1412A.

Index

M